Computer Vision and Internet of Things

Computer Vision and Internet of Things

Technologies and Applications

Edited by

Lavanya Sharma and Mukesh Carpenter

CRC Press
Taylor & Francis Group
Boca Raton London New York

CRC Press is an imprint of the
Taylor & Francis Group, an **informa** business
A CHAPMAN & HALL BOOK

First edition published 2022
by CRC Press
6000 Broken Sound Parkway NW, Suite 300, Boca Raton, FL 33487-2742

and by CRC Press
4 Park Square, Milton Park, Abingdon, Oxon, OX14 4RN

CRC Press is an imprint of Taylor & Francis Group, LLC

Library of Congress Cataloging-in-Publication Data
Names: Sharma, Lavanya, editor. | Carpenter, Mukesh, editor.
Title: Computer vision and internet of things : technologies and applications / Lavanya Sharma, Mukesh Carpenter.
Description: First edition. | Boca Raton : Chapman & Hall/CRC Press, 2022. | Includes bibliographical references and index. | Summary: "The book Computer Vision and Internet of Things: Technologies and Applications explores the utilization of Internet of Things with Computer Vision and its underlying technologies in different applications areas. The text explores a series of present and future applications — business insights, indoor-outdoor securities, smart grids, human detection and tracking, intelligent traffic monitoring, e-health department, medical imaging and many more. The book also focuses on providing a detailed description of the utilization of IoT with computer vision and its underlying technologies in critical application areas, such as smart grids, emergency departments, intelligent traffic cams, insurance, and the automotive industry. This book is primarily aimed at graduates and researchers working in the areas of IoT, computer vision, big data, cloud computing and remote sensing. It will also be an ideal resource for IT professionals and technology developers"— Provided by publisher.
Identifiers: LCCN 2021052518 (print) | LCCN 2021052519 (ebook) |
ISBN 9781032154367 (hbk) | ISBN 9781032154404 (pbk) |
ISBN 9781003244165 (ebk)
Subjects: LCSH: Computer vision. | Internet of things.
Classification: LCC TA1634 .C648933 2022 (print) | LCC TA1634 (ebook) |
DDC 006.3/7—dc23/eng/20211227
LC record available at https://lccn.loc.gov/2021052518
LC ebook record available at https://lccn.loc.gov/2021052519

ISBN: 9781032154367 (hbk)
ISBN: 9781032154404 (pbk)
ISBN: 9781003244165 (ebk)

DOI: 10.1201/9781003244165

Typeset in Palatino
by codeMantra

Dedicated to my Dada Ji (Late Shri Ram Krishan Choudhary Ji)

Ek prerna mayeh Vyaktitavh

Dr. Lavanya Sharma

Contents

Preface

This book explores the utilization of Internet of Things (IoT) with Computer Vision and its underlying technologies in different applications areas. Using a series of present and future applications – business insights, indoor-outdoor securities, smart grids, human detection and tracking, intelligent traffic monitoring, e-health department, medical imaging and many more – this publication will support readers to acquire more deeper knowledge in implementing the IoT with visual surveillance.

This book comprises five parts that provide an overview of basic concept from rising of machines and communication to IoT with computer vision, critical application domains, tools, technologies, and solutions to handle relevant challenges. This book provides a detailed description to the readers with practical ideas of using IoT with visual surveillance (motion-based object data) to deal with human dynamics, challenges involved in surpassing diversified architecture, communications, integrity, and security aspects. IoT in combination with visual surveillance proved to be most advantageous for the companies to efficiently monitor and control their day-to-day processes such as production, transportation, maintenance, implementation, and distribution of their products.

Overall *Computer Vision and Internet of Things: Technologies and Applications* helps the readers to understand the value of Internet of Things with Computer Vision to individuals as well as organizations.

Editors

Dr. Lavanya Sharma completed her M.Tech. (Computer Science and Engineering) in 2013 at Manav Rachna College of Engineering, affiliated with Maharshi Dayanand University, Haryana, India. She completed her Ph.D. at Uttarakhand Technical University, India, as a full-time Ph.D. scholar in the field of digital image processing and computer vision in April 2018, and received a TEQIP scholarship for the same. Her research work is on motion-based object detection using background subtraction technique for smart video surveillance.

She is the recipient of several prestigious awards during her academic career and qualified certification courses from ISRO Dehradun unit, India. She has published 40+ research papers to her credit in reputed journals of Elsevier (SCI Indexed), Inderscience, IGI Global, IEEE Explore, and many more. She has published three books with Taylor & Francis and two books with CRC Press in 2019, 2020, and 2021, respectively. She also has two patents in her account on object detection in visual surveillance. She has also contributed as an Organizing Committee member to Springer's ICACDS 2016, Springer's ICACDS 2018, Springer's ICACDS 2019, Springer's ICACDS 2020, ICRITO 2021, and Springer's ICACDS 2021 conferences. Presently, she is the editorial member/reviewer of various journals of repute and also an active program committee member of various IEEE and Springer conferences. Her primary research interests are digital image processing and computer vision, artificial intelligence, machine learning, deep learning, and IoT. Her vision is to promote teaching and research, providing a highly competitive and productive environment in academic and research areas with tremendous growing opportunities for the society and her country.

Dr. Mukesh Carpenter completed his MBBS at Manipal Academy of Higher Education, Kasturba Medical College, Mangalore, India, in 2011 and Master of Surgery at Mallya Hospital, Vittal Mallya Road, Bangalore, India, in 2014. He is the recipient of several prestigious awards and recognition in his career. He is presently working as a Unit Head and consultant with Alshifa Hospital, Delhi, India, and is having more than 6 years of experience. He has published several articles and research papers, and contributed as hospital committee member and unit head. He has also delivered expert talk in prestigious hospitals and educational institutes of India on topics including breast cancer awareness, cancer awareness drive, Covid-precautions and measures, and also serves as the brand ambassador of Pro Biotic (Expert Talk on Fever 104 FM, January 2021). He is also guiding Ph.D. students in the field of computer science for medical imaging domain.

List of Contributors

Alpana
Department of Computer Science and Technology
Manav Rachna University
Haryana, India

Sunil L. Bangare
Department of I.T.
Sinhgad Academy of Engineering
Pune, India

Thierry Bouwmans
Lab. MIA
University of La Rochelle
La Rochelle, France

Dharmendra Carpenter
Department of Anaesthesiology & Critical Care
Narayana Multispecialty Hospital
Jaipur, India

Mukesh Carpenter
Department of Surgery
Alshifa Multispecialty Hospital
Okhla, India

Nikhil J. Chaudhari
Department of I.T.
Sinhgad Academy of Engineering
Pune, India

Soumitra Das
DYPIT
Pune, India

Belmar Garcia-Garcia
Lab. MIA
University of La Rochelle
La Rochelle, France

Deepa Gupta
Amity Institute of Information Technology
Amity University
Noida, India

Kavish Gupta
Amity Institute of Information Technology
Amity University
Noida, India

Priyanka Gupta
Department of Computer Science and Technology
Manav Rachna University
Haryana, India

Yaman Hooda
Faculty of Engineering and Technology
Manav Rachna International Institute of Research and Studies
Haryana, India

Pramod Jadhav
Department of Computer Engineering
G. H. Raisoni Institute of Engineering and Technology
Pune, India

Vinod Kumar Jangid
Department of Respiratory Medicine
Medical College Kota
Rajasthan, India

Vukoman Jokanović
ALBOS doo, Innovative Company
Belgrade, Serbia
Vinča Institute of Nuclear Sciences
Belgrade, Serbia

Isha S. Kuchangi
Department of I.T.
Sinhgad Academy of Engineering
Pune, India

Nirvikar Lohan
University of Patanjali
Uttarakhand, India

Yash M. Lokhande
Department of I.T.
Sinhgad Academy of Engineering
Pune, India

Sandeep Dwarkanath Pande
DYPIT
Pune, India

Shantanu S. Pathak
DHI Training and Research Consultancy
Pune, India

Amol D. Sawant
DYPIT
Pune, India

Kamal Sehairi
Lab. LTSS
University of Laghouat
Laghouat, Algeria

Sudhriti Sengupta
Department of Computer Science and Engineering
Galgotias University
Greater Noida, India

Lavanya Sharma
Amity Institute of Information Technology
Amity University
Noida, India

Jaskirat Singh
Amity Institute of Information Technology
Amity University
Noida, India

Hardeo Kumar Thakur
Department of Computer Science and Technology
Manav Rachna University
Haryana, India

Aditya N. Varade
Department of I.T.
Sinhgad Academy of Engineering
Pune, India

El-Hadi Zahzah
Lab. L3i
University of La Rochelle
La Rochelle, France

M. Živković
School of Dentistry
University of Belgrade
Belgrade, Serbia

S. Živković
School of Dentistry
University of Belgrade
Belgrade, Serbia

Acknowledgement

I am especially grateful to my dada ji, my parents, my husband, and my beautiful family for their continuous support and blessings. I would like to thank my husband Dr. Mukesh (general and laparoscopic surgeon) for his continuous motivation and support throughout this project. Apart from his busy schedule, he always motivated and supported me. I owe my special thanks to Ms. Samta Choudhary ji and Late Shri Pradeep Choudhary ji for their invaluable contributions, cooperation, and discussions.

Above all, I express my heartiest thanks to God (The One to Whom We Owe Everything) Sai Baba of Shirdi for all blessings, guidance, and help [by you and only you]. I would like to thank God for believing in me and being my defender. Thank you, God Almighty

Lavanya Sharma

Part 1

Introduction to Computer Vision and Internet of Things

1

Rise of Computer Vision and Internet of Things

Lavanya Sharma
Amity Institute of Information Technology, Amity University, Noida, India

CONTENTS

1.1 Introduction

Computer Vision (CV) represents the domain of artificial intelligence (AI) which trains the system to identify and interpret the visual world. Machines detect objects and classify them into various categories as per the vision. To detect objects, digital cameras, video stream, and deep learning (DL) models are used. CV involves various important tasks such as three-dimensional scene modeling, multi-model camera geometry, motion-based, stereo correspondence, point cloud processing, motion estimation, and many more. There are three basic steps involved in this process as shown in Figure 1.1. With the advancements of AI, systems can proceed to the next level and take appropriate actions based

DOI: 10.1201/9781003244165-2

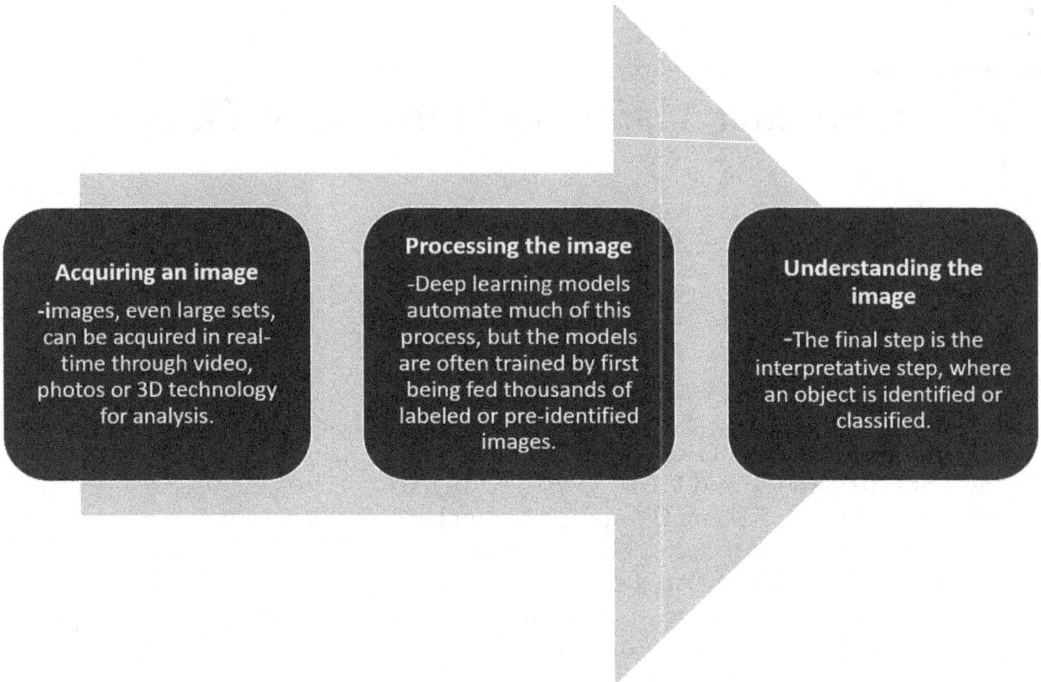

FIGURE 1.1
Working of Computer Vision.

on the first step (Figure 1.1.). In literature, various kinds of CV can be used in a different manner such as segmentation, object detection, face recognition, and edge and pattern detection. CV is an emerging technology that captures and stores an image, or frames and then transforms them into valuable information which can be further acted upon [1–11]. It comprises various technologies working all together such as ML, AI, sensor technology, image processing, and computer graphics. CV, combined with Internet Protocol connectivity, advanced data analytics, and AI, acts as a catalyst for each other and gives rise to revolutionary leaps in the Internet of Things (IoT) innovations and technology [7,11–14].

- **Image segmentation**
  This technique segments a digital image into various smaller segments or set of pixels to be examined separately. These segments correspond to different objects or parts of objects. Every pixel in a frame is allocated to one of these categories [15–17].
- **Object detection**
  This identifies that a particular object from a video stream may be a single object or multi objects in a frame sequence in case of both outdoor and indoor scenes as shown in Figure 1.2. These models use a coordinate system (X, Y) to create bounding boxes and identify all the objects in a frame. In Figure 1.2, object detection is done using background subtraction (BGS) techniques to detect foreground objects by hiding all the background pixels [17–24]. Figure 1.2 shows two different scenarios—outdoor and indoor—along with the ground truth images and output results.

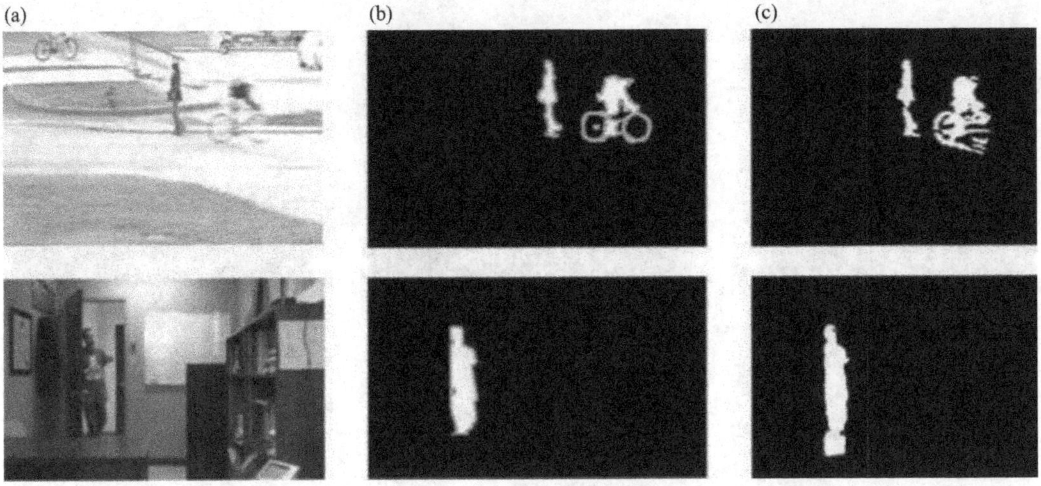

FIGURE 1.2
Detection of moving object in outdoor and indoor video sequence (row wise): (a) video frames, (b) ground truth, and (c) desired output.

- **Facial recognition**
  It is an advanced type of technique which identifies a human face and detects an individual in a frame as shown in Figure 1.3. This technique identifies or verifies the individual's face using their faces, and it also captures, analyzes, and compares the pattern based on the existing details of a person [7,23,25,26].
- **Edge detection**
  This technique is used to detect the outer boundary edge of an object or landscape for clear identification of the content of an image. It identifies points in an image with discontinuities, and these points where brightness changes are known as edges [19,23,24].
- **Pattern detection**
  This technique recognizes repeated shapes, colors, and other visual indicators in the frame. This technique is an automated identification of patterns and regularities in data and is widely used in statistical data analysis, signal processing, image processing, and CV [20,21,26].
- **Image classification**
  This groups images into various categories. The most common image classification technique is supervised and unsupervised image classifications. In literature, various algorithms are present for image classification such as support vector machine, regression, and K map [20,21,23–26].
- **Feature matching**
  It is a kind of pattern detection that matches the similarities in a frame to classify them accurately. It is a part of various CV applications such as image registration, camera calibration, and object detection [26–28].

Simple real-time applications use one of these techniques. It can be advanced ones like driverless cars rely on multiple techniques to achieve their specific goal.

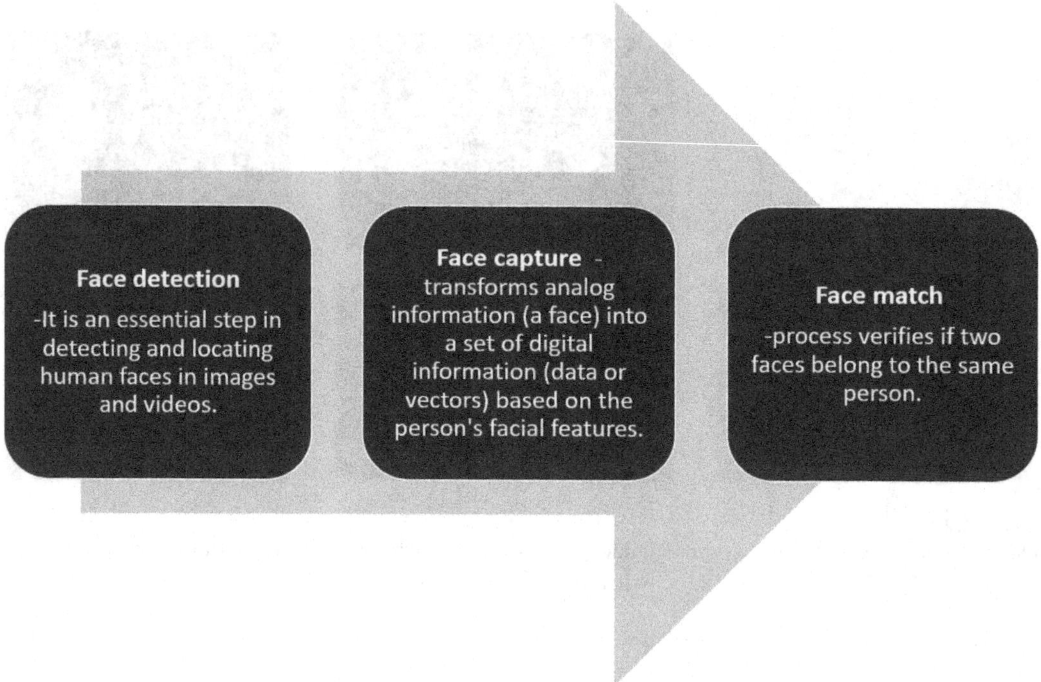

FIGURE 1.3
Facial recognition steps.

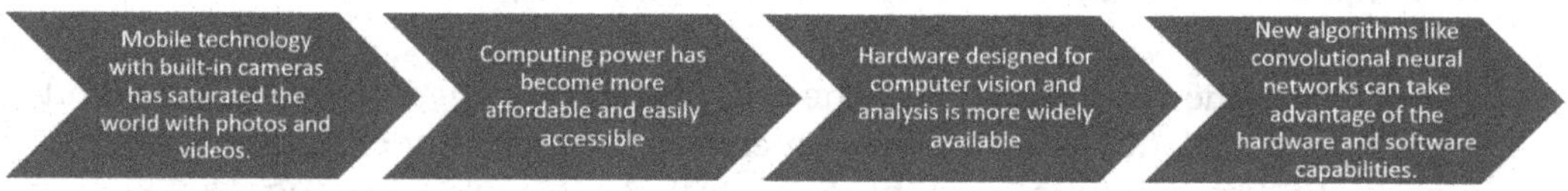

FIGURE 1.4
Factors affecting renaissance in Computer Vision.

1.2 Evolution of CV and IoT

In 1950, early experiments in CV took place with the first neural network to detect an edge of an object and sorting of objects such as a circle or square. Later, in 1970, the first commercial use of CV interpreted typed or handwritten text using optical character recognition. This advancement was used by the visually impaired to interpret written text. In the 1990s, usage of internet was also increased, and large datasets were easily available to developers or researchers for analysis and recognition. With the presence of a large amount of dataset, machines can classify objects from frames or videos [2,23]. Today, several factors have come together to bring about a renaissance in CV as shown in Figure 1.4. There is an outstanding effect of these advancements on CV, and the accuracy rate also increases from 50% to 99%. So, systems can accurately detect and track objects more accurately than humans.

FIGURE 1.5
Computer Vision and innovative technologies.

1.3 Evolving Toward CV and IoT

It is one of the most remarkable technologies to come out of DL and AI domain. The advancements that DL has contributed to the CV have set this domain apart. From face detection to processing the live action of a football game, CV rivals and surpasses humanoid visual abilities in various areas as shown in Figure 1.5. This technology is widely used in industries to enhance the client experience, cost reduction, and security, in manufacturing industries to identify product defects in real time, and in the healthcare system such as MRIs, CAT scans, and X-rays to detect abnormalities as accurately as clinicals.

IoT provides new costs and benefits to CV and has a route toward integrating AI, ML, and DL into an inspection system. The rapid growth of IoT devices has been drastically aided by the availability of several light-weighted internet protocols such as Bluetooth and Zigbee that share low-bandwidth messages. These protocols have good communication connectivity in applications where delays may be acceptable.

Traditional CV analysis deals with identifying defects or pattern matching with an unknown dataset. But AI is trainable and has wide scope in locating, identifying, and segmenting a large number of objects or defects. New techniques can be added to smart frame grabbers to perform better in complex situations with a camera and video data transmitted from the device to CV software. The embedded device offers a direct path to integrate AI into vision applications, and with the help of cloud-based processing it provides data sharing between multiple smart devices. These techniques trained the model to identify objects, defects, matching patterns while supporting a migration toward self-learning robotics systems [5,11,23].

1.4 Enhancement of IoT Using CV and 5G Technology

CV gives rise to new IoT innovation and application when merged with advanced data analytics and AI. In today's world, CV is present everywhere from a smartphone camera to games, robotics, and many more. The ability to detect and match predefined patterns in real-time conditions represents a large amount of opportunity with hundreds of use cases. An improvement such as a face-filtering app can affect our lives. This app can scan the images using face detection, identify facial characteristics, and add an augmented reality concept to it [3,9,15,23].

The performance of IoT depends on the time taken for communication with devices, smartphones, or any other software. With the help of 5G, the speed of data transfer can be increased easily and have a more stable connection between the devices. This is a very important factor for any IoT especially for connected devices such as monitoring system which depends on real-time updates. It also results in the reduction of power consumption that is up to 90% and provides a guarantee of battery for 10 years in low-powered IoT devices.

1.5 Challenging Issue

There are various obstacles to overcome in building the technology more feasible, realistic, and economical for the masses [4,11,23]:

- Embedded platforms require incorporating deep neural design for less power consumption, less cost, more flexibility, and more accuracy.
- There should be a common standard or set of protocols to allow systems and IoT devices to communicate with each other and data sharing.
- There should be less amount of human intervention. The system should be unsupervised learning and improve themselves without human intervention. The complete software update has new significance in ML.
- Systems are no longer passive collectors of data. They need to act upon the data with minimal human intervention. They need to learn and improvise by themselves. The whole software/firmware update process has new values in the ML domain.
- Hackers can also take advantage of new security vulnerabilities in CV and AI. So, developers must take this into account.

1.6 CV and IoT in Real-Time Applications

CV represents AI domain that trains the system to interpret the visual world. Machines can detect, classify, and react accordingly. In real time there are various CV applications with IoT from smart parking to object detection [2–4]. Some of the major applications are listed as follows.

1.6.1 Autonomous Vehicles

As per the WHO report, about 1.25 million persons die every year due to roadside accidents. To overcome this situation, driverless cars are developed using CV to bring this ratio down and enhance transportation. Several companies are using ultrasonic sensors, cameras, and radar to acquire images from the surrounding environment [3,6,29–35]. Autonomous vehicles can easily detect objects, signals, LiDAR, marked lanes, and traffic signs for a safe drive. Autonomous vehicles can be seen in cities next door as shown in Figure 1.6.

- In Boston, a company named "nuTonomy" has recently got approval to test its driverless cars throughout the city.

FIGURE 1.6
Self-driven cars with distance from each other.

- Another example is a grocery chain called Kroger (Scottsdale, Arizona), which has begun to pilot driverless vehicles for grocery delivery services.
- In Phoenix, Google has been actively testing its driverless cab service, and by the end of this year, they will launch it as a full-fledged business.

1.6.2 Healthcare System

In this department, CV technology plays an important role by providing accurate illness or condition prediction that may save patient lives. New CV-based technologies such as X-ray, CT, and mammography help doctors to detect the issues at an early stage [11,23,36,37]. CV technology that can assist health systems is described as follows.

1.6.2.1 Precise Diagnosis

CV systems offer precise diagnosis, which reduces false positives. This technology can potentially eliminate the requirement for unnecessary surgical operations and other high-priced therapies. The algorithms are trained using a large amount of dataset that can easily detect the presence of a condition that was unidentified by doctors due to sensory limitations. In the healthcare system, this technology provides a high level of precision for up to 100% correction.

1.6.2.2 Timely Detection of Illness

Fatal illnesses such as ovarian cancer, lung cancer, throat cancer, gall bladder cancer, and breast cancer must be diagnosed at the early stage. CV enables the detection of these illnesses at an early stage with pattern recognition capability as shown in Figure 1.7. This results in early diagnosis and timely treatment of a patient [11,23].

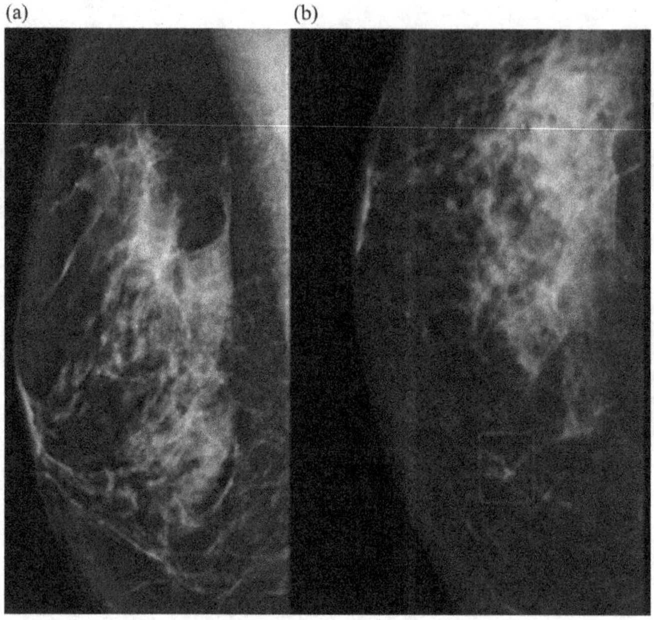

FIGURE 1.7
Detection of the breast using neural network (row wise): (a) input image and (b) resultant image.

1.6.2.3 Facial Recognition

This technique is very useful from both indoor and outdoor security perspectives. In today's world, advanced hacking and cyberattack are increasing day by day, so various companies can be benefited from this technology. It is capable of match-making human faces from an image or video stream to authenticate users via ID verification, checkpoints, and measuring features from the input image. Anand et al. [7] provide an efficient local binary pattern histogram-based technique for facial pattern recognition which is well suited for realistic scenarios using python [7]. A facial recognition using the binary histogram technique is shown in Figure 1.8.

1.6.3 CV-Based Agriculture

Agriculture is among the economy-booster area that makes every country stand out in the international market. In agriculture, CV provides a strong foothold. It provides agriculturalists with an efficient technique to increase profit, cost reduction, and detection of objects [2,8,12,11,23,38,39]. For example, "semi-autonomous combine harvester" can detect the optimal route and analyze the quality of crops. Some of the most significant contributions that exist at present are discussed as follows.

1.6.3.1 Drone-Based Monitoring and Smarter Farming

Drone technology is a very innovative and advanced technology well equipped with AI, ML, and remote-sensing features. Drones have become an essential factor in crop monitoring with their flying capabilities and inbuilt camera. They can cover a significant amount of distance and capture data that can be used to detect unfavorable conditions, pesticide spraying, crop damage information, irrigation monitoring, aerial view of the field, and

FIGURE 1.8
A facial recognition using binary histogram technique (row wise): (a) original frame, (b) grayscale, (c) Haar cascade technique, and (d) local binary classifier-based technique.

soil analysis with its geo-sensing features [11,23,40]. For example, segmentation and image annotation are used for object detection.

1.6.3.2 Yield Analysis

DL-based models are used for yield analysis. The data collected from drones, satellite imaging, nitrogen and moisture level information, climatic condition, and soil analysis are fed as an input smart system which is trained to handle a large amount of data. The overall report helps in prediction beforehand about any natural disruption [1–9,12].

1.6.3.3 Crop Grading and Sorting

There is a high demand for AI-powered CV technologies in farming. Grading and sorting of crops becomes easy with the use of this technology. The smart system can detect crop's longevity and infection in crops which results in less crop damage. Both vegetables and fruits cab be graded as per the quality and packed accordingly [3,6,39].

1.6.3.4 Automated Pesticide Spraying and Phenotyping

CV-based drone technology is used for pesticide spraying, which results in an increase in production and quality of crops. This technology can effectively monitor and detect the infected crops and spray an adequate amount of pesticide as per the requirement. Using this technology, both the intervention of humans and chances of pesticide exposure are less. Phenotyping is an effective approach based on an advanced CV to detect the crop traits for precision farming. The CV algorithms embedded with image processing features

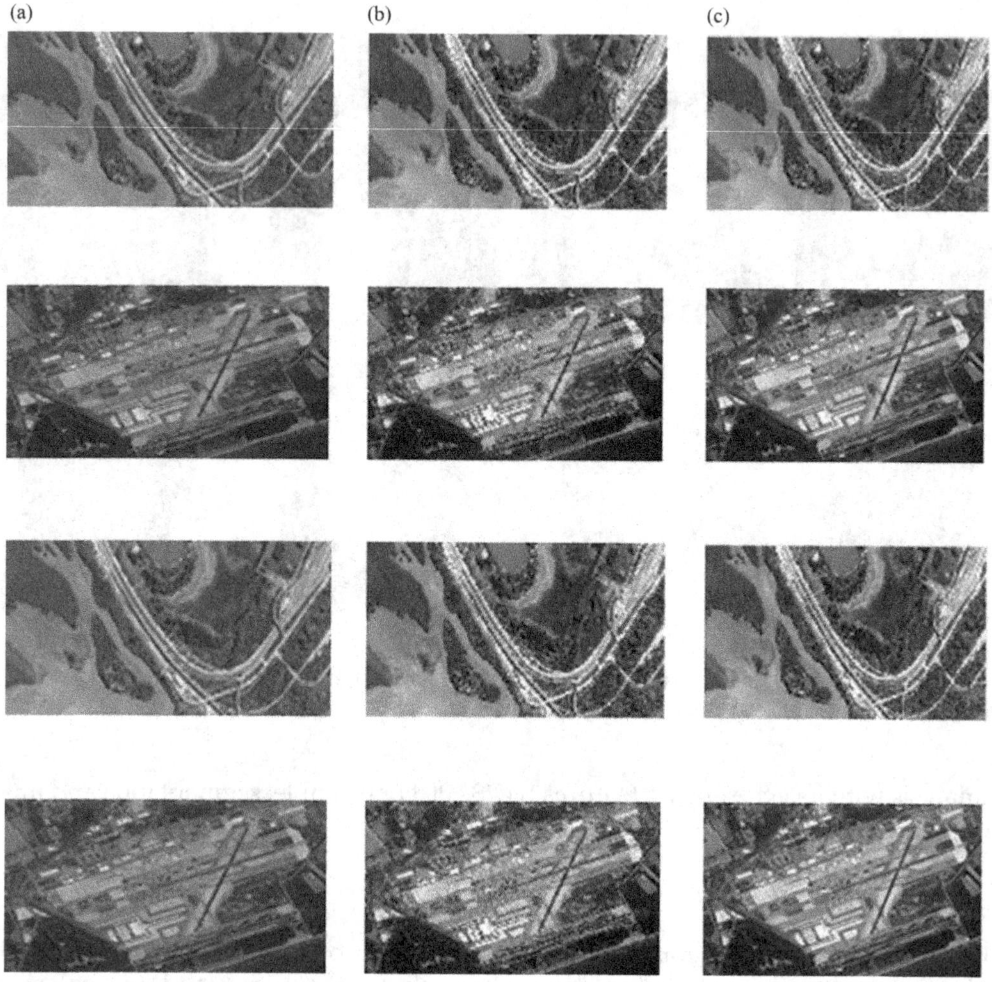

FIGURE 1.9
Satellite imaging (row wise): (a) test frame, (b) histogram-based technique, and (c) conventional technique.

can help in discarding the unwanted or damaged crop data and keep a record of relevant information on precise measurement as shown in Figure 1.9.

1.6.3.5 Forest Information

Drone and CV technologies are used for aerial view and collection of data. The data collected using this technology can be used for object detection in the forest such as yield estimation, tree counting and classification, stem drainage information, unused land and boundary of the forest, and animal counting as shown in Figure 1.10 [23].

1.6.4 Less Traffic Congestion

With the help of AI-powered CV technology, the government can monitor and detect the traffic flow and report abnormalities. It also provides valuable insight for the operation

FIGURE 1.10
Animal detection in the forest using AI-based drone technology in thermal images.

and management of suitable resources such as repairing of roads, snow removal, congestion due to animals, and lifesaving methods. This system can store traffic data to classify various vehicles on a road and also provide information about traffic flows (congestion, damaged road). This information can be further used to avoid that disruption [1,11,23,40].

1.6.5 Smart Parking

In smart cities, IoT-enabled technology is used for parking issues. This technology consists of inbuilt sensors that can detect the vacant parking space. All the data is gathered and transferred to a cloud server which results in easy access to a map of available space. The IoT sensors can detect the distance to its undercarriage in case of an occupied area. To scan the number plate of vehicles smart cameras and parking meters are used. Smart video surveillance is used by the government to mark parking points without the use of parking meters [1,2,3,6,11,23]. These cameras can detect the object or vehicles in the parking space and time duration. In this smart parking system, automatic number plate detection is used. With the help of this, all the free space data can be updated on a real-time basis. As a result, drivers can use the app to allocate the parking space as shown in Figure 1.11. Using this CV technology, driver identification, car model, number plate of a car can be easily scanned and any unwanted activities at parking space detected.

1.6.6 Object Detection and Tracking

Detection of motion-based objects is the first step in ML and image processing applications. This step involves the extraction of the most informative pixel from the video frame. In literature, various algorithms are present that can efficiently detect the moving object from a video frame [5–31]. This is a pivotal and initial step which is applied by several researchers for motion detection, unauthorized or illegal activities, traffic monitoring, medical imaging, industries, defense, and indoor–outdoor security aspects as shown in Figure 1.10. Figure 1.12 presents object detection and tracking using the BGS technique. This technology is often combined with IoT to detect multiple objects in a single frame such as a car, a person, or an animal.

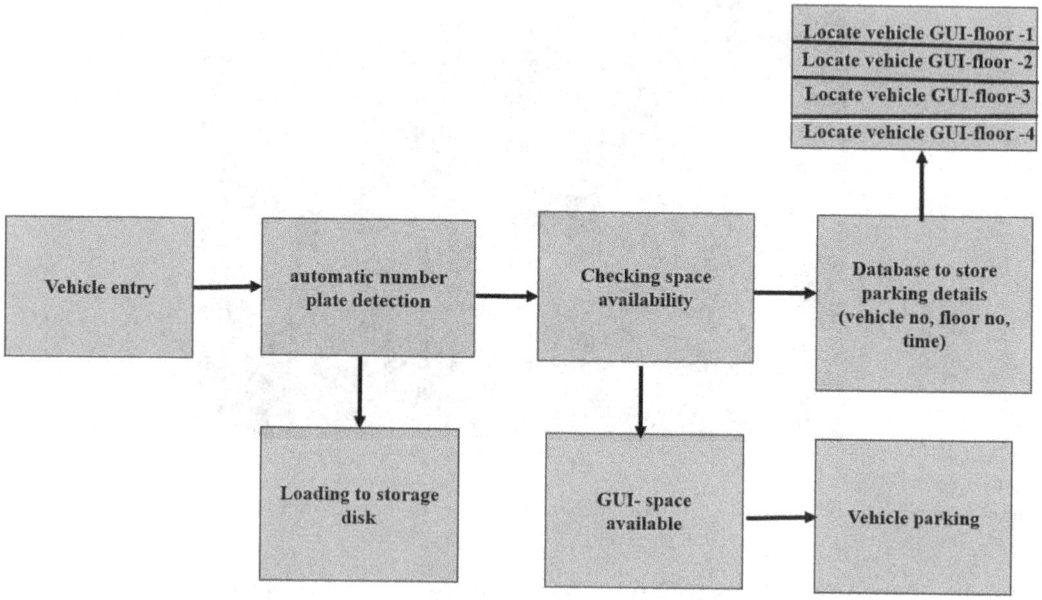

FIGURE 1.11
AI-powered CV technology for smart parking.

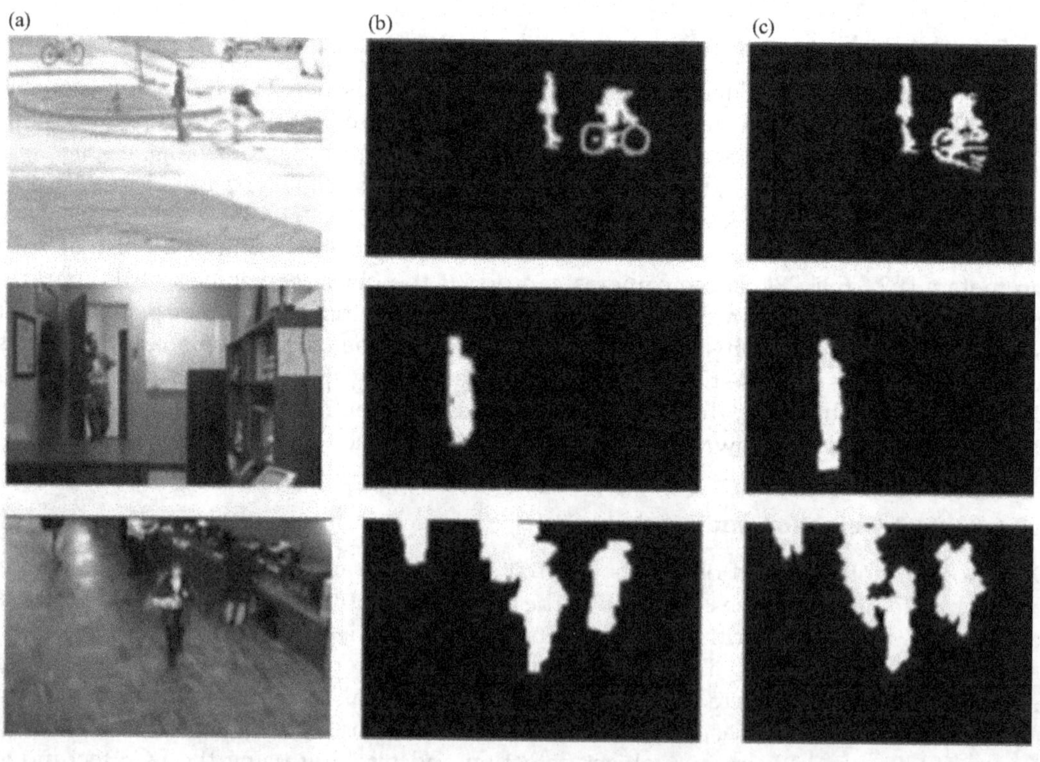

FIGURE 1.12
Detection of a motion-based object in indoor and outdoor sequence (row wise): (a) original image, (b) ground truth, and (c) desired output.

1.7 Conclusion

This chapter gives an overview of CV and its different kinds of techniques such as image segmentation, facial and pattern recognition, object detection & tracking, edge detection, and feature matching. This chapter also provides details of an evolution of CV and IoT, enhancement of IoT using CV and 5G technology, various challenging issues, and real-time applications such as automated cars, early diagnosis and detection of illness in the healthcare system, object detection (cars, animals, and humans) and tracking using BGS technique, drone technology in smart farming, satellite imaging in agriculture, facial recognition smart parking, and forest information.

References

1. Computer vision and IoT. Available at: https://www.wwt.com/article/four-ways-iot-ai-and-computer-vision-make-your-life-better [accessed on 20 June 2021].
2. Computer vision applications. Available at: https://www.sas.com/en_in/insights/analytics/computer-vision.html [accessed on 20 June 2021].
3. Computer vision applications. Available at: https://broutonlab.com/blog/iot-and-computer-vision-in-parkings [accessed on 20 June 2021]
4. Computer vision and IoT. Available at: https://www.iotforall.com/computer-vision-iot [accessed on 20 June 2021].
5. Computer vision applications. Available at: https://www.photonics.com/Articles/Machine_Vision_Makes_the_Move_to_IoT/a65141 [accessed on 20 June 2021].
6. Computer vision applications. Available at: https://www.smartcitiesworld.net/opinions/opinions/driving-autonomous-vehicles-forward-with-intelligent-infrastructure [accessed on 20 June 2021]
7. Akshit Anand, Vikrant Jha, Lavanya Sharma, "An improved local binary patterns histograms techniques for face recognition for real time application", *International Journal of Recent Technology and Engineering*, Volume-8, Issue-2S7, pp. 524–529, July 2019. Indexed in Scopus, ISSN: 2277-3878. DOI: 10.35940/ijrte.B1098.0782S719.
8. https://medium.com/swlh/computer-vision-in-agriculture-d84b69c6858e [accessed on 20 June 2021].
9. Computer vision applications. Available at: https://viso.ai/applications/computer-vision-applications-for-coronavirus-control/ [accessed on 10 July 2021].
10. Drone usage in Covid-19. Available at: https://www.unicef.org/supply/media/5286/file/%20Rapid-guidance-how-can-drones-help-in-COVID-19-response.pdf [accessed on 10 July 2021].
11. Lavanya Sharma, Pradeep K. Garg (Eds.), *From Visual Surveillance to Internet of Things*. Chapman and Hall/CRC, New York, 2020. DOI: 10.1201/9780429297922.
12. https://news.mongabay.com/2019/03/ai-and-drone-based-imagery-improve-survey-power-of-cryptic-animals/ [accessed on 20 June 2021].
13. Lavanya Sharma, Nirvikar Lohan, "Performance analysis of moving object detection using BGS techniques in visual surveillance", *International Journal of Spatiotemporal Data Science*, Volume-1, pp. 22–53, January 2019.
14. Lavanya Sharma, Dileep Kumar Yadav, "Histogram based adaptive learning rate for background modelling and moving object detection in video surveillance", *International Journal of Telemedicine and Clinical Practices*, June, 2016. ISSN: 2052-8442, DOI: 10.1504/IJTMCP.2017.082107.

15. Lavanya Sharma, Nirvikar Lohan, "Internet of things with object detection", In: *Handbook of Research on Big Data and the IoT.* IGI Global, pp. 89–100, March, 2019. ISBN: 9781522574323. DOI: 10.4018/978-1-5225-7432-3.ch006.
16. Lavanya Sharma, "Introduction", In: Lavanya Sharma and Pradeep K. Garg (eds) *From Visual Surveillance to Internet of Things.* Taylor & Francis, CRC Press, Boca Raton, FL, 2018, Vol.1, pp. 14.
17. Lavanya Sharma, Pradeep K. Garg, "Block based adaptive learning rate for moving person detection in video surveillance", In: *From Visual Surveillance to Internet of Things.* Taylor & Francis, CRC Press, Boca Raton, FL, Vol. 1, pp. 201.
18. Steve Makkar, Lavanya Sharma, "A face detection using support vector machine: Challenging issues, recent trend, solutions and proposed framework", In: Singh M., Gupta P., Tyagi V., Flusser J., Ören T., Kashyap R. (eds) *Advances in Computing and Data Sciences.* ICACDS 2019. Communications in Computer and Information Science, vol. 1046. Springer, Singapore, 2019. DOI: 10.1007/978-981-13-9942-8_1.
19. Lavanya Sharma, Pradeep K. Garg, "IoT and its applications", In: *From Visual Surveillance to Internet of Things.* Taylor & Francis, CRC Press, Boca Raton, FL, Vol. 1, pp. 29.
20. Lavanya Sharma, Dileep Kumar Yadav, Sunil Kumar Bharti, "An improved method for visual surveillance using background subtraction technique," *2015 2nd International Conference on Signal Processing and Integrated Networks (SPIN),* Amity University, Noida, India, Feb. 19–20, 2015, pp. 421–426. DOI: 10.1109/SPIN.2015.7095253.
21. Dileep Kumar Yadav, Lavanya Sharma, Sunil Kumar Bharti, "Moving object detection in real-time visual surveillance using background subtraction technique", *2014 14th International Conference on Hybrid Intelligent Systems,* Gulf University for Science and Technology, Kuwait, December 14–16, 2014, pp. 79–84. DOI: 10.1109/HIS.2014.7086176.
22. Lavanya Sharma, Annapurna Singh, Dileep Kumar Yadav, "Fisher's linear discriminant ratio based threshold for moving human detection in thermal video", *Infrared Physics and Technology,* Volume-78, pp. 118–128, Elsevier, March, 2016.
23. Lavanya Sharma (Ed.), *Towards Smart World.* Chapman and Hall/CRC, New York, 2021. DOI: 10.1201/9781003056751.
24. Lavanya Sharma, "Human detection and tracking using background subtraction in visual surveillance", In: *Towards Smart World.* Chapman and Hall/CRC. New York, December 2020, pp. 317–328. DOI: 10.1201/9781003056751.
25. Lavanya Sharma, Dileep Kumar Yadav, Manoj Kumar, "A Morphological Approach for Human Skin Detection in Color Images", *2nd national conference on "Emerging Trends in Intelligent Computing & Communication",* GCET, Gr. Noida, India, 26–27 April 2013.
26. Lavanya Sharma, Sudhriti Sengupta, Birendra Kumar, "An improved technique for enhancement of satellite images", *Journal of Physics: Conference Series.* Volume-1714, p. 012051, January 2021.
27. Supreet Singh, Lavanya Sharma, Birendra Kumar, "A machine learning based predictive model for coronavirus pandemic scenario", *Journal of Physics: Conference Series.* Volume-1714, p. 012023, January 2021.
28. Gouri Jha, Lavanya Sharma, Shailja Gupta, Future of Augmented Reality in Healthcare Department. In: Singh P.K., Wierzchoń S.T., Tanwar S., Ganzha M., Rodrigues J.J.P.C. (eds) *Proceedings of Second International Conference on Computing, Communications, and Cyber-Security.* Lecture Notes in Networks and Systems, vol. 203. Springer, Singapore, 2021. DOI: 10.1007/978-981-16-0733-2_47.
29. Gouri Jha, Lavanya Sharma, Shailja Gupta, E-health in Internet of Things (IoT) in Real-Time Scenario. In: Singh P.K., Wierzchoń S.T., Tanwar S., Ganzha M., Rodrigues J.J.P.C. (eds) *Proceedings of Second International Conference on Computing, Communications, and Cyber-Security.* Lecture Notes in Networks and Systems, vol. 203. Springer, Singapore, 2021. DOI: 10.1007/978-981-16-0733-2_48.
30. Sanjay Kumar, Priyanka Gupta, Sachin Lakra, Lavanya Sharma, Ram Chatterjee, "The zeitgeist juncture of "Big Data" and its future trends", *2019 International Conference on Machine Learning, Big Data, Cloud and Parallel Computing (COMITCon),* 2019, pp. 465–469. DOI: 10.1109/COMITCon.2019.8862433.

31. Shubham Sharma, Shubhankar Verma, Mohit Kumar, Lavanya Sharma, "Use of motion capture in 3D animation: Motion capture systems, challenges, and recent trends", *2019 International Conference on Machine Learning, Big Data, Cloud and Parallel Computing (COMITCon)*, 2019, pp. 289–294. DOI: 10.1109/COMITCon.2019.8862448.
32. Thierry Bouwmans, Fatih Porikli, Benjamin Horferlin, Antoine Vacavant, *Handbook on Background Modeling and Foreground Detection for Video Surveillance*. CRC Press, Taylor and Francis Group, Boca Raton, FL, 2014.
33. T. Bouwmans, N. Aybat, E. Zahzah, *Handbook on Robust Low-Rank and Sparse Matrix Decomposition: Applications in Image and Video Processing*. Taylor and Francis Group, Boca Raton, FL, 2016.
34. T. Bouwmans, B. Hoferlin, F. Porikli, A. Vacavant, "Traditional approaches in background modeling for video surveillance", In: Thierry Bouwmans, ,Necdet Serhat Aybat and El-hadi Zahzah (eds) *Handbook Background Modeling and Foreground Detection for Video Surveillance*. Taylor and Francis Group, Boca Raton, FL, July 2014.
35. T. Bouwmans, B. Hoferlin, F. Porikli, A. Vacavant, "Recent approaches in background modeling for video surveillance", In: *Handbook Background Modeling and Foreground Detection for Video Surveillance*. Taylor and Francis Group, Boca Raton, FL, July 2014.
36. J. Giraldo, T. Bouwmans, "GraphBGS: Background subtraction via recovery of graph signals", *International Conference on Pattern Recognition, ICPR 2020*, Milan, Italy, January 2021.
37. J. Giraldo, T. Bouwmans, "Semi-supervised background subtraction of unseen videos: Minimization of the total variation of graph signals", *IEEE International Conference on Image Processing, ICIP 2020*, Abu Dhabi, UAE, October 2020.
38. A. Griesser, S. De Roeck, A. Neubeck, "GPU-based foreground-background segmentation using an extended colinearity criterion", Vision, Modeling, and Visualization, VMV 2005, 2005.
39. A. Griesser, "Real-time GPU-based foreground-background segmentation", Technical Report 269, 2005.
40. Motion Based Object Detection using Background Subtraction Technique for Smart Video Surveillance. Available at: http://hdl.handle.net/10603/204721 [accessed on 20 June 2021].

2

IoE: An Innovative Technology for Future Enhancement

Sudhriti Sengupta
Department of Computer Science and Engineering, Galgotias University, Greater Noida, India

CONTENTS

2.1 Introduction

The Internet of Things (IoT) is the interconnection of physical devices or objects known as "things" for communication and data sharing. The concept of IoT is further extended into the Internet of Everything (IoE), which encompasses people, processes, data, and things in the network. This concept is used in not only large industrial tools but also small household appliances. Many devices such as kitchen appliances, cars, thermostats, baby-monitoring systems are connected to the internet via sensors, actuators, or other devices to produce better communication and decision-making capabilities. Development of low-cost and highly efficient sensors, connective medium, cloud and fog infrastructures, and data analytics tools helped in developing the IoE technology rapidly. It is estimated that by 2025, 22 billion devices will be connected by IoT [1,11,15–19]. In this chapter, we will discuss the overview of IoT, its architectural framework, its role, and security issues in IoT. The difference and similarity between IoT and IoE are also discussed. The issues concerning the security aspects of IoE along with the role of IoE in various domains are discussed. A case study of smart city development in France is mentioned to show the usage and impact of IoE.

2.2 IoT and Its Present Perspectives

The interconnection between computing devices, machines, objects, and human beings is brought together to facilitate the transfer of data and communication by a network ID

DOI: 10.1201/9781003244165-3

called IoT. The primary point in IoT is called "things," which are sensors, software, and allied technology to facilitate communication. IoT spans from mundane household objects to large or highly complex industrial objects. IoT has widely evolved because of multidisciplinary fields such as machine learning, sensors, and embedded system. IoT products have opened the field of smart homes and smart cities. Home automation and smart homes are IoT device-implemented systems which can provide automatic lighting, security management, room temperature automation, etc. A smart home is an interface that controls and uses devices in a general household for controlling different aspects of livelihood leading to a better lifestyle. An example can be taken from Homekit produced by Apple Inc., which controls different applications like Siri. Similarly, IoT has paved the way for almost every aspect such as agriculture, healthcare, industry, and transportation. A smart city means a city where technology is induced in various aspects such as monitoring of traffic, collection of garbage, lighting of street lamps, and providing helplines. Telensa is one of the major agencies, which pivots the development of smart cities [2,17–19].

The architecture framework of IoT consists of four general modules: sensors/actuators, data acquisition system, Edge IT, and data cloud center. The information and data from the environment are collected by the sensors or actuators. Actuators manipulate the external environment and send the data to different gateways. These constitute the end point of IoT (Figure 2.1).

Sensors or actuators generally collect the data in analog form. Data acquisition system converts these analog data into digital form. This system also performs aggregation and conversion of data. It supplies the internet gateways with these converted and summarized data. Data are generally saved in cloud infrastructure by using the Edge IT aspects, which is decentralized and hence more secure. The data are sent to the data cloud center for processing and storage.

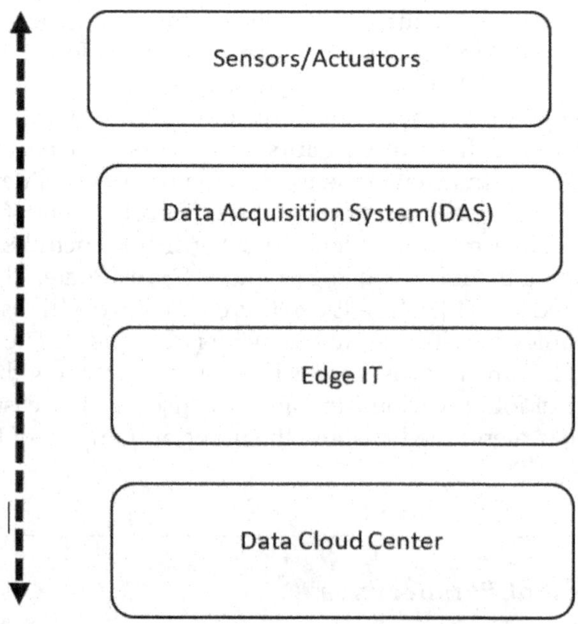

FIGURE 2.1
Architectural framework of IoT.

IoT was introduced in 1999, and since then it has paved the way to make life easier, simpler, and automatic. IoT is used in security, healthcare, energy-saving facilities, water supplies, traffic control, transportation, etc. Along with the immense benefit of IoT there is a need to have an enhanced and elaborate system of security in IoT as well. There are some differences between security in IoT and security in Traditional IT systems. Traditional IT accesses the data by using defined links, whereas in IoT data are produced by machines forming machine-to-machine communications. Human users are the source of usage and consumption of data in IT [16,17,19]. Some of the challenges in IoT security are as follows:

- In IoT, the variety and quantity of devices are significantly larger.
- Volume of data generated and consumed by IoT is comparatively larger.
- Failure in IoT security may lead to very severe conditions even if it supports a system like healthcare.

Though there are challenges in IoT infrastructure such as providing security, the benefit and role of IoT in the modern world toward creating smart cities, smarter lifestyles, and other things have led the researchers in paving the way for an extension of IoT into a bigger domain which is called IoE. The next section discusses the concept of IoE [3,4].

2.3 IoE—Its Role and Responsibility

IoE brings together people, things, and data focusing on networking devices equipped with sensors. It refers to the devices, products, and people using them as an interconnected module which facilitates in data collection, analysis, storage, and transmission. It is a concept of information interchange focusing on the association of things to one another and their uses. The main components of IoE are people, things, data, and processes. People are the users of the internet or the needs used for sharing data, such as social networking sites and sensors. Physical devices such as gadgets, sensors, and actuators are used for collecting and transmitting data. Data encompass both raw and processed data, which are useful in providing efficient decision-making [5,6].

The main goal of IoE is to increase the efficiency of operation, create new business scope, and improve the lifestyle of the human being. For instance, a person is not sure whether he has closed his gas valve at home before leaving his house. An IoE solution enables the user to automatically enquire about the gas valve status remotely. Some of the applications of IoE are checking of remote health care, usage of smart grid for increased network connection, building better experience to enhance relations between consumers and marketers/retailers, etc. Despite the rewarding scope and opportunities associated with IoE, it could produce significant security risks and threats. Cybercrimes or cyberattacks on IoE devices and their framework produce significant risk or potential damage to the domain of IoE including healthcare and security. One of the noteworthy attacks on IoE was the Mirai botnet which caused "DDoS attack on Dyn" and even brought down the whole system. This resulted in the malfunctioning of sites like CNN and Netflix [7]. IoE unites individuals, cycles, information, and things to make arranged associations more significant and important than any other time-transforming data into activities that make new capacities, more extravagant encounters, and phenomenal financial chances for organizations, people, and nations [8,9].

IoE was recorded as one of the top patterns in 2015 by Gartner. Cisco characterizes IoE as the organized association of individuals, cycles, information, and things. Getting advantage from the compound effect of interfacing individuals, cycles, information, and things, IoE-expanded connectedness makes "everything" possible on the web. IoE brings remarkable opportunities for associations, people, networks, and nations to acknowledge drastically more noteworthy incentives from organized associations among individuals, cycles, information, and things. IoE is the forthcoming imaginative and ubiquitous innovation progression which will make organized associations more applicable and significant. Transforming data right into it makes new abilities, more extravagant encounters, and uncommon monetary freedoms for organizations, people, and nations.

IoE alludes to billions of gadgets and purchaser items associated with the web in a wise organized climate with extended advanced highlights. It is fundamentally a way of thinking in which our innovation future is undermined by various kinds of machines, gadgets, and things associated with the worldwide web. As of now the web association is simply limited to phones, tablets, PCs, and a small bunch of different gadgets [10]. In more straightforward terms, IoE is the astute association of individuals, cycles, information, and things that will change our reality so that there will be billions of associated gadgets having sensors to recognize, measure, and access their status which will all be associated over open or private organization worked over standard conventions like TCP/IP. IoE carries with it the organization's insight to tie these ideas together into a firm framework. There are four pillars of IoE: individuals, things, information, and cycle. Figure 2.2 depicts the pillars of IoE.

People generate data which are analyzed by suitable processes. These analyzed data are presented to things which produces applications to be used in a cohesive manner.

A. **Individuals**

Individuals will be associated with the web in more important ways, will produce information, and will be cooperating with gadgets by using mobiles, tablets, PCs, and social networks. In addition, through sensors set on human skins or

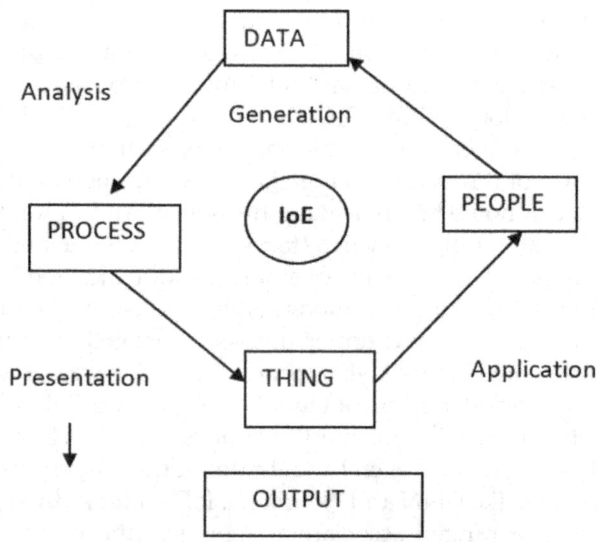

FIGURE 2.2
Pillars of IoE.

sewn into garments individual's crucial signs can be collected. Along these lines, individuals will themselves become nodes on the web. A genuine model may be Nike's wearable wellness bands which read an individual's fundamental signs.

B. **Things**

Things and actual things, for example, sensors, industry gadgets, customer items, undertaking resources will be associated with the web or to one another, additionally getting data from their surroundings, will be mindful of the settings, more intellectual, more astute, frequently purported to be web of things.

C. **Information**

Instead of collecting essentially raw information, these associated gadgets will send more significant level, more prepared information back to separate workers for quicker evaluation or more astute assessment. Here the information is more about sagacious data and activity plans than simply irregular pieces. Classifying out an approach to translate the correct progression of data is the way for utilizing the Big Data and as the kinds of information and sources increment, to draw helpful knowledge there will be a need to arrange data and examine them.

D. **Cycle**

Compared with the IoE process, the correct data will be conveyed to the right individual at the right time in a proper manner. Information will help innovations in business to settle on additional choices and advance their work process cycles and techniques and will be along these lines contending to use the information quicker than their rivals for a lithe and quicker dynamic.

2.4 Interplay between IoE and IoT

Though the terms IoT and IoE are similar, they differ in their basic definition. IoT is the connection of physical entity between things, whereas IoE comprises people, things, data, and processes. IoT is the network of devices used to collect and estimate data with human interference [7,8]. IoE is the next-generation IoT, connecting people, processes, data, things for providing decision-making capabilities by using information and communication. IoT can be defined as a subset of IoE. In IoT, the communication occurs between machine and machine, whereas in IoE, communication occurs between machine and machine, machine and people, and technology and people. It is more complex than IoT. Examples of IoT usage are wearable monitor sensors for health checking, smart services, etc. Some examples of IoE applications are connecting food and people to the supply chain and monitoring of traffic activity to ensure no congestion or delay in case of availing emergency services such as ambulance and fire engines [10–14] (Figure 2.3).

The main components of IoT are networks and things without any people intervention (Figure 2.4).

People, things, processes, data, and networks are the main components of IoE.

Although IoT and IoE are different terms, there are similarities:

- **Decentralization**

  The two frameworks are dispersed and don't have a solitary community; every hub fills in as a little administration place and can play out specific errands autonomously.

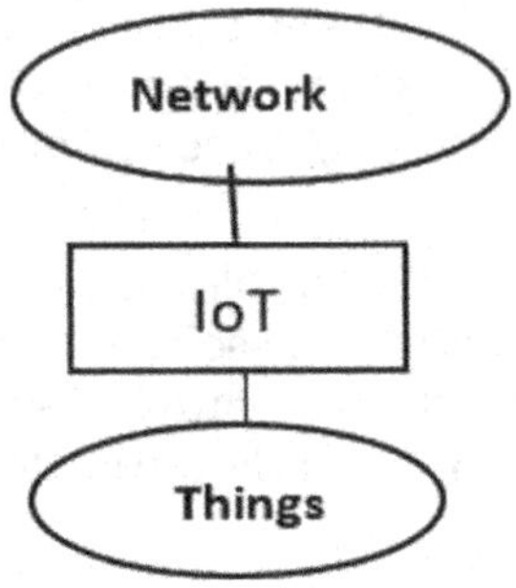

FIGURE 2.3
Components of IoT.

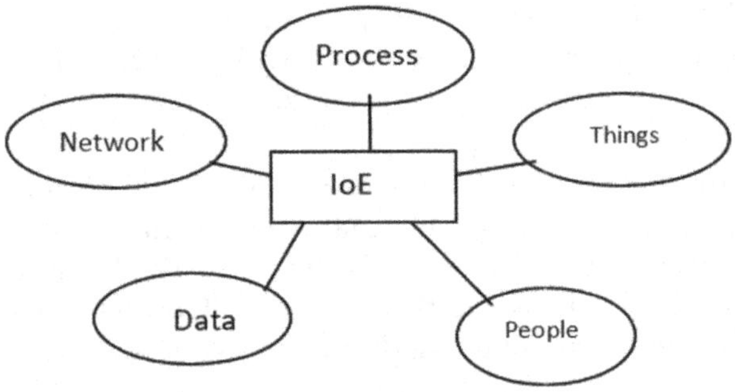

FIGURE 2.4
Components of IoE.

- **Security Issues**

  Disseminated frameworks are still exceptionally helpless against infiltration and cyberattacks; the more gadgets are associated with the organization, the higher the vulnerability to breaks.

On the one hand, decentralization is one of the IoE and IoT focal points, since the entire framework doesn't disseminate regardless of whether there are issues in two or three hubs. On the other hand, such an appropriation causes inconveniences and might be dangerous to information security and individual protection.

2.5 Security in IoE

The correct model for IoE security will empower associations to appreciate the advantages of IoE while keeping a significant level of information protection and insurance, and guaranteeing dependable, continuous help. The model comprises three columns that interface with each other—deceivability, danger mindfulness, and activity.

Danger mindfulness works with the undefined edge, assuming trade off and sharpening our capacity to distinguish dangers depending on getting typical and irregular conduct, recognize markers of a bargain, decide, and react quickly. This requires conquering unpredictability and discontinuity in our surroundings. When we distinguish a danger or odd conduct we need to make a move. It will lead to correct advances, cycles, and individuals cooperating—and quickly—to be successful [10–19].

Moving toward a predictable goal led to adjustments fully expecting potential dangers isn't simple, yet it's fundamental. So, security groups need to get innovative. As of now, it's excessively costly and too clumsy to even think about monitoring each and every organization association. Security groups are also reliant on gadgets that radiate information that can be devoured by another gadget. The objective is to implant security observable and control into as many gadgets under IT's influence as could be allowed and join this with current organization approaches, making the organization a tremendous, extensible sensor. The role and application of IoE are evident in the current world with the rise of internet users and subsequently IoE devices. There is a significant risk in IoE security. This section aims to discuss certain risks and also possible steps to mitigate the risks. One of the risks associated with IoE is to find any kind of data which is sensitive but not encrypted. The attackers will have the opportunity to modify or delete the data leading to resource hampering and loss.

Another risk associated with IoE is a denial-of-service attack for slowing down or preventing data access. Compromised key attacks and password-based attacks are also risks to IoE. In the first case, the private key of the encryption algorithm is stolen making the cryptographic algorithm a failure. Password-based attacks are used to break into the network on a device by guessing or applying algorithms like a brute force to guess the password. Another risk associated with IoE transmission is a man-in-the-middle attack, where a third party accesses the data transmitted between two groups. There are several steps to reduce this attack. Some of them are as follows:

1. Use of all security features present in the device.
2. Updating IoE-based products regularly to install patches which will make the system more secure.
3. Use of a strong password consisting of a complex sequence of alphabets, numbers, etc.

2.6 Role and Importance of IoE

These days, IoE assumes a significant part in different areas such as home automation, smart cities, education, industry, healthcare, business, innovation, and agriculture. The usage of IoE innovation in the agricultural framework makes the farmers screen their agricultural fields and yields and controls things distantly from their mobile phones. The different wireless sensor network can detect the parameters and send the deliberate and observed information to the ranchers through IoE organization. Similarly, things can be controlled adroitly. This aids in applications such as soil dampness and supplements. Sensing, reporting climatic conditions, and custom compost profiles upheld soil science, controlling the use of water for ideal plant development. It additionally incorporates ranch vehicles, stockpiling, and so on. In the next subsection, the usage of IoE in the area of healthcare, retail, transportation, energy management, and manufacturing is discussed. According to Cisco, IoE will have a direct effect on sectors such as healthcare, retail, and transport. These sectors embrace the likelihood of connecting the pillars of IoE, viz., people, processes, data, and things [10–17].

- **IoE in Healthcare**
  Healthcare industry requires reliable, robust, and effective technology for communication so that patients will be monitored easily and remotely. Also, diagnosis and treatment of disease by specialist doctors can be enabled by this technique. Medical and healthcare providers will be able to monitor the prognosis of the disease remotely and constantly [11,17–19].
- **IoE in Retail**
  IoE has enabled applications such as dynamic pricing, smart shelves, inventory management, and customer-oriented advertising to usher to the new age of retailing both in online and offline modes. IoE promises to play a key role in developing retail facilities in various sizes and domains of retail outlets from a supermarket to small shops to a chain of shopping brands. This technology also plays an important role in the optimization of logistics in retail.
- **IoE in Transportation**: IoE aims to bring together modes of transportation such as railways, vehicles, and water vehicles to create an autonomous network to enable easier transport over a greater distance. IoE enables smart cities to remotely observe the physical condition of roads, tracks, highways, etc. The journey planning application can utilize the real-time data collected and analyzed by IoE. Smart devices can be used for measuring the flow of traffic, predicting congestion, and adaptively controlling traffic paths.
- **IoE in Energy Management**
  As per Cisco, a smarter device can converse more energy which will make energy conversion optimal. Constant monitoring of energy-consuming devices will lead to energy saving. This technology will lead to the development of an interconnection system of energy infrastructure parts used for loading, storing, and distribution of energy. The utilization of IoE in the energy sector and smart grid brings forward the development of Internet of Energy paradigm.
- **IoE in Manufacturing**
  In manufacturing firms and industries an intertwining of hardware and software with the internet creates a smarter environment for the creation and innovation of products. In manufacturing sectors, optimized production and quality of products are very crucial for successful investment. IoE can process smarter production lines for monitoring the production cycle and point deficit in the process.

The above points are some of the areas where IoE can be used. IoE can also be used in various other fields such as logistics, agriculture, car parking, smart cars, and home appliances. IoE has an impact not only on large-scale industries but also on the smaller households.

2.7 France: IoE Smart City Pilot

Cisco is joining forces with the city of Nice, France, and a few nearby and other industry accomplices to construct a smart city for additional development through the introduction of the effects of IoE for urban communities. The undertaking's fundamental goals are to test and approve an IP-empowered innovation design and financial model, just as to decide the social advantages of IoE. The task depends on a shared stage intended to be more adaptable, granular, and versatile to create metropolitan working frameworks. The shared stage is proposed to make it simpler to set up the new associations that are basic for Nice to turn into a smart city. Also, the undertaking will fill in as an impetus for joining key revelations from this and other smart city activities. The aim is to share what Nice has realized with other hopeful urban areas so that they can make their own smart city structure. The undertaking incorporates four city benefits that can quickly show the advantages and estimation of IoE for the two inhabitants and city authority. As these arrangements are actualized, Cisco and the city of Nice are surveying how accumulated information can be utilized to make data setting explicit and helpful across various administrations. For example, information caught by sensors for traffic designs can help to impart traffic signals automatically [10–19]. The ramifications of information "crossfertilization" and cross-joint effort go past mechanical possibility because they additionally sway the choices of city supervisors, cross-departmental coordinated effort, and back-office activities.

2.8 Conclusion

The concept and introduction of IoE enable connecting people, processes, data, and physical things to usher the development of a lifestyle of people and advancement in various fields such as manufacturing, retail, and healthcare. IoE has preceded its precursor IoT in its impact in real-time application. A case study of IoE usage in France is discussed. Also, the security aspects of IoE are mentioned as an area which requires significant attention.

References

1. https://www.oracle.com/in/internet-of-things/what-is-iot/.
2. D. Kohli, S. Sengupta, "Recent trends of IoT in smart city development", in *Proceedings of the International Conference on Computer Networks*, 49, 275–280, 2020.
3. S. Sen Gupta, M. Shad Khan and T. Sethi, "Latest trends in security, privacy and trust in IOT", in *2019 3rd International conference on Electronics, Communication and Aerospace Technology (ICECA)*, 2019, pp. 382–385, doi: 10.1109/ICECA.2019.8822178.
4. S. Tyagi and S. Sengupta, Role of AI in gaming and animation. Lecture Notes on Data Engineering and Communication Technology, Springer, vol. 49. pp. 259–267, 2020.
5. S. Higginbotham and M. Pesce, "Internet of Everything: Macro & Micro", *IEEE Spectrum*, vol. 58, no. 2, pp. 20–21, February 2021, doi: 10.1109/MSPEC.2021.9340118.
6. I. Bandara and F. Ioras, "The evolving challenges of internet of everything: Enhancing student performance and employability in higher education", in *10th annual International Technology, Education and Development Conference*. Vol. 10, pp. 121–131, 2016.
7. What is Mirai Botnet? Accessed at https://www.cloudflare.com/engb/learning/ddos/glossary/mirai-botnet/ on 04th Feb 2021
8. M. Miraz, M. Ali, P. Excell, and P. Rich, "A review on Internet of Things (IoT), Internet of Everything (IoE) and Internet of Nano Things (IoNT)", in *Proceeding of Internet Technologies and Applications*, Wrexham, UK, 2015.
9. A. Rustagi, C. Manchanda, and N. Sharma, "IoE: A boon & threat to the mankind", in *2020 IEEE 9th International Conference on Communication Systems and Network Technologies (CSNT)*, Gwalior, India, 2020, pp. 114–119, doi: 10.1109/CSNT48778.2020.9115748.

10. S. Sengupta and A. Rana, "Role of bloom filter in analysis of big data", in *2020 8th International Conference on Reliability, Infocom Technologies and Optimization (Trends and Future Directions) (ICRITO)*, 2020, pp. 6–9, doi: 10.1109/ICRITO48877.2020.9197859.
11. L. Sharma, "Introduction", *From Visual Surveillance to Internet of Things*, Taylor & Francis, CRC Press, Vol.1, pp. 14.
12. L. Sharma and P. K Garg, "Block based adaptive learning rate for moving person detection in video surveillance", *From Visual Surveillance to Internet of Things*, Taylor & Francis, CRC Press, Boca Raton, FL, Vol. 1, pp. 201, 2018.
13. L. Sharma and P. K Garg, "IoT and its applications", *From Visual Surveillance to Internet of Things*, Taylor & Francis, CRC Press, Boca Raton, FL, Vol. 1, pp. 29, 2019.
14. L. Sharma (Ed.), *Towards Smart World*, Chapman and Hall/CRC, New York, 2021. doi: 10.1201/9781003056751.
15. Object detection. Available at: http://hdl.handle.net/10603/204721 [accessed on 20 august 2021].
16. L. Sharma and N. Lohan, "Internet of things with object detection", in *Handbook of Research on Big Data and the IoT*, IGI Global, pp. 89–100, March, 2019. (ISBN: 9781522574323, DOI: 10.4018/978-1-5225-7432-3.ch006)
17. S. Singh, L. Sharma, and B. Kumar, "A machine learning based predictive model for coronavirus pandemic scenario", *Journal of Physics: Conference Series*. Volume-1714, p. 012023, January 2021.
18. G. Jha, L. Sharma, and S. Gupta, Future of Augmented Reality in Healthcare Department. In: Singh P.K., Wierzchoń S.T., Tanwar S., Ganzha M., Rodrigues J.J.P.C. (eds) *Proceedings of Second International Conference on Computing, Communications, and Cyber-Security*. Lecture Notes in Networks and Systems, vol. 203. Springer, Singapore, 2021. DOI: 10.1007/978-981-16-0733-2_47.
19. G. Jha, L. Sharma, and S Gupta, E-health in Internet of Things (IoT) in real-time scenario. In: Singh P.K., Wierzchoń S.T., Tanwar S., Ganzha M., Rodrigues J.J.P.C. (eds) *Proceedings of Second International Conference on Computing, Communications, and Cyber-Security*. Lecture Notes in Networks and Systems, vol. 203. Springer, Singapore, 2021. DOI: 10.1007/978-981-16-0733-2_48.

3

An Overview of Security Issues of Internet of Things

Lavanya Sharma
Amity Institute of Information Technology, Amity University, Noida, India

Sudhriti Sengupta
Department of Computer Science and Engineering, Galgotias University, Greater Noida, India

Nirvikar Lohan
University of Patanjali, Uttarakhand, India

CONTENTS

3.1 Introduction

The Internet of Things (IoT) is a collective term for the field of study of multiple devices connected through the internet. It also comprises devices which are not part of the conventional internet. A wide range of services provided by IoT devices makes lives easier by helping us in many aspects of modern lifestyles, such as energy management, medical and healthcare, business, education, manufacturing industry, and personal mobile device. With the expansion of IoT in various spheres of our lives, our data are now more vulnerable and susceptible to fall into undesirable hands. It is no longer restricted to leakage of personal data but more deeply connected to our non-virtual lives. The biggest challenges to our security are to protect the system from data loss, theft, unauthorized access, and physical damage (mishandling and attaching malicious devices). Overcoming these threats, we must maintain data confidentiality and integrity of the stored information [1,2,5,7,8,34,35]. Linux Foundation is also working on an operating system project for IoT and will be supported by Intel and other companies in the industry. With this project IoT will achieve a new level. According to the most recent analytics, the valuation of various IoT industries would be around $1.9 trillion by 2020 [35–37,39] (Figure 3.1).

DOI: 10.1201/9781003244165-4

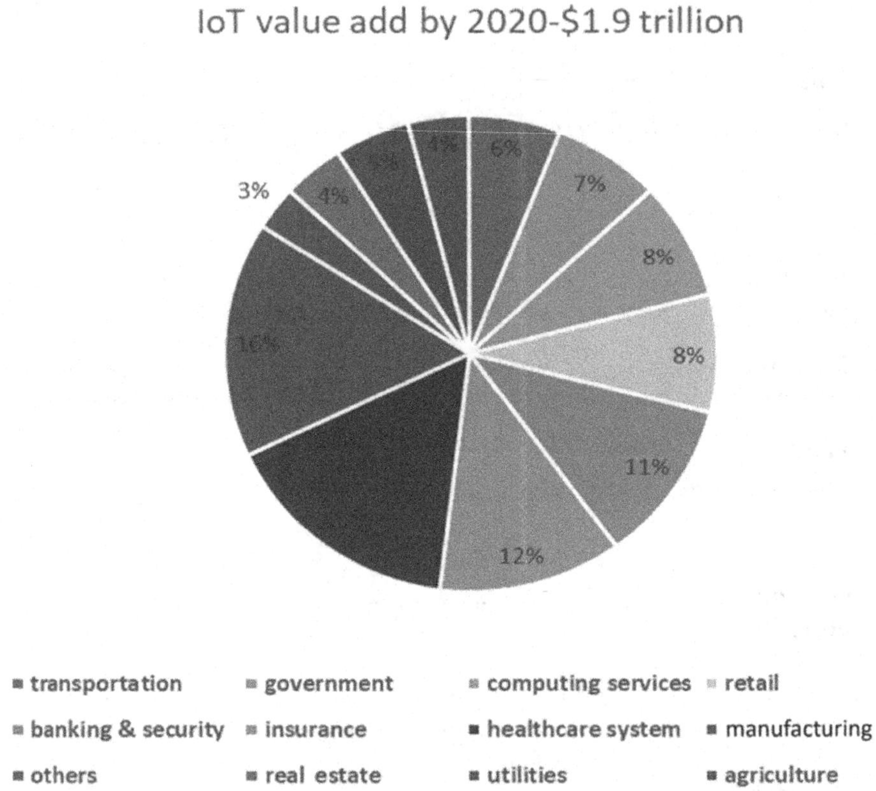

FIGURE 3.1
Valuation of various IoT industries in 2020.

Depending on the type and size of an organization or the position of the person, the threat can be of a less or more severe type. These threats are connected not only to our online profile or social life but also to our daily non-virtual life. The data transmitted over the internet can contain our personal/private data [38,39,41]. Some of those examples are footage from personal devices (mobile, camera, and laptop), CCTV, data from our fitness or medical devices such as fitness bands, medical aid devices, government records about public (Aadhaar card data, PAN card data, and bank details), or even military data [1,2] as shown in Figure 3.2.

This chapter is categorized into seven sections. Section 3.1 deals with the introductory part of IoT and its security aspects, whereas Section 3.2 discusses related work. In Section 3.3, we define IoT devices. In Section 3.4, we define what security means for IoT devices. In Sections 3.5 and 3.6, threats to the security of IoT devices and purposes of attacks are described. In Section 3.7, the classification of types of intruders is given. In the last section, the conclusion of the work is discussed.

3.2 Literature Review

Gubbi et al. [3] discussed IoT with a cloud-centric vision. Some application areas and key technologies help paving a way for research in the field of IoT. Lee et al. [4] discussed essential technologies for the deployment of IoT. Also, some IoT fields are useful in the

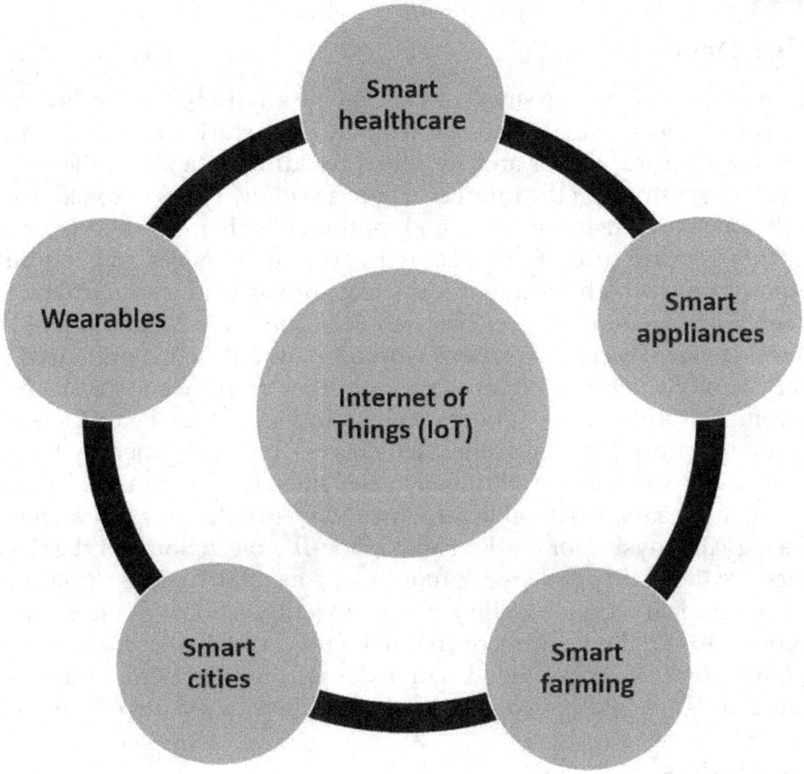

FIGURE 3.2
Sectors of Internet of Things (IoT).

commercial sector and customer relations. Dieter et al. [5] discussed the implementation of IoT used for monitoring domestic conditions with a low-cost sensing system. They also described the network architecture and mechanisms for the measurement of parameters by sensors. Xu et al. [6] described IoT CAD security techniques. Farooq et al. [7] analyzed security issues and provided an architecture of security to be adopted by a larger section of people. Mahmoud et al. [8] presented a survey and a detailed analysis of IoT security concerns. Riahi et al. [9] presented a new approach for designing new security mechanisms and their deployment. The author gives a complete outline of the approaches and attempts to find compatible ways of deploying them. Wurm et al. [10] discussed a detailed security analysis procedure on home automation systems for diagnosing security vulnerabilities. Many security mitigations and solutions are also discussed. Hwang et al. [11] described concerns and threats for privacy and security in services of IoT. They also provided an approach to solve these issues in the industrial field. Nawir et al. [12] presented security matters of network health care, transportation, and healthcare domains. Chaabouni et al. [13] discussed classifications of security threats and challenges to IoT. Ahmad Khan et al. [14] surveyed security issues in IoT and reviewed popular security issues. Al-Garadi et al. surveyed different methods to enhance security in IoT. Miettinen et al. [15] discussed a new system for the identification of devices. Minoli et al. [16] discussed challenges in the deployment of IoT. Blythe et al. [17] discussed the information value of the consumer security index.

3.3 Smart IoT Devices

An IoT device, also referred to as a smart device, can be anything such as home appliances, medical healthcare devices, vehicles, homes, workshops, factories, and cities. Anything can be attached with a microprocessor and sensors, providing data about the real world and transferring those data through the internet. There are many types of sensors (e.g., temperature, humidity, pressure, distance, light, and motion) which are embedded in the device. An IoT device can be configured to interact with other IoT devices and computers. These devices communicate through various means (e.g., broadband, cellular data, and Wi-Fi) [5,6]. Power supply to these IoT devices plays an important role in mobility or rigidity. For example, a small device which is capable of working without a constant wired power supply can be very handy. Such devices are generally preferred by consumers because they are more convenient. Other types of bigger things in IoT include healthcare devices (e.g., CT scanner, monitor), buildings, and cities, which are rigid and generally have a constant wired power supply. Also, there are things which constantly move and also have a wired power supply such as cars, bikes, and airplanes [5,7]. IoT devices can also be classified whether it is a logical/physical or an IP-enabled/non-IP object. Some of the characteristics of an IoT device are the ability to sense, actuate, and control the energy/power and its connection with the physical world, mobility, and connectivity. Some devices are required to be fast and robust [40–42]. Some are required to be precise, while some are required to be long-lasting. Some devices are provided with external security factors (cases, covers, and triggers), while others are totally exposed. Some examples of IoT devices are as follows:

- **Wearable Devices**
  Fitness bands like Google Home [18] and smartwatches like Apple watch [19] and Samsung Galaxy Gear [20].
- **Amazon Echo**
  It is a hands-free speaker which is connected to a cloud-based voice service [21].
- **Philips Hue**
  It is a smart home lighting system which can be controlled remotely and can sense time and day to adjust lights accordingly [22] (Table 3.1).

3.4 Major Security Issues of IoT Devices

Security in IoT devices includes protecting information and data, hardware components, and services of the device from unauthorized access. Both the data and information stored in the device and those in transit should be protected [16,18]. The major problems with IoT devices are identified as follows:

- **Data Integrity**
  The integrity of data is defined as the assurance of maintaining data accuracy and consistency throughout the storage lifecycle [23].
- **System Security**
  This issue mainly focuses on the overall security of IoT systems to detect various security issues to design various security frameworks and offer appropriate security guidelines to maintain the security of a network.

TABLE 3.1
Major Organizations for IoT Devices

Organization	Smart Devices	Characteristics
TP-Link [32]	Smart Wi-Fi routers, Wi-Fi LED bulbs	It supports a whole range of IoT devices such as Google Home and Amazon Alexa, has a touch screen display, and works on both 5 GHz and 2.4 GHz.
Apple [19]	HomeKit, Smart Watch	Users can communicate with all connected devices with their app. Smart Watch can track activities and connect with your phone.
Samsung [20]	Galaxy Gear Watch, Samsung Connect Home, SmartThings Hub	Gear series smartwatches started the smartwatch trend. Samsung Connect Home and SmartThings Hub are smart home devices and can connect with multiple devices without a wire.
Google [18]	Google Home, Chromecast	Voice command functions, video wireless casting. Cast videos from various platforms such as Android, Windows, Mac, and Linux.
Philips [22]	Philips Hue wireless LED lighting solutions	Philips wireless lighting connects with your phone with an app to control the light according to your choice and conditions.
August [33]	Smart Lock, Doorbell Cam	It provides features for home security and IoT like DoorSense.

- **Authorization**
  The process of granting privilege and specifying access rights is known as authorization [24,25].
- **Application Security**
  This security works for the application to manage security challenges or issues as per situation constraints. In general, security evaluation at the application level prevents data hijacking within the app such as hardware or software that minimizes security vulnerabilities.
- **Data Confidentiality**
  The practice of keeping private data secret is known as data confidentiality [23–25].
- **System Vulnerabilities**
  A lot of work is done by researchers in software vulnerability. Various IoT devices have low-quality software susceptibility to different types of vulnerabilities which are common in the early 2000 and late 1990s. These devices are vulnerable to weak usage of cryptography, authentication, deployment issue, system software (s/s) exploits, and so on.
- **Network Security**
  This security handles communication attacks on the data which can be transmitted between servers and IoT devices.
- **Lack of Common Standard**
  There are various standards for IoT device-manufacturing companies. Therefore, it becomes a major challenging issue to differentiate between authorized and non-authorized devices connected to the internet.

This defines some fundamental problems in IoT devices. The user accessing the device and its services should be properly authorized in order to view, modify, or add any kind of data to the device storage. An IoT device should be able to authorize the person to access the device. Hence, access and authorization control become necessary factors for establishing a secure connection between multiple devices and their services. Privacy protection is an imperative issue in IoT gadgets and administration because of the universal character of the IoT condition [10,17,19]. Elements are associated, and information is conveyed and transmitted over the web, making client protection a delicate subject in many research works. Protection in information accumulation, just as information sharing and its management, and information security matters stay important issues to be updated.

3.5 Threats to Security

A threat is an action that takes advantage of security weaknesses in a system and has a negative impact on it [26,31,43]. There are two types of primary sources: nature and humans. Threats by humans are the kind of threats which are caused by humans. These can be either internal (people with authorized access) or external (people without authorization). It involves people with malicious intent such as stealing, interrupting, hijacking, deleting, modifying, and changing the data that can be stored in a device, in transit, or on a cloud as shown in Figure 3.3. Natural threats include damage to the system caused by natural calamities such as earthquakes, lightning and thunder, flood, fire, and hurricanes [19,20,22].

3.5.1 Vulnerabilities

Vulnerabilities are the weak spots or backdoors in the devices. These are the weaknesses in the system or its design. Weaknesses in the system can be found in multiple areas of the

FIGURE 3.3
Use of early warning system in an El Salvador village.

system. These vulnerabilities are often those areas which are not properly tested or areas with a backdoor. For example, a poorly written code can provide access to unauthorized personal details without alerting the system [28,29]. In 2014, a security vulnerability in St. Jude's pacemakers was reported and confirmed by FDA. The vulnerability gave intruders direct access to the device without a wire. Figure 3.4 shows the pacemaker device and its implementation.

3.5.2 Attack

An attack is an action which has a malicious intent to harm or disrupt the system by exploiting its vulnerabilities using tools and techniques (Figure 3.5). An attacker or intruder can, in many ways, be harmful to an individual device or a whole system of connected devices [20,26,30]. An attack can happen in many forms. Below are some of the common forms of cyberattacks:

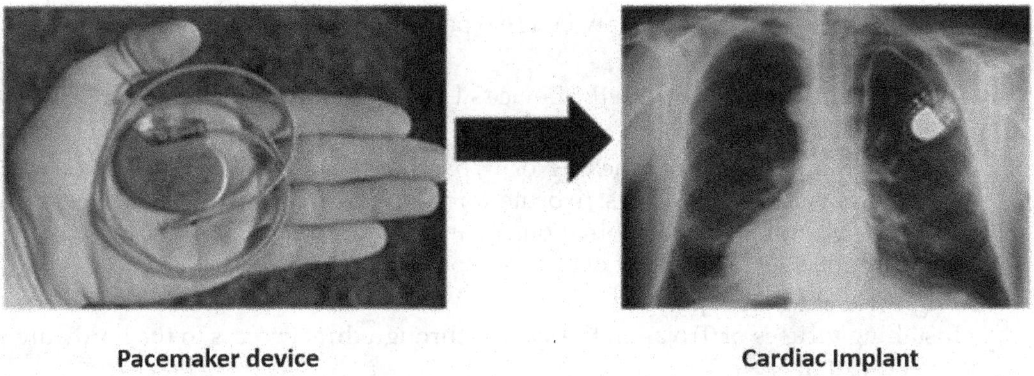

FIGURE 3.4
Pacemaker device of St. Jude.

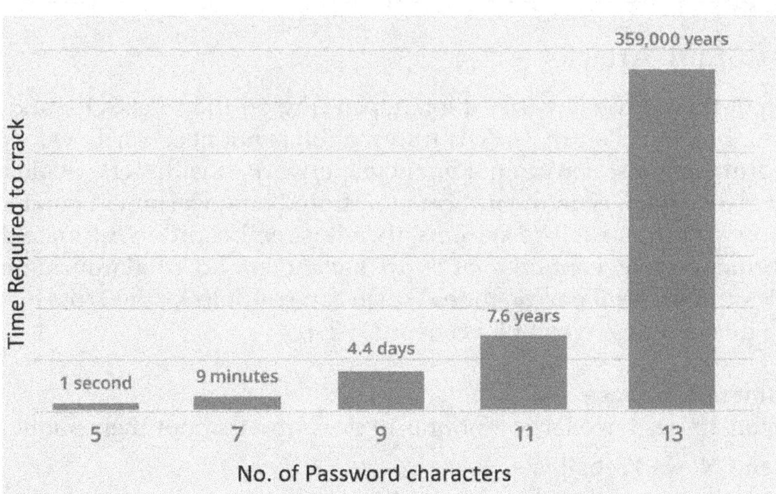

FIGURE 3.5
Brute-force computing time for password cracking (15 million keys per second is assumed).

- **Physical Attacks**
  The attacker tampers with the hardware components of the device. Because of the vast area of implementations of IoT devices many devices are left unattended for a long interval of time. Most of these devices are operated outdoors and hence are vulnerable to such attacks [21]. Some of them are listed below:
- **Reconnaissance Attack**
  Finding and discovering services, systems, and their vulnerabilities. For example, scanning network ports, traffic analysis, sending queries [26,27].
- **Denial of Service**
  In this kind of attack, the attacker or intruder aims to disrupt the services of any host on the network by flooding the targeted resources [26].
- **Access Attacks**
  Gaining unauthorized access to a device by either physically accessing the device or remotely gaining access with an IP address [21].
- **Eavesdropping**
  Listening to a conversation between two parties [29].
- **Tracking**
  Tracking a user's device with the device's Unique Identification Number [26].
- **Password-Based Attacks**
  Attacks in an attempt of cracking or bypassing the password. There are two ways of password-based attacks: (i) brute-force attacks— using tools which generate all possible number of combinations in an attempt to crack user password; (ii) dictionary attack— trying a set of common passwords [26].
- **Using Viruses and Trojan**
  Installing viruses or Trojan in the system through direct access to the hardware or infecting a file [29].

3.6 Purpose of IoT Attacks

The valuation of these targets varies a lot, depending on the sensitivity and value of the information as shown in Figure 3.6. The information is not necessarily valued in terms of money. If information is stolen from a financial network, it is directly evaluated in terms of capital, but if the attack is on the information stored on government databases, military networks, or public infrastructure systems, then it is really hard to estimate the real value since the information may contain a lot of private and confidential information about the general public or influential personalities which can result in loss of trust by the public in government agencies or even complete chaos [26,29].

- **Government Database**
  These databases have large amounts of data stored about their public [6,34].
- **Media and News Websites**
  Sensitive data are related to different cases. These media websites may contain sensitive information which, if released, can create chaos [34].

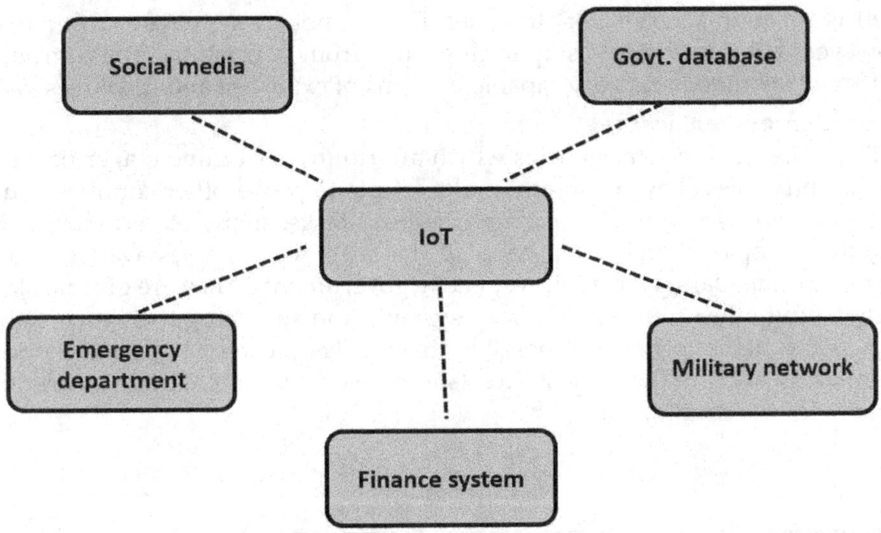

FIGURE 3.6
Major targets of IoT attacks.

- **Military Storage Facilities and Military Networks (Security and Defense)**
  Every country has its military network and respective storage facilities. These facilities can contain important information on multiple sensitive topics [4,34].
- **Financial System Networks**
  Networks of banks which are responsible for large fund transfers can prove very damaging if put in the wrong hands [8,11,34].

3.7 Classification of Intruders

Intruders can be individuals, a group of people, or an agency, and the people may belong to an internal or external area. An internal intruder has proper authorization and access but has malicious intents. An external intruder is a person who does not have authorization but has malicious intents of harming the system. These intruders can belong to any one of these following categories [29,34]:

- **Individuals**
  Hackers, professionals, or even people not having any prior knowledge of hacking can use available tools and techniques for their malicious intent. It is very common in youngsters who try to use these tools to either achieve fame for themselves or do it just for fun or revenge [34].
- **Organized Group of Persons**
  Groups of people with criminal intents are becoming more and more common over time. These groups are well organized, keep their original identity unknown, and use an alias as their group name. These groups have some professionals as

well as amateurs who all work together. They do not always have a criminal intent; however, it is important to stop such groups from flourishing. These groups are sufficiently funded and very capable in terms of expertise and resources [34].

- **Intelligence Agencies**

  These are intelligence agencies which are run by government agencies in most of the countries. They constantly make efforts to probe other country's military networks and systems. To accomplish these tasks, many experts are working together. People of this group have all the latest technologies available to them and are funded largely by their respective governments. They are given tasks such as invading other country's military systems and searching their own country's network systems to find out possible threats. They use strong surveillance and monitoring and are the biggest threats to networks but are treated as prime safeguards for the country [34].

3.8 Conclusion

IoT devices have several threats and security issues. This chapter provides information about probable threats to IoT and documents, vulnerabilities and attacks faced by IoT, and types of intruders and their categories and capabilities. Keeping in mind the importance of data integrity, privacy, and confidentiality, this chapter concludes that various security updates such as a better authorization system, better data encryption, stronger network security, and intrusion detection system are required to overcome challenges posed by threats. These measures are also important to keep IoT a field of choice for the public, organizations, and government agencies. In the future, we will develop a new security algorithm that will help in identifying and restricting possible threats and intrusions.

References

1. Samsung Galaxy Gear. Available at https://www.samsung.com/in/wearables/ [accessed on 11 March 2019].
2. IoT Introduction. Available at https://www.slideshare.net/xinoe/iot-introduction-55189395 [accessed at 9 March 2019].
3. J. Gubbi et al., "Internet of Things (IoT): A vision, architectural elements, and future directions", *Future Generation Computer Systems*, Volume 29, Issue 7, Pages 1645–1660, September 2013.
4. I. Lee, "The Internet of Things (IoT): Applications, investments, and challenges for enterprises", *Business Horizons*, Volume 58, Issue 4, Pages 431–440, July–August 2015.
5. S. Dieter et al., "Towards the implementation of IoT for environmental condition monitoring in homes", *IEEE Sensors Journal*, Volume 13, Issue 10, Pages 3846–3853, October 2013.
6. T. Xu et al., "Security of IoT systems: Design challenges and opportunities", *ICCAD '14 Proceedings of the 2014 IEEE/ACM International Conference on Computer-Aided Design*, San Jose, CA, Pages 417–423, November, 2014.
7. M.U. Farooq et al., "A critical analysis on the security concerns of Internet of Things (IoT)", *International Journal of Computer Applications*, Volume 111, Issue 7, February 2015.

8. R. Mahmoud et al., "Internet of Things (IoT) security: Current status, challenges and prospective measures", *2015 10th International Conference for Internet Technology and Secured Transactions (ICITST)*, London, December 2015.
9. A. Riahi et al., "A systemic approach for IoT security", *IEEE International Conference on Distributed Computing in Sensor Systems*, Cambridge, MA, Pages 351–355, May 2013.
10. J. Wurm et al., "Security analysis on consumer and industrial IoT devices", *2016 21st Asia and South Pacific Design Automation Conference (ASP-DAC)*, Macao, China, 25–28 January 2016.
11. Y.H. Hwang, "IoT security & privacy: Threats and challenges", *IoTPTS '15 Proceedings of the 1st ACM Workshop on IoT Privacy, Trust, and Security*, New York, Page 1, 2015.
12. M. Nawir et al., "Internet of Things (IoT): Taxonomy of security attacks", *2016 3rd International Conference on Electronic Design (ICED)*, Phuket, 11–12 August 2016.
13. N. Chaabouni et al., "Network intrusion detection for IoT security based on learning techniques", *IEEE Communications Surveys & Tutorials*, Volume 21, Pages 2671–2701, 30 January 2019.
14. M. Ahmad Khan et al., "IoT security: Review, blockchain solutions, and open challenges", *Future Generation Computer Systems*, Volume 82, Pages 395–411, May 2018,
15. M. Miettinen et al., "IoT SENTINEL: Automated device-type identification for security enforcement in IoT", *2017 IEEE 37th International Conference on Distributed Computing Systems (ICDCS)*, Atlanta, GA, 5–8 June 2017.
16. D. Minoli et al., "IoT security (IoTSec) considerations, requirements, and architectures", *2017 14th IEEE Annual Consumer Communications & Networking Conference (CCNC)*, Las Vegas, NV, 8–11 January 2017.
17. J.M. Blythe et al., "The consumer security index for IoT: A protocol for developing an index to improve consumer decision making and to incentivize greater security provision in IoT devices", *Living in the Internet of Things: Cybersecurity of the IoT – 2018*, London, Page 7, 28–29 March 2018.
18. Google Home. Available at https://store.google.com/product/google_home [accessed on 9 March 2019].
19. Apple Watch. Available at https://www.apple.com/in/watch/ [accessed on 9 March 2019].
20. Samsung Galaxy Gear. Available at https://www.samsung.com/in/wearables/ [accessed on 9 March 2019].
21. Amazon Echo. Available at https://www.amazon.in/Amazon-Echo-Smart-speaker-Powered/-dp/B0725W7Q38 [accessed on 9 March 2019].
22. Philip Hue. Available at https://www2.meethue.com/en-in [accessed on 9 March 2019].
23. Data Integrity. Available at https://en.wikipedia.org/wiki/Data_integrity [accessed on 8 March 2019] [last edited on 1 March 2022].
24. Authorization. Available at https://en.wikipedia.org/wiki/Authorization [accessed on 8 March 2019] [last edited on 8 February 2022].
25. Authorization. Available at https://tools.ietf.org/html/rfc2196 [accessed on 8 March 2019].
26. H.G. Brauch, "Concepts of security threats, challenges, vulnerabilities and risks", in Brauch H. et al., *Coping with Global Environmental Change, Disasters and Security*, Pages 61–106. Berlin: Springer, 2011.
27. IoT Early Warning System. Available at http://iot.salon/wp-content/uploads/2017/04/-Internet-of-Things-is-Saving-El-Salvador-min.jpg [accessed on 9 March 2019].
28. Pacemaker Device. Available at https://zdnet4.cbsistatic.com/hub/i/r/2017/02/08/9a4c2af1-4c38-41ac-a386-400813acd577/resize/770xauto/c688e2f11187aecd084dfe4359ded4ef/screen-shot-2017-02-08-at-09-10-01.jpg [accessed on 9 March 2019].
29. I. Naumann and G. Hogben, "Privacy features of European Eid card specifications," *Network Security*, Volume 2008, Issue 8, Pages 9–13, 2008.
30. Brute Force. Available at https://www.cloudflare.com/img/learning/security/threats/brute-force-attack/brute-force-cracking-time.png [accessed on 11 March 2019].
31. TP-Link. Available at https://www.tp-link.com/us/home-networking/smart-home-router/ [accessed on 9 March 2019].
32. TP-Link. Available at https://august.com/products [accessed on 9 March 2019].

33. M. Abomhara et al., "Cyber security and the Internet of Things: Vulnerabilities, threats, intruders and attacks", *Journal of Cyber Security and Mobility*, Volume 4, Pages 65–88, January 2015.
34. L. Sharma, N. Lohan, "Performance analysis of moving object detection using BGS techniques in visual surveillance", *International Journal of SpatioTemporal Data Science*, Volume 1, Issue 1, Pages 22–53, January 2019.
35. IoT. Available at https://iotworm.com/internet-things-goes-open-source-linux-foundation/ [accessed on 20 May 2021].
36. L. Sharma, P. Garg (Eds.), *From Visual Surveillance to Internet of Things*. New York: Chapman and Hall/CRC, 2020. https://doi.org/10.1201/9780429297922.
37. L. Sharma, P.K. Garg, "IoT and its applications", *From Visual Surveillance to Internet of Things*, Volume 1, Page 29. Taylor & Francis, CRC Press, 2018.
38. L. Sharma, D.K. Yadav, S.K. Bharti, "An improved method for visual surveillance using background subtraction technique", *2015 2nd International Conference on Signal Processing and Integrated Networks (SPIN)*, Pages 421–426, 2015. https://doi.org/10.1109/SPIN.2015.7095253.
39. D.K. Yadav, L. Sharma, S.K. Bharti, "Moving object detection in real-time visual surveillance using background subtraction technique", *2014 14th International Conference on Hybrid Intelligent Systems*, 2014, Pages 79–84. https://doi.org/10.1109/HIS.2014.7086176.
40. L. Sharma, A. Singh, D.K. Yadav, "Fisher's linear discriminant ratio based threshold for moving human detection in thermal video", *Infrared Physics and Technology*, Volume 76, Pages 118–128, Elsevier, March 2016.
41. L. Sharma (Ed.), *Towards Smart World*. New York: Chapman and Hall/CRC, 2021. https://doi.org/10.1201/9781003056751.
42. L. Sharma, "Human detection and tracking using background subtraction in visual surveillance", in *Towards Smart World*, Pages 317–328. New York: Chapman and Hall/CRC, December 2020. https://doi.org/10.1201/9781003056751.
43. IoT Framework. Available at https://www.netburner.com/learn/architectural-frameworks-in-the-iot-civilization/ [accessed on 30 May 2021].

4

Use of Robotics in Real-Time Applications

Lavanya Sharma
Amity Institute of Information Technology, Amity University, Noida, India

Mukesh Carpenter
Department of Surgery, Alshifa Multispecialty Hospital, Okhla, India

CONTENTS

4.1 Introduction

In today's world, everyone is aware of a huge range of thinking to expand robotic technology after its use in industry technology. It is the part of the engineering and technology that comprise electrical engineering and information technology. This branch manages the structure, development, usage to control robots, tangible input, and data preparing [1]. Those robots are intended to be utilized for many reasons; however, these will be utilized in delicate conditions such as bomb recognition, deactivating procedure of different bombs, and so on. Robots would make any frame; however, a significant number of these robots possess human behavior and nature.

These robots will look like humans and have the ability to walk, discourse, and moreover all of the things a human can do [1,2,10–36]. The greater part of these robots is propelled commonly and is called bio-enlivened robots.

There are a few kinds of robots as follows:

(i) **Explained**

Components in these robots are rotating connections and the scope is from 2 to 10 connections. Here, the arm is associated with a revolving connection, and each connection is called the pivot which will give scope for developments [3,4].

(ii) **Cartesian**

These are otherwise called gantry robots. There are three connections that utilize the Cartesian framework: *a, b, c*. These types of robots are furnished with appended wrists to give rotatory movement.

DOI: 10.1201/9781003244165-5

(iii) **Round and hollow**

These sorts of robots have no less than one revolving connection and one kaleidoscopic joint which are utilized to interface the connections. The utilization of rotatory connection is to pivot with the hub and kaleidoscopic connections are used to give straight movement.

(iv) **Polar**

These are otherwise called round robots. Here, the arm is associated with a bending connection and has a mix of two revolving connections and one straight joint. These robots are mostly utilized in gathering uses. Its arm is tube-shaped in the plan. They have the two connections at an equal distance which are utilized to give consistency in one chosen plane.

(v) **Delta**

The formation of these robots resembles arachnid. They work by connecting a trapezium that is associated with the base [4,6]. The trapezium moves in an arch-formed working region. These are utilized for the most part in sustenance and electrical ventures. The World Technology Evaluation Center is an American association that surveys the condition of innovations around the globe. Their investigations can be subsidized by different American government bodies, for example, the National Aeronautics and Space Administration, Defense Advanced Research Projects Agency, and so on. In our endeavors to fire up a mechanical technology organization, I inspected one of their reports distributed in 2006, titled "Universal Assessment of Research and Development in Robotics." This report was composed by researchers gaining practical experience in the field [6–8,39–53]. They visited and talked with researchers from organizations, and the research focuses on various nations: the USA, Japan, South Korea, Australia, and Europe [9]. The report portrays the present condition of mechanical technology, contrasts the USA and whatever remains of the world (as per the report), and talks about future difficulties in applying autonomy, which is of unique enthusiasm tome:

- Mechanical vehicles, space apply autonomy, humanoids.
- Apply autonomy is that part of a building that manages origination, structure, task, and assembling of robots. There was a creator named Isaac Asimov. He said that he was the main individual to give a name to apply autonomy in a short story made in the 1940s. In that story, Isaac proposed three standards about how to direct these kinds of mechanical machines. Later, these three standards were given the name of Isaac's three laws of robotics [10]. These three laws are as follows:
- Robots will never hurt individuals.
- Robots will adhere to directions given by people with infringing upon law one.
- Robots will secure themselves without defying different guidelines.

This chapter is categorized into four sections. Section 4.1 deals with the introductory part of robotics and its types. Section 4.2 describes related work. Section 4.3 deals with challenging issues in robotics. Section 4.4 discusses different areas of robotics in real-time applications. In Section 4.5, the conclusion of the work is discussed.

4.2 Related Work

Greczek et al. [1] discussed how to standardize and replace the computer technologies by socially assistive robotics as it has the potential to do what has the edge of knowing in tangible context, plan to structure mechanical frameworks that are convincing, help youngsters in accomplishing instructive objectives. Rischet [2] discussed the progression in apply autonomy innovation. Unmanned rural apply autonomy is broadly utilized in exactness agribusiness. Architects outfitted with mechanical technology information are exceedingly requested by the present high effectiveness, high-creating rural industry. Shin et al. [3] investigated if the students could learn this core computer science concept while enjoying themselves in the robotic context. A visual questionnaire was developed based upon the combined Bloom and SOLO taxonomies, although it proved to be difficult to construct a questionnaire appropriate for a young student. Jovanović [4] discussed that the structure of present-day automated gadgets faces various necessities and impediments which are identified with enhancement and power. Therefore, these stringent necessities have caused enhancements in many building territories and led to the improvement of new streamlining techniques which better handle new complex items intended for application in modern robots. Ayushnarula et al. [5] discussed that step-by-step instructions to do work-savvy robots in surgeries have been totally determined by the kind of medical procedure. Similarly, the main job of creating savvy robots is as of now being taken up by the private area. Bhattacharyya [6] discussed the utilization of electroencephalogram (EEG) signals for controlling in the field of mechanical autonomy and utilizing a reasonable mapping process known as a brain–computer interface. Different deterioration strategies of the EEG motion for highlight extraction have been proposed by numerous analysts. Joshi et al. [8] discussed the neural circuits that control getting a handle on and perform related visual handling have been examined broadly in macaque monkeys. We are building up a computational model of this framework so as to comprehend its capacity and investigate applications to mechanical technology. Subramanian et al. [9] discussed numerous mechanical spots on the planet from multiple points of view to recognize essential jobs in numerous businesses for some reasons. Liang [10] demonstrated that show-based, probabilistic reverse fortification learning Intelligent Robotics Lab Facilities (IRL) is attainable in high-measurement, state-activity spaces with just a solitary master exhibition. By executing the IRL max edge calculation with a probabilistic model-based fortification learning calculation named PILCO, we can join the calculations to make the IRL/PILCO calculation, which is equipped for replicating master directions by picking appropriate highlights.

4.3 Current Challenging Issues

There are various challenging issues faced by robotics in the recent past. Some of them are listed below:

- **New Materials, Creation Techniques**
  Apparatuses, engines, and actuators are central to the present robots. So, huge works are being carried out with fake muscles, delicate mechanical autonomy, and

techniques that will help build up the upcoming age of self-ruling robots to do many functions at the same time [13].

- **Making Eco-Friendly-Enlivened Robots**
  Naturally enlivened robotics are doing their work progressively basic in automations autonomy labs. The primary aim is to create robots that function more like the effective frameworks found in the atmosphere. The investigation says that the significant difficulties required with this territory have remained, to a great extent, unaltered for a long time—high power cells to coordinate metabolic transformation, muscle-like actuators, self-mending parts that's been used in robotics, independence in every condition, human beings-like recognition, with calculation while thinking accordingly. Materials which are being used together in detection, activation, calculation, with correspondence should be created and discussed and connected with each other. This advancement will prompt robots with highlights, for example, physical support, force decrease, sway security, physiological calculation, and versatility [15].
- **Good Resources in Force**
  Improvising the battery life is a noteworthy case in automatons and portable robotics, in particular. Fortunately, expanded selection in these frameworks is prompting unused or best battery advancements which are moderate, protected, and enduring. The task is given and completed so as to make the segments of a robot more efficient. So, the examination refers to robots that need to work remotely in unstructured situations and will in the long run concentrate vitality for some lightening, oscillations, and mechanical development [16,48].
- **Connections in Robots in Swarms**
  Discernment activity circles are basic to making self-ruling robotic work in unpatented conditions. Robot swarms need correspondence capacity to insert in this input circle. Consequently, recognition activity openness circles are of utmost importance to structuring robot swarms. There are no efficient methodologies for doing this crosswise over expansive gatherings [17].
- **Navigate Untracked Environment**
  Step-by-step instructions to reason about new ideas and their semantic portrayals and find new articles or classes on the earth through learning and dynamic associations [18]. Per the examination:

  > For route, the great test is to deal with disappointments and having the capacity to adjust, learn, and recoup. For investigation, it is building up the natural capacities to make and perceive new disclosures. From a framework viewpoint, this requires the physical heartiness to withstand cruel, alterable conditions, harsh dealing with, and complex control. The robots need huge dimensions of self-rule prompting complex self-checking, self-reconfiguration, and fix with the end goal that there is no single purpose of complete disappointment but instead elegant framework debasement. Whenever possible, arrangements need to include control of different heterogeneous robots; adaptively organize, interface, and utilize various resources; and offer data from numerous information wellsprings of variable unwavering quality and exactness.

- **Artificial Intelligence**
  The investigation considers computer-based intelligence as the "supporting innovation for mechanical autonomy," yet recognizes that "regardless we have far to go to recreate and surpass every one of the aspects of knowledge that we find in people." The key is to consolidate propelled design acknowledgment and

model-based thinking to create artificial intelligence that can reason and has the presence of mind [19]. The examination mentions that Deep Mind's AlphaGo Zero framework is a genuine case; however, it says, "we don't yet have frameworks that can do this effectively crosswise over heterogeneous errands and areas." [42,46].

- **Brain–Computer Interface**
  It empowers some gadgets and tools constrained by your psyche. It would be helpful in expanding human capacity; however, constructing up the invention is the most expensive selection in the test [20,43,12].
- **Therapeutic Mechanical Autonomy with More Self-Sufficiency**
  From the negligibly obtrusive medical procedure, clinic streamlining, to crisis reaction, prosthetics, and home help, medicinal mechanical technology is one of the quickest developing areas. The test is building dependable frameworks with more prominent dimensions of self-sufficiency [21,39]. A long-haul test is to empower one specialist to oversee a lot of robots that can perform the whole set methodology self-sufficiently and just approach specialists amid basic, quiet explicit advances. The investigation says, "Maybe the most noteworthy test of computerizing any clinical undertaking is to almost certainly envision, identify, and react to all conceivable disappointment modes. Restorative gadget guideline of self-governing robots will probably need to create in a way that adjusts the prerequisites for provably safe calculations with consistence costs."

4.4 Different Areas of Robotics in Real-Time Applications

A. **Robotics in Space**
Mechanical innovation deals with the improvement of sharp robots for extraterrestrial examination focusing on [23,50] the following:

- Improvement of robot systems for unattorned, unevenly reliant on naturally spurred innovative speed thoughts.
- Improvising of many utilizable robot's bunches usable for different endeavors running from in-situ examinations to the affiliation and upkeep of the system.
- Re-customizable pre-field frame in the whole investigation.
- Simulated insight in basic methods for a self-ruling course and task organizing in a darkened scene.
- Pic appraisal, object affirmation, and region showing simulated knowledge-based steady systems for legitimate tests.

B. **Marine Robotics**
Robots perform several task such as raising fish to analyzing shipwrecks. It also help water engineers, police, marine biologist to do their jobs efficiently [25] but still there are some major issues which are as follows:

- Structure of systems for controlling and planning of robot in partially merged applications, especially with the best-level sensor progression, for example, "Visual Serving."
- Image evaluation and article assertion.

- Mapping of control systems for cutting-edge, self-choice submerged vehicles [34,37].
- Advancement of normally invigorated and vitality gainful ways in transporting partially merged vehicles, for instance, impacting preplanned tasks.

C. **Electric Vehicles**

Electrical vehicles are controlled by robots resulting in improvement in inventive motors ways which are as follows:

- Structure of a unique way of having mobility and traffic immersion, application support, innovative incorporation.
- Data accumulatation in armada tests firms with innovative electricity motors (see eco-versatility armada in Exploration Offices) [27].
- Makes an opening for new sources in activity and traffic ways in the data recently collected.

D. **Robotics in Agricultural**

These robots automate moderate, monotonous and tiresome tasks for farmers and allow them to focus on overall production improvement. Some of the most common robots in agriculture which are used for harvesting and picking and weed control. Few examples of autonomous robots used for agriculture are:

- Spraying and weeding robots.
- Crop harvesting robots.
- Seeding and planting robots.
- Soil analysis robots.

E. **Logistics, Production, and Consumer**

It improves intellectuality for adaptable assembling by upgraded capacities of robots. Other uses are as follows:

- Canny human–robot coordinated effort utilizing half and half groups for creation situations.
- Novel and safe robots for human–robot joint effort.
- Independent versatile control for intra-logistics and assembling situations.
- Inventive mechanical technology answers for investigation (for example, balance water tanks, transport structures, or passage exhausting machines) [31].

F. **Search and Rescue & Security Robotics**

Robots can also be used for the search and rescue [52] as well as security purposes. Some of them are listed below:

- Advancement of exceedingly portable steps in hardware and software apps.
- Advancement of preplanned tasks that will differentiate potential exploited people or interloper guards (security) [33].
- Advancement with taking and doing a collection of sensors dependent on radars, supersonic scanners, and supervision to distinguish articles and humans responsible [48].

G. **Assistance and Rehabilitation Systems**

Center themes include the following:

- Idea improvement, structure, and development.

- Clever equipment framework models, programming structures [39,42,44].
- Combination of different pulsate sensors [35,45,51].

H. **Army Robotics**

Military or army robots are autonomous robots or sometimes referred to as remote-controlled robots that are specifically designed for army applications such as for transportation to explore different routes and rescue and attack. These systems are currently in use, and more are under development.

4.5 Conclusion

In this chapter, we provide a brief overview of the uses of robotics in real-time applications in space, marine, electric vehicles, agriculture, logistic production and consumer, search and rescue as well as security policies, assistant in rehabilitation system, and army. The main aim of this work is to study the uses of robotics and some of the recent advancements of robotics in real-time applications. This book also presents various challenging issues faced by robotics and provides some current solutions and observations. It can be concluded that robotics in a real-time application is one of the technologies which can enhance lives. It will become a ubiquitous innovation in the future.

References

1. C. Clabaugh, G. Ragusa, F. Sha and M. Matarić, "Designing a socially assistive robot for personalized number concepts learning in preschool children". Joint IEEE International Conference on Development and Learning and Epigenetic Robotics (ICDL-EpiRob), 2015, pp. 314–319, doi: 10.1109/DEVLRN.2015.7346164.
2. Yonghui Wang, Suxia Cui, Eric Risch, Yubin Lan, Jian-ao Lian, and Kevin Lee "Enhance Multi-Disciplinary Experience for Agriculture and Engineering Students with Agriculture Robotics Project". Mechanical Engineering Department, Prairie View A&M University, 2014.
3. Stéphane Magnenat, Jiwon Shin, Fanny Riedo, Roland Siegwart, and Morderchai Ben-Ari. 2014. Teaching a core CS concept through robotics. In *Proceedings of the 2014 conference on Innovation & technology in computer science education (ITiCSE '14)*. Association for Computing Machinery, New York, NY, USA, 315–320. doi:10.1145/2591708.2591714.
4. Velibor Karanović, Mitar Jocanović "Review of Development Stages in the Conceptual Design of an Electro-Hydrostatic Actuator for Robotics". Vol. 11, No. 5, University of Novi Sad Faculty of Technical Sciences Trg Dositeja Obradovića 6, 21000 Novi Sad, Serbia, Mechanical Engineering Technology 111-D Kaufman Hall Norfolk, VA 23529, USA, January 2014..
5. Ayush Narula, N.K. Narula, Satyam Khanna, Ruchi Narula, Jyoti Narula, and Arpi Narula "Future Prospects of Artificial Intelligence in Robotics Software, A Healthcare Perspective". *International Journal of Applied Engineering Research*, Vol. 9, No. 22, pp. 10271–10280, 2014..
6. Arora, A., & Bhattacharyya, S. (2014). An Approach towards Brain Actuated Control in the Field Of Robotics Using Eeg Signals: A Review.
7. Jessica Swenson, Ethan Danahy "Examining Influences on the Evolution of Design Ideas in a First-Year Robotics Project". Tufts University Department of Computer Science, Medford, MA, 5th International Conference Robotics in Education Padova (Italy), pp. 84–92, July 18, 2014.

8. Ashley Kleinhans, Serge Thill, Benjamin Rosman, Renaud Detry, and Bryan Tripp , "Modelling Primate Control of Grasping for Robotics Applications". CSIR, South Africa, University of Skövde, Sweden, University of Liège, Belgium, University of Waterloo, Canada, Computer Vision - ECCV 2014 Workshops.
9. Bin He, Meng Xia, Xinguo Yu, Pengpeng Jian, Hao Meng, Zhanwen Chen, "An educational robot system of visual question answering for preschoolers". *Robotics and Automation Engineering (ICRAE) 2017 2nd International Conference on*, pp. 441–445, 2017.
10. Yuan Gao, Wolmet Barendregt, Mohammad Obaid, Ginevra Castellano, "When Robot Personalisation Does Not Help: Insights from a Robot-Supported Learning Study". *Robot and Human Interactive Communication (RO-MAN) 2018 27th IEEE International Symposium on*, pp. 705–712, 2018.
11. Repairing of Damaged Car Parts with the Help of Robotics. Available at: https://fortunedotcom.files.wordpress.com/2016/07/fut-07-01-16-future-of-workautomotive.jpg [accessed on 08-March-2019].
12. User Interaction with Robotic Library. Available at: https://i.ytimg.com/vi/1H5tCIWVzFE/maxresdefault.jpg [accessed on 08-March-2019].
13. New Materials, Creation Techniques. Available at: https://www.azorobotics.com/Article.aspx?ArticleID=255 [accessed on 09-March-2020].
14. Challenging Issues in Robotics. Available at: https://20kh6h3g46l33ivuea3rxuyuwpengine.netdna-ssl.com/wp-content/uploads/2020/02/robotics-challenges-1024x453.jpg [accessed on 09-March-2020].
15. Making Eco-Friendly-Enlivened Robots. Available at: https://link.springer.com/article/10.1186/s40648-016-0060-4 [accessed on 09-March-2020].
16. Good Resources in Force. Available at: https://www.theguardian.com/science/politicalscience/2017/oct/01/will-robots-bring-about-the-end-of-work [accessed on 09-March-2020].
17. Connections in Robots in Swarns. Available at: https://www.frontiersin.org/articles/10.3389/frobt.2017.00009/full [accessed on 09-March-2020].
18. Robot Navigation in a Dynamic Environment. Available at: https://www.quora.com/Whatare-the-problems-in-robot-navigation-in-a-DynamicEnvironment [accessed on 09-March-2020].
19. The Reason can be Artificial Intelligence. Available at: https://science.howstuffworks.com/robot6.htm [accessed on 09-March-2020].
20. Mind PC Interfaces. Available at: https://jobs.theconversation.com/?utm_source=theconversation.com&utm_medium=we bsite&utm_campaign=topbar [accessed on 09-March-2020].
21. Autonomous-Weapons. Available at: https://spectrum.ieee.org/automaton/robotics/military robots/why-should-we-ban-autonomous-weapons-to-survive [Accessed on 09-March-2020].
22. Information from the 2006 WTEC. Available at: https://cdn2.hubspot.net/hub/13401/file 13223192-gif/images/robotics_challenge[1].gif [accessed on 10-March-2020].
23. Space Robotics. Available at: https://www.robotics.org/blog-article.cfm/Robotics-In-23 Space/10 [Accessed on 11-March-2020].
24. Robotics in Space. Available at: https://robotik.dfkibremen.de/fileadmin/_processed_/7/0/csm_Space2_5b2f4e1cae.jpg [accessed on 11-March-2020].
25. Robots for Marine Uses. Available at: https://www.whoi.edu/marinerobotics [Accessed on 11-March-2020].
26. Marine Robotics. Available at: https://www.ieee-ras.org/marine-robotics#:~:text=Recently%2C%20Marine%20Robotics%20has%20grown,to%20dive%20beyond%206000%20meters.&text=Marine%20Robotics%20as%20a%20field,understanding%20large%20scale%20societal%20problems [Accessed on 11-March-2020].
27. Electric Vehicles. Available at: https://www.sae.org/publications/technicalpapers/content/2019-01-0869/ [accessed on 11-March-2020].
28. Testing of Go-Anywhere Robot Car. Available at: https://i.ytimg.com/vi/6JLJjVSvEV8/maxresdefault.jpg [Accessed on 11-march-2020]

29. Agricultural Robots. Available at: https://www.postscapes.com/agriculture-robots/ [accessed on 11-March-2020]; Sprinkling Water Using Robots. Available at: https://www.governmenteuropa.eu/wpcontent/uploads/2020/06/%C2%A9-iStock-baranozdemir-696x392.jpg [accessed on 11-March-2020].
30. Robots for Logistics and Manufacturing. Available at: https://www.i-scoop.eu/robotscobots-logistics-4-0/ [accessed on 12-March-2020].
31. Logistics, Production and Consumer. Available at: https://robotik.dfkibremen.de/fileadmin/_processed_/1/9/csm_Aila_0023_01cf25cfcf.jpg [accessed on 12-March-2020].
32. Robots for Rescue. Available at: https://www.intechopen.com/books/search-and-rescue robotics-from-theory-to-practice/introduction-to-the-use-of-robotic-tools-for-search-andrescue [accessed on 12-March-2020].
33. Advanced Security Robots. Available at: https://robotik.dfkibremen.de/fileadmin/_processed_/2/9/csm_SAR_a26895106b.jpg [accessed on 12-March-2020].
34. Robots in Rehabilitation. Available at: https://www.britannica.com/technology/rehabilitation-robot [Accessed on 12-March-2020].
35. Robotic Therapy Shown Effective for Stroke Rehab. Available at: https://www.medgadget.com/wp-content/uploads/2014/01/armin-robot.jpg [Accessed on 13-March-2020].
36. Military Robots. Available at: https://www.engineersgarage.com/articles/military-robotics [accessed on 13-March-2020].
37. Lavanya Sharma, Nirvikar Lohan "Internet of Things with Object Detection: Challenges, Applications, and Solutions", *Handbook of Research on Big Data and the IoT*, IGI Global, pp. 89–100, March 2019.
38. Lavanya Sharma, Dileep Kumar Yadav, and Annapurna Singh "Fisher's Linear Discriminant Ratio Based Threshold for Moving Human Detection in Thermal Video", *Infrared Physics & Technology*, Elsevier, Vol. 78, pp. 118–128, September 2016.
39. Lavanya Sharma, Dileep Kumar Yadav "Histogram-Based Adaptive Learning for Background Modelling: 25 Moving Object Detection in Video Surveillance", *International Journal of Telemedicine and Clinical Practices*, Inderscience, Vol. 2, No. 1, pp. 74–92, 2017.
40. Thais Oliveira Almeida "Adaptation Content in Robotic Systems: A Systematic Mapping Study", *Frontiers in Education Conference (FIE) 2018 IEEE*, pp. 1–9, 2018.
41. Roxanna Pakkar, Caitlyn Clabaugh, Rhianna Lee, Eric Deng, Maja J Mataricć, "Designing a Socially Assistive Robot for Long-Term In-Home Use for Children with Autism Spectrum Disorders". *Robot and Human Interactive Communication (RO-MAN) 2019 28th IEEE International Conference on*, pp. 1–7, 2019.
42. Lavanya Sharma, Nirvikar Lohan "Performance Analysis of Moving Object Detection Using BGS Techniques in Visual Surveillance". *International Journal of SpatioTemporal Data Science*, Vol. 1, No. 1, pp. 22–53, January 2019.
43. Drone Usage in Covid-19. Available at: https://www.unicef.org/supply/media/5286/file/%20 Rapid-guidance-how-can-drones-help-in-COVID-19-response.pdf [accessed on 10 July 2021].
44. Lavanya Sharma, Pradeep K. Garg (Eds.) *From Visual Surveillance to Internet of Things*. New York: Chapman and Hall/CRC, 2020. https://doi.org/10.1201/9780429297922.
45. Lavanya Sharma (Ed.) *Towards Smart World*. New York: Chapman and Hall/CRC, 2021. https://doi.org/10.1201/9781003056751.
46. Lavanya Sharma, "Human Detection and Tracking Using Background Subtraction in Visual Surveillance", *Towards Smart World*. New York: Chapman and Hall/CRC, pp. 317–328, December 2020. https://doi.org/10.1201/9781003056751.
47. Lavanya Sharma, Sudhriti Sengupta, and Birendra Kumar "An Improved Technique for Enhancement of Satellite Images". *Journal of Physics: Conference Series*, Vol. 1714, p. 012051, January 2021.
48. Supreet Singh, Lavanya Sharma, and Birendra Kumar, "A Machine Learning Based Predictive Model for Coronavirus Pandemic Scenario". *Journal of Physics: Conference Series*, Vol. 1714, p. 012051, January 2021.

49. Gourv Jha, Lavanya Sharma, and Shailja Gupta "Future of Augmented Reality in Healthcare Department". In: Singh P.K., Wierzchoń S.T., Tanwar S., Ganzha M., Rodrigues J.J.P.C. (eds) *Proceedings of Second International Conference on Computing, Communications, and Cyber-Security*. Lecture Notes in Networks and Systems, Vol. 203. Singapore: Springer, 2021. https://doi.org/10.1007/978-981-16-0733-2_47.
50. Gourv Jha, Lavanya Sharma, and Shailja Gupta "E-health in Internet of Things (IoT) in Real-Time Scenario". In: Singh P.K., Wierzchoń S.T., Tanwar S., Ganzha M., Rodrigues J.J.P.C. (eds) *Proceedings of Second International Conference on Computing, Communications, and Cyber-Security*. Lecture Notes in Networks and Systems, Vol. 203. Singapore: Springer, 2021. https://doi.org/10.1007/978-981-16-0733-2_48.
51. Sanjay Kumar, Priyanka Gupta, Sachin Lakra, Lavanya Sharma, and Ram Chatterjee "The Zeitgeist Juncture of "Big Data" and Its Future Trends", *2019 International Conference on Machine Learning, Big Data, Cloud and Parallel Computing (COMITCon)*, 2019, pp. 465–469. https://doi.org/10.1109/COMITCon.2019.8862433.
52. Shubham Sharma, Shubhankar Verma, Mohit Kumar, and Lavanya Sharma "Use of Motion Capture in 3D Animation: Motion Capture Systems, Challenges, and Recent Trends", *2019 International Conference on Machine Learning, Big Data, Cloud and Parallel Computing (COMITCon)*, 2019, pp. 289–294. https://doi.org/10.1109/COMITCon.2019.8862448.
53. Thierry Bouwmans, Fatih Porikli, Benjamin Horferlin, and Antoine Vacavant *Handbook on Background Modeling and Foreground Detection for Video Surveillance*. Boca Raton, FL: CRC Press, Taylor and Francis Group, 2014.

Part 2

Tools and Technologies of IoT with Computer Vision

5

Preventing Security Breach in Social Media: Threats and Prevention Techniques

Lavanya Sharma
Amity Institute of Information Technology, Amity University, Noida, India

CONTENTS

5.1 Introduction

Social media networks such as Facebook, Instagram, Twitter, and WhatsApp have become the prime source of sharing personal information, thoughts, news, photos, videos, messages with our friends on a daily basis. While using social media in the present it seems very easy to share data within seconds but before 1970 it was not that easy. In 1971, the first email was sent between two computers which were sitting next to each other; it was

DOI: 10.1201/9781003244165-7

the beginning of the era of sharing information within two different devices in real time. Bulletin Board System was introduced in 1987 to share data over phone lines all around the world. The first social media site namely Geocities.ws was founded by David Bohnett and John Razner in November 1994 which was called Beverly Hills Internet; it was further occupied by Yahoo in 1999.

Social media networks deal with a huge amount of data that need to be provided to the productive analyzing companies in a secure way keeping in mind the difference between the private and the public data of the users. Handling that much becomes difficult which leads to breach of user privacy. These breaches would also lead to physical damage as social media requires the permission of accessing contact, sharing location, etc., which is the credential data of a victim. Social media users User Interface Description Language to develop multilingual, multi-platform, and multimodal user interfaces so that the user could operate in the social network from anywhere and any type of device easily.

5.2 Related Research Work

William et al. [1] discussed the behavior of different types of users and groups in a social media website and describe how these types of activity affect privacy in different social media applications. Hashimoto et al. [2] discussed the benefits of using social media sites in a safe way. The author showcases the key components affecting the user's privacy in social media and how to overcome those components without changing the functionality of the application or keeping the information sharing concept as it is. Kumar et al. [3] described the privacy issues, security issues, identity theft, profiling risk present in these types of applications. It also discusses different types of attacks which can be performed on the social network to hack the privacy of the user which also includes the new attack strategy occupied by the hackers. It also provides us with the step to prevent the above attacks. Fire et al. [4] discussed a detailed review of the different privacy and security risks, which threaten the well-being of online social network users in general, and children in particular. It also presents an overview of existing solutions that can provide better protection, security, and privacy for online social network users. Senthil Kumar et al. [5] describe social media as an essential part of humans in day-to-day life. While enjoying the social media platforms user need to understand which information is needed to be kept private and undisclosed so that no one could use that information in the wrong way. Kumari and Singh [6] described privacy as the main concern in social media sites such as Facebook and Twitter. The main purpose of these sites is to provide the user with the facility to share information. Lack of attention on these sites can lead to a big violation to overcome these various methods for securing the data being covered. Kumar et al. [7] discuss how the prevention of different social media vulnerabilities can be integrated into the design of the social sites to facilitate interaction while enhancing privacy and security. Kumar et al. [8] discussed the present privacy and security issues of the social network and also discussed each of the popular sites with some of the precaution which needs to be taken up.

5.3 Importance of Social Media

Usage of social media websites has increased to the extent that almost every person on earth is using the social network to share, communicate, and discuss different things with colleagues, family members, friends, etc., in a formal or an informal way. Social networking sites are performing well on the internet by receiving more than 10,000 hits per second and this is because these sites have many advantages which help their users daily. Here are a few of them:

- **Communication**
  The main role of social media is it helps its audience to communicate with anyone all around the world at no cost which keeps them connected and up to date with other people, society, organizations, etc.
- **Exchanging Data**
  Information sharing is very easy and quick, we can share our information with a single person at a time or we have the choice to share it with multiple persons at a time using groups.
- **Discloser of Inactivity**
  It also helps in the discloser doing inappropriate activities like harming animals. This could be a great advantage for police to detect the person doing inactivity across the nation.
- **Enhance Business**
  Helps in creating your community of people through which you can earn by sharing your links, blog posts. On YouTube we can share our videos and earn per view, and on Facebook we can share our posts, photos, and videos on our pages.
- **Cost-Effective Advertising**
  Online advertising has become the greatest source of getting the right customer but there are very few sites which provide effective advertising. So, the role of social network sites provides the facility to promote our business with ads in a cost-effective way.
- **Doubt Solving**
  As a student or professional we come across difficult questions to answer. The role of social media sites like Reddit and Quora helps users to get answers to their questions in any field.
- **Entertainment**
  These sites also help users to refresh themselves through different sources of entertainment such as videos, jokes, and photos. Which are being posted by the other you're and you too have the right to present you entertaining stuff on the media sites and become popular through it.
- **Quick Response**
  The information posted on social media spreads quickly which could be a good point for correct news but it also helps the rumors to spread quickly. This reach also depends on the number of members associated with the person posting the content. The quicker the information spread, the quicker we will get a response.

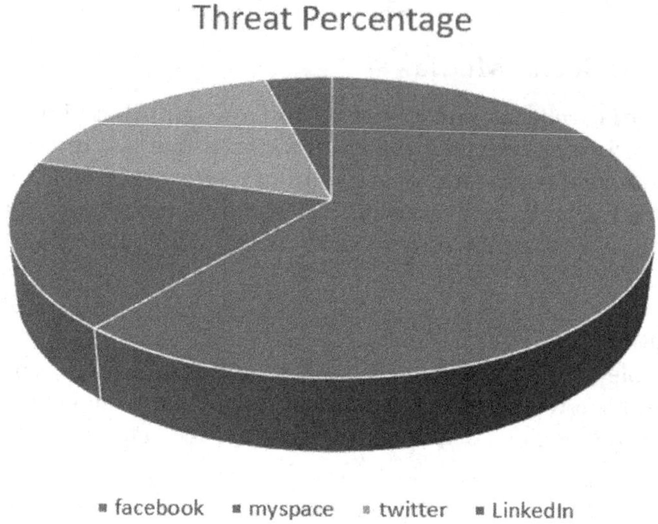

FIGURE 5.1
Threats percentage.

5.4 Privacy and Network Threats in Social Media

With the increase in the number of users, it becomes more dangerous to keep information private and protected. Hence, many research papers have been published discussing the security problems, risks, and privacy threats in using this type of network. Some of them came up with precautions which need to be taken while using these online sites. In a social network, threats are the point of risk or methods which can be used to exploit the user's privacy as shown in Figure 5.1. These are mainly divided into four categories.

5.4.1 Classic Threats

In these types of threats, user's data is being exploited by using its published public information visible to everyone which also includes the attack on his friends too [9–20]. Below are the threats which come under this category:

- **Malware**
  In a view to the computer system malware could be software, an application, or a program which is used to gain unauthorized access to a secure network or to harm it. So, here it means gaining access to the secure user profile by using malware applications.
- **Phishing Attack**
  This attack was first performed in 2007. The attacker creates a fake social media login page which looks exactly like the original and sends that page to the user in an attempt that the user may try to login through the fake login page and the attacker could easily gain the user's confidential data such as username and password. This attack is most used today.

- **Cross-Site Scripting**
  This is an attack on the web application or browser which a victim uses to login into social media. In this attack the attacker injects some malicious script on the web browser of a victim's computer which means whenever the victim tries to visit the website the script loads up and the private credentials are being stolen.

5.4.2 Modern Threats

These types of attacks have been used by most of the attackers as this threat includes the creation of a fake profile by using the personal information of the victim user on the profile page and after the creating of the fake ID sending a friend request to the other target user.

- **Clickjacking**
  Click is performing some action on the screen and jacking means to like a picture, video, or page so meaning after combining both is forcing a victim to click a malicious link, post, or video.
- **Fake Profile**
  It is also known as Profiling Risk or Profile Cloning. It is nothing but creating a duplicate profile page as the victim page with some name and photo of the victim and then using that profile ID for the wrong purpose.
- **Discloser of Sensitive Information**
  Many of us share our profile phone public as every social media website has a feature to share it so that other people on the website could identify us easily and communicate to us but there is a risk involved which is disclosing of user's identity and the content present in the image as the image could reveal details such as identification of image location and object present in the image.
- **Fake Links**
  While browsing social media we often get to message or a post that offers a huge discount for a product which encourage us to click it but when we click on that link, we are sending our profile information to the attacker in the form of giving our name, address, phone number and sometimes critical bank information too such as credit and debit card numbers.
- **Watering Hole Attack**
  This attack is not that common. Considering the make it came from the predators near the watering holes like lakes looking for the opportunity to attack the animals drinking the water. So, this type of attack exploits the security by using the less secure medium in the organization such as workers and employees and infecting the main application with the malware that makes the website or application vulnerable.
- **Interference Attack**
  It is also called a Jamming attack. This attack includes disruption of one's network by accessing the network in an unauthorized way. This attack is a bit difficult to perform because while performing the attack the network signals are being disrupted very often. This attack can be performed using Bluetooth, phones, or microwave ovens.
- **Eavesdropping**
  Most of the social site users use mobile as media to connect to these applications. As the social site requires an internet connection, many users connect to

the internet from public places through free wireless connection through Wi-Fi. Hence, connecting themselves to a vulnerable network source leads to the risk of capturing the user's data or the network using different sniffing tools available in the market.

5.4.3 Combination Threats

With the upcoming technology where the world is coming up with new prevention techniques, the attackers combine classic and modern threats to create a more sophisticated attack. The prevention technique of both threats is different and so a combination attack can create a big problem. This attack can be performed in many ways for example an attacker can hack your account by performing the phishing attack and then can post in your profile with a link containing the clickjacking attack.

5.4.4 Threats Targeting Children

Younger children are the prime victim as classic and modern attacks are not that much specific to an age group, but the below threats are intentionally for the young youth.

- **Cyberbullying**: This is done using electronic devices such as mobile phones and PCs in which these devices are used as media to communicate, share, and post negative content about someone such as sending text messages and communicating through social media websites, online blogs, online forums, or groups, etc.
- **Online Predators**: It relates to the personal information safety of children. Livingstone and Haddon of EU Kids Online defined a typology to understand the risk and harm related to the following online activities: harm from content (a child's exposure to pornography or harmful sexual content), harm from contact (a child who is contacted by an adult or another child for sexual abuse), and harm from conduct (the child as an active initiator of abusive or risky behaviors) [4].

5.5 Prevention Techniques and Strategies

5.5.1 Never Connect to Open Wi-Fi Networks

In most of the cases the user is not concerned about the forms of attacks which can be performed using the free network as such. But keep in mind that the network may not be secured and may use the information shared by you using that internet connection. The prevention of this is too possible use your mobile data or else use a VPN service like 1.1.1.1 by Cloudflare which makes your data secured and faster free of cost.

5.5.2 Check Before You Post

Always remember that any information of data you share on the internet is always available to everyone if you make that information public and even the deleted data from your side also remain on the web in form of cache in the other user's browser who viewed your data. So, you have to be careful while posting anything which you are comfortable with anyone viewing that data. Another thing to keep in mind is to limit the amount of information we post, we should never disclose our credentials such as phone number and home address to the public.

5.5.3 Limit the Number of People

With the increase in the number of users, it is easier for someone to communicate with different people and misuse this feature by pretending as someone else. To prevent this we should keep the friends we know and be careful about sharing the level of information with everyone. It is better to avoid interaction with a stranger.

5.5.4 Avoid Rumors

Never believe in information shared on the web as already discussed. Social media could be a great platform to spread information quickly but the shared information has no confirmation that it is true as it is being posted by an individual and hence can be a rumor. To avoid these, an individual should use common sense to figure out the authenticity of the information.

5.5.5 Use New Authentication Techniques

As we know with the use of phishing attacks, we can find out the username and password of the user but what if we have a two-factor authentication which means a user must go through two verification processes to be signed into the profile. So, the attacker may know the username and password but fails while trying the second authentication process. Almost all the sites have adopted this technique. The new method also includes multi-factor authentication such as face and voice recognition and fingerprint scanning.

5.5.6 Use of Antivirus

It is always better to be on the safer side and the best possible way to do so is by installing an antivirus which will minimize your concern and take care of data. Using a free antivirus could be good for students not having money to afford but it is always recommended to buy a subscription to any popular antivirus as paid stuff will always perform better than a free version.

5.5.7 Make Password Strong

Password is the first and the main authentication keyword for a user's data, hence it becomes very important to make it strong. There are possible ways to crack the weak password using the possible word phrases of your date of birth, phone number, etc. To make the password strong you must always use a random combination of uppercase, capital, and lowercase letter with different numbers and symbols. Using a different password for different websites and accounts is always recommended.

5.5.8 Keep Software Update

Updating the application every time a new update is available is important so that the application could have the latest algorithms which would block the attacker to attack on the breaches present in the previous versions. The updating is also useful to come up with the new features and to make the latest operating system compatible with that version of the app.

5.5.9 Analyze the Setting

This is true that without your permission or any action, no one can get access to your account data. So, it is essential to stay updated with the latest privacy settings. The default settings may allow everyone to view your profile data, but they can be customized. You can restrict access to only a few people whom you are comfortable with when seeing your data. Check the privacy setting regularly as new privacy settings are being added very frequently to make the account safer.

5.5.10 Observe Your Children

Children are prime victims of threats as they are not familiar with the vulnerabilities. For example, a child can try to misrepresent his age to bypass the age restrictions on different sites so If you are a parent, you should apply all the abovementioned techniques to your children's social profile. Activity monitoring software can be used to keep an eye on the children.

5.6 Future Scope

There are several existing and emerging attacks in the field of social network privacy and security which could lead to the creation of new researchers who can provide a much considerable solution to these problems. To prevent the strategies adopted by the new era of hackers we need to create a comprehensive mixture of different solutions available. For example, using different algorithms experts can predict private traits of almost all users such as age and gender, based on the number of likes they give to a certain post and combine this technique to find out the fake profiles in a social network. Other techniques involve flashing warning messages on the user to check the posted information as it could be sensitive and therefore algorithm advice not to post that information. Eventually, these methods could effectively help in detecting malware, miscellaneous activity, fake profile, fake URLs, phishing attacks, and all other threats. Research should come up with new techniques to educate users about privacy and policies by providing them with the appropriate settings for their profiles.

5.7 Conclusion

The usage of social media websites is increasing to a great extent and on average a person spends more time on social media than any of the other online activities. So, privacy is a big concern and the only way to resolve the privacy and security issues is to make the audience aware of gathering details about how someone can be easily tricked through these applications. We have recommended ten solutions and precautions which can be taken care of while using social networks. We also request users to educate the surroundings regarding online threats. Users must be careful while sharing stuff, adding a new friend on social media as there is no way to identify the other user's real identity. A great way of preventing these threats is keeping your systems clean—having no malware or virus (use antivirus) and software up to date. As far as future research is concerned social networks and other organizations are coming up with applications having an improved version of security features. In the end, our security is in our hands only; nevertheless having the state-of-the-art security solution could only help if we pay attention. Thus while using the social network correct precautions need to be taken by the user to preserve his security.

References

1. Kaven William, Andrew Boyd, Scott Densten, Ron Chin, Diana Diamond, Chris Morgenthaler, "Social Networking Privacy Behaviors and Risks", Seidenberg School of CSIS, Pace University, White Plains, NY, January 2009.
2. Gilberto Tadayoshi Hashimoto, Pedro Frosi Rosa, Edmo Lopes Filho, Jayme Tadeu Machado, "A Security Framework to Protect Against Social Networks Services Threats", *2010 Fifth International Conference on Systems and Networks Communications*, IEEE, Nice, France, 2010.
3. Abhishek Kumar, Subham Kumar Gupta, Animesh Kumar Rai, Sapna Sinha, "Social Networking Sites and Their Security Issues", *International Journal of Scientific and Research Publications*, Vol. 3, No. 4, pp. 1–5, April 2013.
4. Michael Fire, Roy Goldschmidt, Yuval Elovici, "Online Social Networks: Threats and Solutions", *IEEE Communication Surveys and Tutorials*, Vol. 16. No. 4, pp. 2019–2036, 2014.
5. N. Senthil Kumar, K. Saravanakumar, K. Deepa, "On Privacy and Security in Social Media – A Comprehensive Study", *International Conference on Information Security & Privacy (ICISP2015)*, Nagpur, India, 11–12 December 2015.
6. Sangeeta Kumari, Shailendra Singh, "A Critical Analysis of Privacy and Security on Social Media", *Fifth International Conference on Communication Systems and Network Technologies*, Gwalior, India, 2015.
7. Mukesh Kumar, Nupur Sharma, Shreesh Kumar Shrivastava, "Online Social Networks: Privacy Challenges and Proposed Security Framework for Facebook", *International Journal of Soft Computing and Engineering*, Vol. 4, No. 1, pp. 129–133, March 2014.
8. Damera Vijay Kumar, P. Satya Shekar Varma, Shyam Sunder Pabboju, "Security Issues in Social Networking", *International Journal of Computer Science and Network Security*, Vol. 13, No. 6, June 2013.
9. Matthew Smith, Christian Szongott, Benjamin Henne, Gabriele von Voigt, "Big Data Privacy Issues in Public Social Media", IEEE, Campione d'Italia, Italy, 2013.
10. Leucio Antonio Cutillo, Refik Molva, Thorsten Strufe, "Privacy Preserving Social Networking through Decentralization", *WONS L 2009 6th International Conference on Wireless On-Demand Network Systems and Services*, ACM, New York, 2007, pp. 357–366.
11. Abdullah Al Hasib, "Threats of Online Social Networks", *International Journal of Computer Science and Network Security*, Vol. 9, No. 11, pp. 288–293, November 2009.
12. Anchises M. G. de Paula, "Security Aspects and Future Trends of Social Networks", *International Journal of Forensic Computer Science*, Vol. 1, pp. 60–79, 2010.
13. Number of Users in Social Media. Available at: https://www.statista.com/graphic/1/278414/number-of-worldwide-social-network-users.jpg [accessed on 20-Jan-2019].

14. Threats Percentage-Pose on Social Network Sites. Available at: https://www.researchgate.net/profile/Wajeb_Gharibi/publication/221663523/figure/fig2/AS:669097822023690@1536536760666/Threats-percentage-pose-on-social-networks-Sophos-2010-Security-Threat-Report.jpg [accessed on 11-Feb-2019].
15. Privacy Setting on Facebook. Available at: https://www.facebook.com [accessed on 11-Feb-2019].
16. Qiang Cao, Michael Sirivianos, Xiaowei Yang, Tiago Pregueiro, "Aiding the Detection of Fake Accounts in Large Scale Social Online Services," *Proceedings of 9th USENIX Conference, NSDI*, 2012, p. 15. [Online]. Available at: http://dl.acm.org/citation.cfm?id=2228298.2228319.
17. Gianluca Stringhini et al., "Follow the Green: Growth and dynamics in twitter follower markets", *Proceedings of the 2013 Conference Internet Measurement Conference*, 2013, pp. 163–176.
18. http://mobile.scmagazineuk.com/a-lack-of-security-on-social-networkingsites-causes-problems-for-businesses/marticle/173602/.
19. Anna C. Squicciarini, Mohamed Shehab, "Privacy policies for shared content in social network sites," Proceedings of the CHI'10, ACM Press, Boston, MA, 30 June 2010.
20. Yabing Liu, Krishna P. Gummadi, "Analyzing Facebook Privacy Settings: User Expectations vs. Reality", *Proceedings of the 10th International Conference on Privacy Enhancing Technologies*, New York, 2011.

6

Role of Image Processing in Artificial Intelligence and Internet of Things

Sudhriti Sengupta
Department of Computer Science and Engineering, Galgotias University, Greater Noida, India

CONTENTS

6.1 Introduction

The Internet of Things (IoT) denotes the interconnection of interrelated devices, called "thing," by using communication technology to collect and transfer data without intervention from a human. IoT promises to transform the business model and lifestyle rapidly. Sensors, actuators, and other devices can collect and analyze data in real time to effect different aspects of different domains [1,2]. Along with IoT, another branch impacting the development of automatized modern society is image processing. This method performs a variety of operations on an image to receive an enhanced image or to extract useful information from it. Both IoT and image processing have wide applications in a variety of industries independently. The individual applications of this field have been successful in many areas. The integration of these techniques is also viable and yields optimum outcomes [1–12]. The recent explosion of "huge records" has ushered in a new generation age of artificial intelligence (AI) algorithms in a lot of fields of technological interest. In this chapter, the applications and general architecture of IoT, steps in image processing, the importance of image processing in IoT, and the role of AI in image processing and IoT are discussed.

DOI: 10.1201/9781003244165-8

6.2 Image Processing

Image processing is the process of applying some procedures to an image so that we can get a more desirable output in the form of an image or extract some beneficial data from it. Image processing is a category of signal processing in which the image is the input data. The output may be an image or attributes associated with that image. Effectively making use of unstructured facts from large amounts of image and voice statistics proves to be pretty challenging for data mining professionals [13]. The processing of unstructured statistics usually includes the usage of deep learning algorithms and those algorithms can be daunting for novices. In addition, processing unstructured records usually calls for powerful graphics processing units and a large number of computing sources. The advancement of AI and big facts technology has helped gas improvements within the retail enterprise. Cutting-part products, inclusive of Alibaba Cloud's Image Search and machine learning (ML) platform for AI, have changed the means of customers' interaction and shopping for products. Customers now do not want to queue up in brick-and-mortar stores; they can also conveniently search for products by means of performing a short photograph seek [5,14–28].

However, the use of image-based search generation is not confined to an easy product search on e-commerce platforms in the retail enterprise. In latest years, we've witnessed an increasing number of projects to offer new clients easy content-based image search. This consists of using image look for inventory retaining unit, and additionally for matching merchandise with supply substances and or styles.

There are two major categories of image processing namely, analog and digital image processing. Analog image processing generally uses hard copies like printouts. Researchers use different operations of interpretation while the usage of those visible strategies. Digital image processing methods enable to manipulate digital photos by way of the use of computers. The three well-known phases that information needs to go through at the same time as the use of the digital approach are pre-processing, enhancement, and output [29–36].

6.3 AI Solutions for Image Processing

Researchers have achieved extraordinary advances in unleashing the power of AI on applications related to image processing. Techniques of image processing collaborated with AI are extensively utilized in different domains, such as medicinal drugs, law enforcement, cybersecurity, and retail. AI integrated with image processing has paved the way for computing devices to acquire new functionalities by using the method of ML [6–28,33–42]. Some of these areas are as follows:

- Image category
- Object recognition
- Object tracking
- Image generation
- Image retrieval

For utilizing the different image processing functionality with AI, some tools and resources such as datasets, libraries, and frameworks are also required. Researchers can utilize loads of open-source services to use to make develop AI-based image processing applications.

The following frameworks will assist in better knowledge or function of AI in image processing:

6.3.1 OpenCV

An important open-source library for image processing and computer vision application is the Open Source Computer Vision Library or OpenCV. OpenCV was created to give general architecture for computer vision packages. OpenCV is a Berkeley Software Distribution-certified product and it is easier to use in different areas or programs. It's a free prescient library which allows users to carry out different methods of image processing such as:

- Input
- Compression and decompression
- Enhancement and recovery
- Denoising
- Segmentation
- Data extraction.

This library contains more than 2,500 optimized algorithms, which incorporates a comprehensive set of traditional image processing techniques along with the latest trends like ML.

6.3.2 TensorFlow

TensorFlow is an open-source software program library for various applications. It is utilized in a huge range of obligations but is specifically focused on training and inference of deep neural networks. We can use this to process the veracity of information, related to image processing, such as detection of an object, feature extraction, recognition, and feature extraction. TensorFlow consists of libraries to develop and train customized deep learning and neural networks. One of the main advantages of this framework is that it supports Jupyter Notebook and is an easy coding platform. TensorFlow is compatible with many famous languages, which include C++, Python, Java, Rust, and Go.

6.3.3 Keras

Keras is an application programming interface designed for an artificial neural network. Keras follows practices for reducing cognitive load while giving regular and simple application programming interfaces. It minimizes the range of consumer actions required for common use instances, and it affords clear and actionable mistake messages. It also has great documentation and developer publications.

It is widely popular for those who simply start to use machine gaining knowledge of algorithms of their tasks as it clarifies the making of a deep getting to know the model from scratch. Keras is simple to control and a prototype of different styles of neural networks. The library became built on TensorFlow and is presently absolutely integrated into the framework. This approach that one may write its deep studying model in Keras, because it has a much greater user-friendly interface, and then effortlessly put in force a selected functionality or feature from TensorFlow on this model. It can also be located on top of other popular AI frameworks consisting of Microsoft Cognitive Toolkit and Theano.

6.3.4 VXL

The VXL is fixed for libraries of computer vision and image processing. It implements some of the popular computing imaging and prescient technology algorithms and associated functionalities.

6.3.5 AForge.NET

AForge.NET is a computer programming library with a couple of libraries that may be used for picture or image processing and computer vision to utilize neural networks and fuzzy computations [37].

6.4 Advantages of the Use of AI in Image Processing

The implementation of image processing strategies has had a massive effect on many application domains and industries [10–28,33–42]. Following are most of the most beneficial blessings of image processing, regardless of the sector of operation:

- **Processing Is Less Difficult and Fast**
  It allows the customers and industries to feedback for their merchandise and service. It's a feasible way to the life of computer vision in rapid computers.
- **Reliability**
  Computer systems and cameras don't have the human aspect of tiredness which is eliminated from it. Efficiency commonly stays equal; it doesn't rely on outside elements together with illness or sentimental repute.
- **Accuracy**
  The precision of computer imaging and computer vision will make sure a higher accuracy on the final product.
- **An Extensive Variety of Use**
  We can see the identical laptop machine in numerous special fields and activities. We can also find this in factories with warehouse monitoring and transport of elements and in the clinical industry via scanned pics, among different alternatives.
- **The Reduction of Prices**
  Time and mistakes fees are reduced inside the system of computer imaging. The computer vision techniques will require the help of experts in imaging and designing variety of tools and applications.

(i) The virtual picture can be made to be had in any desired format (improve photo, X-Ray, bad photograph, and many others).
(ii) It allows improving pictures for human interpretation.
(iii) Information may be processed and extracted from pictures for device interpretation.

(iv) The pixels in the photograph can be manipulated to any desired density and contrast.

(v) Images can be stored and retrieved without difficulty.

(vi) It allows for easy electronic transmission of photographs to 0.33-birthday party companies.

6.5 Challenges of AI in Image Processing

There is no doubt about the importance and role of AI in image processing in the modern-day application domain [12–28,36–42]. However, these approaches have certain challenges as well.

- **Lack of Specialists**: The computer vision era entails the use of AI and ML. To train a computer vision application powered by means of AI and ML, corporations want to have a group of professionals with technical understanding. Without them, building a machine which could analyze the surrounding details isn't always feasible.
- **Need for Constant Monitoring**: Being a machine, the devices using this application can break down or have a technical glitch. To avoid this, agencies must get a devoted group onboard for everyday monitoring and assessment.
- **Necessity for Experts**: There's a massive necessity for professionals related to the sector of ML and AI. A large number of people are benefitted by this decrease in training rate and other expenditure.
- **Delaying with Failure**: Eliminating the human factor may be suitable in a few cases. But whilst the device or tool fails, it doesn't announce or count on that problem, whereas a human individual can inform in advance when the person won't come. When the device fails due to a glitch, cyberattack, or other software problems, it's quite possible that computer vision and image processing will fail. But if we do not resolve the hassle, the capabilities of the device can disappear. It can completely freeze the production in the case of warehouses.

Despite their current barriers, image processing systems can convey corporations' tremendous possibilities to increase revenue streams, meet productivity dreams, and streamline the decision-making approach.

6.6 Image Processing in IoT

In this section, the role of image processing in IoT-based applications is discussed. There are numerous IoT-based projects which focus on or utilize image processing techniques. Image processing provides methods of image acquiring, enhancement, restoration, filtering, etc., which helps to facilitate the development of IoT-ready applications. In this section, image processing-enabled IoT-based services related to healthcare, agriculture, solutions, smart city, security, etc., are discussed. Image processing in IoT has prominent usage in healthcare sectors, such as clinical monitoring, remote patient surveillance system, and preventive and controlling system. Different types of images related to physical modalities such as X-ray, CT, MRI, and dermascopy images can be integrated with IoT applications for having real-time and distance diagnostics and treatment options. Agricultural technological integrations lead to better production of crops which is a fundamental requirement for life [10–28,33–42]. Greenhouse farming, small to large scale farming uses the IoT technology coupled with image processing applications to do many things such as checking crop quality, growth cycle of crops, presence of pests in the crops, and monitoring of soil or water or other resources necessary for the crop to grow. This led to the core concept of smart farming and agriculture to produce improved crops at lower prices. Moreover, communication with the sellers and retailers of the crops also increases profit for the farmers because they can send the crops to the market at the correct time. The consumers are also benefitted as they get the crops in the best condition of ripeness and quality. Another promising utilization of IoT integrated with image processing techniques is the development of smart cities. Smart cities use information technology methods to optimize operation methods by sharing information in a real-time environment. IoT applications are inevitable in smart city development and control. Many of these IoT applications rely on image processing techniques. Some of the areas which use image processing techniques along with IoT applications are automated street lights, smart parking, smart traffic management, smart security system, etc. In a smart security system, image processing plays an important role in acquiring, enhancing, and detecting modalities from web cameras or IoT sensors on a 24*7 basis [33,35,36,38,39,41]. This includes surveillance of sensitive areas, government infrastructures, monuments, roads, and many more. Some other role of image processing in IoT includes biometric identification system, smart home development, production in industries, etc. With this, it can be comprehended that image processing integrated with IoT applications has many roles in the development of the lifestyle of humans [43–45]. In this chapter, the work of one of the important and popular image processing techniques, that is, filtering is shown practically to display the impact of this in an image which leads to better development of IoT applications. A test image is taken in Figure 6.1.

Figure 6.1 represents the test image taken by the author. The entropy of the image in Figure 6.1 is 7.1553 and the mean of the image is 86.390. Then two popular filtering techniques are applied to the test image.

FIGURE 6.1
Test Image 1.

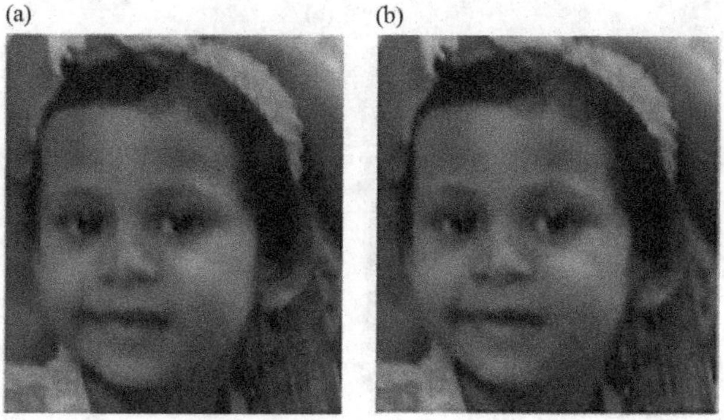

FIGURE 6.2
(a) Test Image 1 after application of median filtering (row wise). (b) Test Image 1 after application of Weiner filtering.

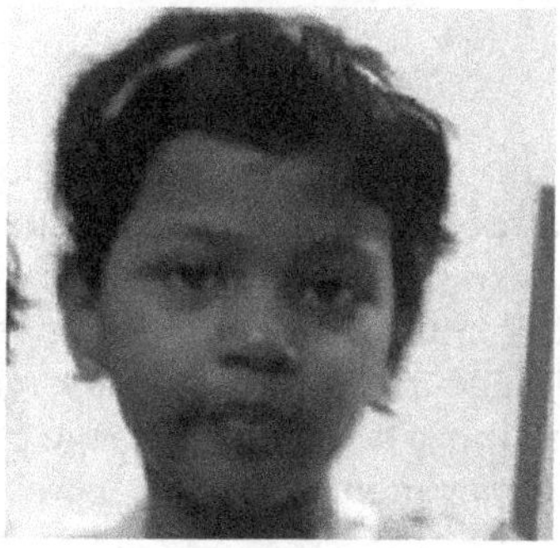

FIGURE 6.3
Test Image 2.

In Figure 6.2a, the entropy and mean are 7.1721 and 86.846, respectively. The image in Figure 6.2a is obtained by applying median filtering in Figure 6.1. Similarly, the entropy and mean of the image in Figure 6.2b after the application of median filtering are 7.2781 and 86.368, respectively. Entropy and mean are two image quality measuring parameters. In both cases of filtering, the resultant images show better quality as per entropy and mean.

To prove the efficiency of image processing techniques in improving the quality of images helping in the development of IoT applications, contrast enhancement techniques are also introduced. Contrast enhancement techniques are used to improve the contrast or visual image. Figure 6.3 represents the test image used to illustrate the contrast enhancement techniques.

In Test Image 2, we have applied two contrast enhancement techniques such as intensity-based and histogram-based enhancement techniques. This is given in Figure 6.4a and b, respectively.

From the above images, the visual cues for perception are better. The image processing techniques show that improved image quality can be used in various IoT applications to get optimal results. Many other image processing techniques are present which is integrated with IoT applications to have efficient and robust results. Image processing techniques create the basic ground for application developers to deploy the power of IoT for better experiences.

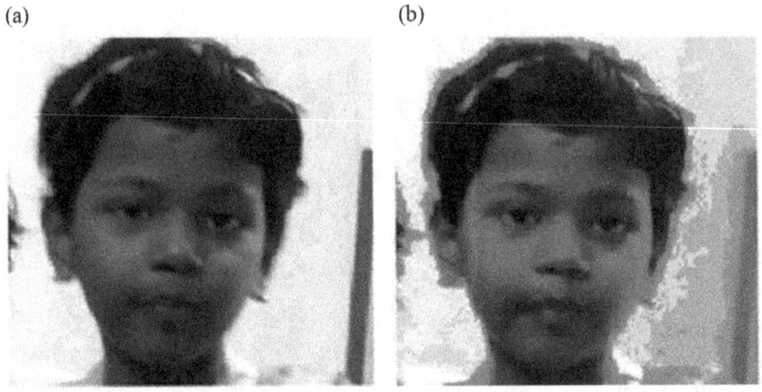

FIGURE 6.4
(a) Test Image 2 after application of contrast enhancement technique based on intensity (row wise). (b) Test Image 2 after application of histogram equalization.

6.7 Conclusion

Image processing techniques are essential in the development of various modern-day applications, such as health monitoring system, CCTV surveillance system, natural disaster management, and incident response team. These are often integrated with IoT in fracture to usher the development of automation and enhanced lifestyle. The tools which are popularly used for AI-based image processing are discussed along with the advantages and limitations in image processing. The usage of image processing methods in IoT is also discussed with a comparative experimental result.

References

1. S. Chen, H. Xu, D. Liu, B. Hu and H. Wang, "A Vision of IoT: Applications, Challenges, and Opportunities With China Perspective," *IEEE Internet of Things Journal*, vol. 1, no. 4, pp. 349–359, August 2014, doi: 10.1109/JIOT.2014.2337336.
2. A. Zanella, N. Bui, A. Castellani, L. Vangelista and M. Zorzi, "Internet of Things for Smart Cities," *IEEE Internet of Things Journal*, vol. 1, no. 1, pp. 22–32, February 2014, doi: 10.1109/JIOT.2014.2306328.
3. M. Dinesh and K. Sudhaman, "Real Time Intelligent Image Processing System with High Speed Secured Internet of Things: Image Processor with IOT," *2016 International Conference on Information Communication and Embedded Systems (ICICES)*, 2016, pp. 1–5, doi: 10.1109/ICICES.2016.7518840.
4. A. Kapoor, S. I. Bhat, S. Shidnal and A. Mehra, "Implementation of IoT (Internet of Things) and Image Processing in Smart Agriculture," *2016 International Conference on Computation System and Information Technology for Sustainable Solutions (CSITSS)*, 2016, pp. 21–26, doi: 10.1109/CSITSS.2016.7779434.
5. L. Sharma and N. Lohan, "Internet of Things with Object Detection: Challenges, Applications, and Solutions," *Handbook of Research on Big Data and the IoT*, 2019, https://www.igi-global.com/chapter/internet-of-things-with-object-detection/224265.
6. L. Sharma and N. Lohan, "Performance Analysis of Moving Object Detection Using BGS Techniques in Visual Surveillance", *International Journal of Spatiotemporal Data Science*, Inderscience, vol. 1, pp. 22–53, January 2019.

7. A. Anand, V. Jha and L. Sharma, "An Improved Local Binary Patterns Histograms Techniques for Face Recognition for Real Time Application," *International Journal of Recent Technology and Engineering*, vol. 8, no. 2S7, pp. 524–529, July 2019.
8. L. Sharma and D. K. Yadav, "Histogram Based Adaptive Learning Rate for Background Modelling and Moving Object Detection in Video Surveillance," *International Journal of Telemedicine and Clinical Practices*, Inderscience, June 2016, ISSN: 2052-8442, doi: 10.1504/IJTMCP.2017.082107.
9. L. Sharma and N. Lohan, "Performance Analysis of Moving Object Detection Using BGS Techniques," *International Journal of Spatio-Temporal Data Science*, Inderscience, Vol. 1, pp. 22–53, February 2019.
10. L. Sharma and N. Lohan, "Internet of Things with Object Detection," *Handbook of Research on Big Data and the IoT*, IGI Global, pp. 89–100, March, 2019, ISBN: 9781522574323, doi: 10.4018/978-1-5225-7432-3.ch006.
11. L. Sharma, "Introduction: From Visual Surveillance to Internet of Things," *From Visual Surveillance to Internet of Things*, Taylor & Francis, CRC Press, Boca Raton, FL, vol. 1, p. 14, 2018.
12. L. Sharma and P. K. Garg, "Block based Adaptive Learning Rate for Moving Person Detection in Video Surveillance," *From Visual Surveillance to Internet of Things*, Taylor & Francis, CRC Press, Boca Raton, FL, vol. 1, p. 201.
13. J. Roe, "A review of applications of artificial intelligence techniques to naval ESM signal processing," *IEE Colloquium on Application of Artificial Intelligence Techniques to Signal Processing*, 1989, pp. 5/1–5/5.
14. A. Wajhal and S. Sengupta, "Analysis of Biometric Modalities," *Proceeding of the International Conference on Computer Networks, Big Data and IoT (ICCBI – 2019). ICCBI 2019.* Lecture Notes on Data Engineering and Communication Technology, Springer, Cham, vol. 49. pp. 281–290, 2020.
15. S. Makkar and L. Sharma, "A Face Detection Using Support Vector Machine: Challenging Issues, Recent Trend, Solutions and Proposed Framework," In: Singh M., Gupta P., Tyagi V., Flusser J., Ören T., Kashyap R. (eds) *Advances in Computing and Data Sciences. ICACDS 2019.* Communications in Computer and Information Science, Springer, Singapore, vol. 1046, 2019. doi: 10.1007/978-981-13-9942-8_1.
16. L. Sharma and P. K. Garg, "IoT and Its Applications," *From Visual Surveillance to Internet of Things*, Taylor & Francis, CRC Press, Boca Raton, FL, vol. 1, p. 29.
17. L. Sharma, D. K. Yadav and S. K. Bharti, "An Improved Method for Visual Surveillance Using Background Subtraction Technique," *2015 2nd International Conference on Signal Processing and Integrated Networks (SPIN)*, 2015, pp. 421–426, doi: 10.1109/SPIN.2015.7095253.
18. D. K. Yadav, L. Sharma and S. K. Bharti, "Moving Object Detection in Real-Time Visual Surveillance Using Background Subtraction Technique," *2014 14th International Conference on Hybrid Intelligent Systems*, 2014, pp. 79–84, doi: 10.1109/HIS.2014.7086176.
19. L. Sharma, A. Singh and D. K. Yadav, "Fisher's Linear Discriminant Ratio based Threshold for Moving Human Detection in Thermal Video," *Infrared Physics and Technology*, Elsevier, vol. 78, pp. 118–128, March 2016.
20. L. Sharma (Ed.), *Towards Smart World*, Chapman and Hall/CRC, New York, 2021, doi: 10.1201/9781003056751.
21. L. Sharma, "Human Detection and Tracking Using Background Subtraction in Visual Surveillance," *Towards Smart World*, Chapman and Hall/CRC, New York, pp. 317–328, December 2020, doi: 10.1201/9781003056751.
22. L. Sharma, D. K. Yadav and S. K. Bharti, "An Improved Method for Visual Surveillance using Background Subtraction Technique," *IEEE, 2nd International Conference on Signal Processing and Integrated Networks (SPIN-2015)*, Amity University, Noida, India, February 19–20, 2015.
23. D. K. Yadav, L. Sharma and S. K. Bharti, "Moving Object Detection in Real-Time Visual Surveillance using Background Subtraction Technique," *14th International Conference in Hybrid Intelligent Computing (HIS-2014)*, IEEE, Gulf University for Science and Technology, Kuwait, December 14–16, 2014.

24. L. Sharma, D. K. Yadav and M. Kumar, "A Morphological Approach for Human Skin Detection in Color Images", *2nd National Conference on "Emerging Trends in Intelligent Computing & Communication", GCET*, Gr. Noida, India, 26–27 April 2013.
25. L. Sharma, S. Sengupta and B. Kumar, "An Improved Technique for Enhancement of Satellite Images," *Journal of Physics: Conference Series*, vol. 1714, p. 012051, January 2021.
26. S. Singh, L. Sharma and B. Kumar, "A Machine Learning Based Predictive Model for Coronavirus Pandemic Scenario," *Journal of Physics: Conference Series*, vol. 1714, p. 012023, January 2021.
27. G. Jha, L. Sharma and S. Gupta, "Future of Augmented Reality in Healthcare Department," In: Singh P.K., Wierzchoń S.T., Tanwar S., Ganzha M., Rodrigues J.J.P.C. (eds) *Proceedings of Second International Conference on Computing, Communications, and Cyber-Security*. Lecture Notes in Networks and Systems, vol. 203, Springer, Singapore, 2021, doi: 10.1007/978-981-16-0733-2_47.
28. G. Jha, L. Sharma and S. Gupta, "E-Health in Internet of Things (IoT) in Real-Time Scenario," In: Singh P.K., Wierzchoń S.T., Tanwar S., Ganzha M., Rodrigues J.J.P.C. (eds) *Proceedings of Second International Conference on Computing, Communications, and Cyber-Security*. Lecture Notes in Networks and Systems, vol. 203, Springer, Singapore, 2021, doi: 10.1007/978-981-16-0733-2_48.
29. R.C. Gonzalez and R.E. Woods, *Digital Image Processing*, 4th Edition, Pearson, New York, 2018.
30. S. Tyagi and S. Sengupta, "Role of AI in Gaming and Animation," *Proceeding of the International Conference on Computer Networks, Big Data and IoT (ICCBI – 2019). ICCBI 2019*. Lecture Notes on Data Engineering and Communication Technology, Springer, Cham, vol. 49. pp. 259–267, 2020.
31. S. Sengupta and N. Astha, "Comparative Analysis of Contrast Enhancement Techniques for MRI Images," *Proceeding of the International Conference on Computer Networks, Big Data and IoT (ICCBI – 2019). ICCBI 2019*. Lecture Notes on Data Engineering and Communication Technology, Springer, Cham, vol. 49. pp. 290–297, 2020.
32. V. Giri and S. Sengupta, "Satellite Image Restoration and Enhancement," *Proceeding of the International Conference on Computer Networks, Big Data and IoT (ICCBI – 2019). ICCBI 2019*. Lecture Notes on Data Engineering and Communication Technology, Springer, Cham, vol. 49. pp. 252–259, 2020.
33. S. Kumar, P. Gupta, S. Lakra, L. Sharma and R. Chatterjee, "The Zeitgeist Juncture of "Big Data" and its Future Trends," *2019 International Conference on Machine Learning, Big Data, Cloud and Parallel Computing (COMITCon)*, 2019, pp. 465–469, doi: 10.1109/COMITCon.2019.8862433.
34. S. Sharma, S. Verma, M. Kumar and L. Sharma, "Use of Motion Capture in 3D Animation: Motion Capture Systems, Challenges, and Recent Trends," *2019 International Conference on Machine Learning, Big Data, Cloud and Parallel Computing (COMITCon)*, 2019, pp. 289–294, doi: 10.1109/COMITCon.2019.8862448.
35. T. Bouwmans, F. Porikli, B. Horferlin and A. Vacavant, *Handbook on Background Modeling and Foreground Detection for Video Surveillance*, CRC Press, Taylor and Francis Group, Boca Raton, FL, 2014.
36. T. Bouwmans, N. Aybat and E. Zahzah, *Handbook on Robust Low-Rank and Sparse Matrix Decomposition : Applications in Image and Video Processing*, Taylor and Francis Group, Boca Raton, FL, 2016.
37. S. Sengupta, "Programming Languages in Artificial Intelligence," *Artificial Intelligence: Technologies, Applications, and Challenges*, 2021.
38. T. Bouwmans, B. Hofer-lin, F. Porikli and A. Vacavant, "Traditional Approaches in Background Modeling for Video Surveillance," *Handbook Background Modeling and Foreground Detection for Video Surveillance*, Taylor and Francis Group, USAJuly 2014.
39. T. Bouwmans, B. Hoferlin, F. Porikli and A. Vacavant, "Recent Approaches in Background Modeling for Video Surveillance," *Handbook Background Modeling and Foreground Detection for Video Surveillance*, Taylor and Francis Group, July 2014.
40. J. Giraldo and T. Bouwmans, "GraphBGS: Background Subtraction via Recovery of Graph Signals," *International Conference on Pattern Recognition, ICPR 2020*, Milan, Italy, January 2021.
41. J. Giraldo and T. Bouwmans, "Semi-Supervised Background Subtraction of Unseen Videos: Minimization of the Total Variation of Graph Signals," *IEEE International Conference on Image Processing, ICIP 2020*, Abu Dhabi, UAE, October 2020.

42. A. Griesser, S. De Roeck and A. Neubeck, "GPU-Based Foreground-Background Segmentation using an Extended Colinearity Criterion", *Vision, Modeling, and Visualization, VMV 2005*, Erlangen, Germany, 2005.
43. D. Kohli and S. Sengupta, "Recent Trends of IoT in Smart City Development," *Proceeding of the International Conference on Computer Networks, Big Data and IoT (ICCBI – 2019). ICCBI 2019.* Lecture Notes on Data Engineering and Communication Technology, Springer, Cham, vol. 49. pp. 376–380, 2020.
44. S. Sen Gupta, M. Shad Khan and T. Sethi, "Latest Trends in Security, Privacy and Trust in IOT," *2019 3rd International conference on Electronics, Communication and Aerospace Technology (ICECA)*, IEEE, Coimbatore, India, 2019, pp. 382–385, doi: 10.1109/ICECA.2019.8822178.
45. S. Sengupta, "A Study of Bluetooth Smart for Medical Application," *2015 4th International Conference on Reliability, Infocom Technologies and Optimization (ICRITO) (Trends and Future Directions)*, IEEE, Noida, 2015, pp. 1–3, doi: 10.1109/ICRITO.2015.7359250.

7

Computer Vision in Surgical Operating Theatre and Medical Imaging

Mukesh Carpenter
Department of Surgery, Alshifa Multispecialty Hospital, Okhla, India

Dharmendra Carpenter
Department of Anaesthesiology & Critical Care, Narayana Multispecialty Hospital, Jaipur, India

Vinod Kumar Jangid
Department of Respiratory Medicine, Medical College Kota, Rajasthan, India

Lavanya Sharma
Amity Institute of Information Technology, Amity University, Noida, India

CONTENTS

DOI: 10.1201/9781003244165-9

7.1 Introduction

Computer vision (CV) is a branch of artificial intelligence (AI) which aims to mimic the tasks done by human vision. The processing of images is taken from digital cameras or video streams to detect an object of interest and classify them accordingly. In the medical domain, radiologists find a pattern of irregularity in the tissues or bones of a person. The work done by radiologists makes CV an ideal choice to find the irregularity in a more accurate manner and overall results in productivity improvement. CV technology provides potential support in various applications to deliver lifesaving functionalities for patients [1–6]. This emerging technology focuses on training a system to replicate human insight and to understand the object. In healthcare, it helps in early diagnosis and treatment of illness, medical imaging, clinical trial reduction, surgeries (laparoscopic and robotics surgery). In the last few decades, both CV and AI play a vital role in changing the healthcare system. CV algorithms detect anomalies and patterns in medical images to obtain a diagnosis [7–20]. In combination with machine learning (ML) and neural networks, CV detects, evaluates, and interprets the images. The main aim of CV in healthcare care is to provide an accurate diagnosis, timely detection, and clinicals in critical situations (see Figure 7.1). The major benefits of AI-driven CV technology are listed below:

- Increase in productivity of clinicals.
- Screening and sortation of an image from the queue.
- Automatic updating of records.
- Lessen mistakes in manual checking of an image.
- Monitoring and quantification of the infected region across a time frame.

The state of art in AI and CV has various applications which process a huge amount of data to detect the pattern that may evade human eyes. In many situations, minutes of detail have great importance and can be the difference between life and death [7]. The visual pattern

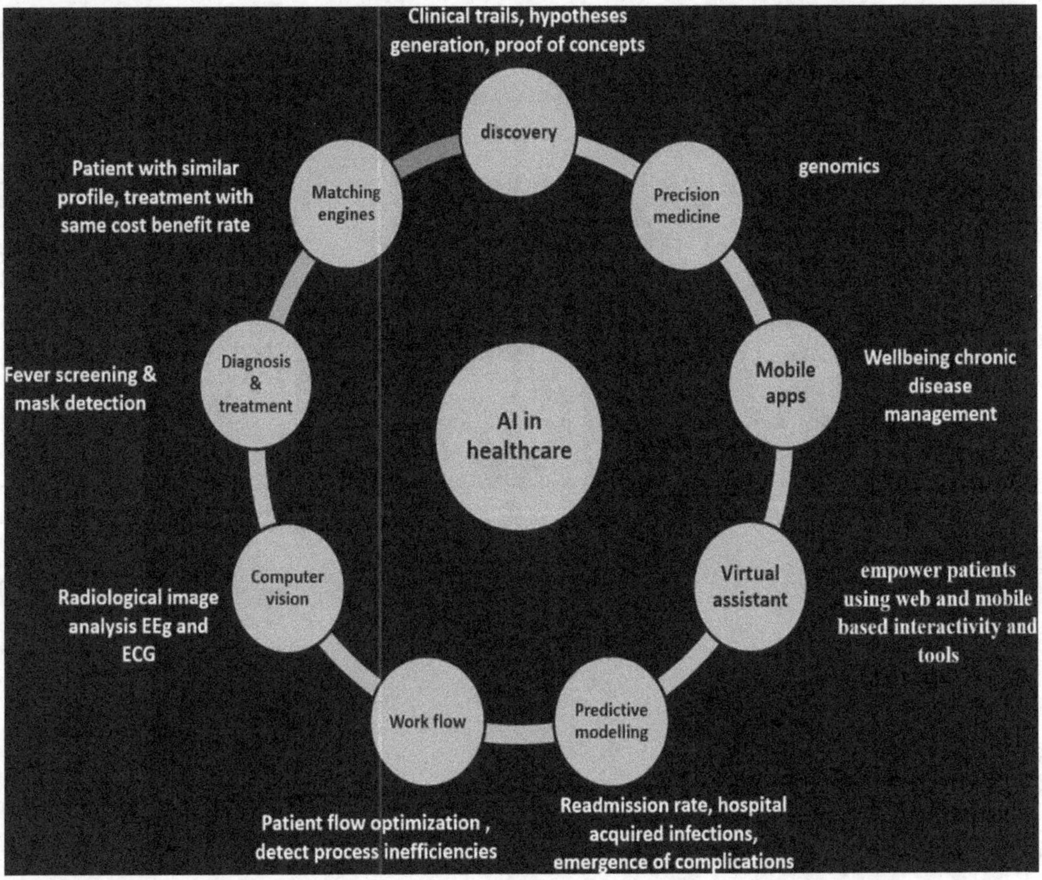

FIGURE 7.1
Applications of AI and CV in healthcare.

recognition capability of AI-driven CV technology transforms the healthcare industry in many ways such as improved diagnosis (X-Ray, CT scan, MRI, EEG, ECG), timely treatment of illness, and increased efficiency in medical procedures (see Figure 7.2).

7.2 Evolution of CV

This technology exists for a decade and initial techniques were based on mathematical tools and image processing that perform edge detection, segmentation in the images, identifying the object from the background. But transformation begins in 2010 with a combination of deep learning (DL). DL algorithms more accurately detect the object in complex situations without explicitly defining the pattern. Recently, a lot of advancement in DL-based CV technology can be seen with a 90% accuracy rate as shown in Figure 7.3. Some of the notable use cases are face detection, video surveillance, automatic vehicles, gaming, robotics, healthcare (radiology, oncology, lab test, surgery, cardiology, dermatology), and defense [10,11,13,21,22].

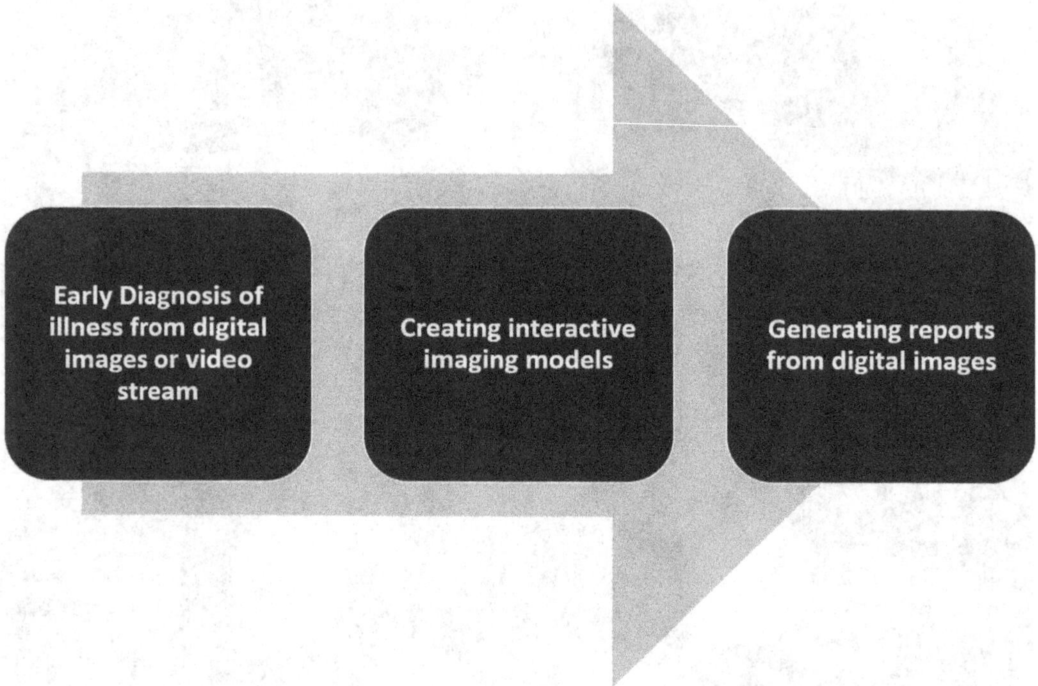

FIGURE 7.2
AI-driven CV technology in healthcare.

7.3 Medical Imaging Techniques

Medical imaging (MI) is a noninvasive technique of imaging inside of a patient's body for medical analysis. This technique is most widely used for bone and tissue examination for any anomalies [22–33]. A radiologist can do the interpretation of the images. MI techniques are as follows.

7.3.1 X-Ray

This technique uses ionized radiations which penetrate various parts of the patient body to produce digital images of bones and tissues. In the medical field, this technique is widely used for the diagnosis of illness or abnormality. This MI technique comprises digitizing devices such as image plates or flat panel detectors [29–31,34]. Recently, this technique is widely used to detect Covid-19 (a deadly virus) infection in the lungs as shown in Figure 7.4. Early symptoms of this virus are fever, breathing difficulty, dropping of saturation level, sore throat, and headache that can be detected using the X-Ray technique.

7.3.2 Computed Tomography (CT)

It is a diagnostic imaging technique which is used to generate detailed information of internal body organs, bones, soft tissue, and blood vessels. It is a useful diagnostic tool for detecting illness and internal injuries. It comprises a series of X-Rays taken from different

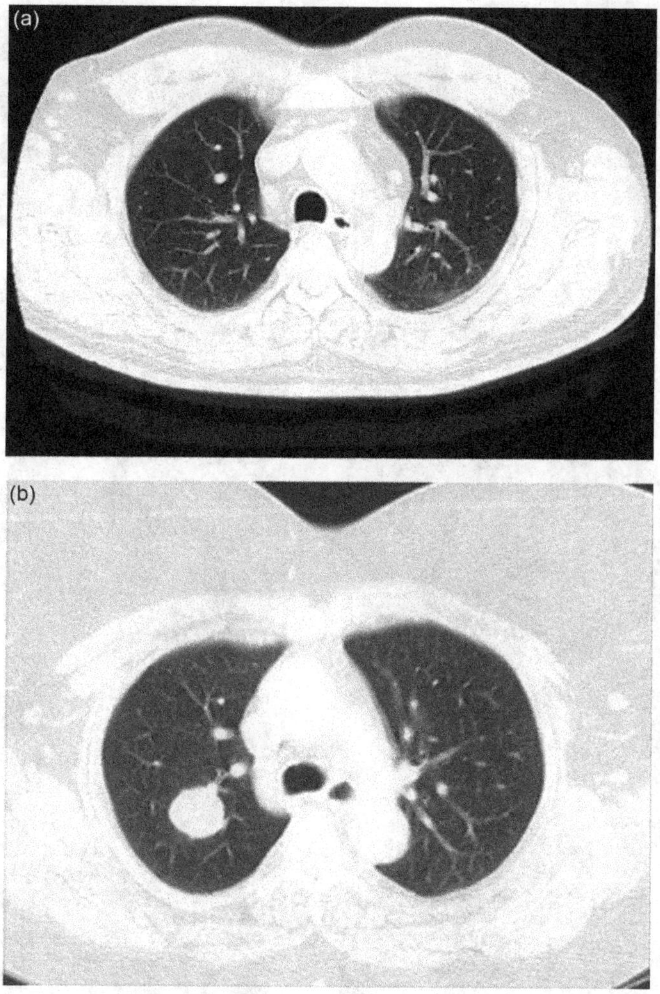

FIGURE 7.3
Lung cancer detection in CT images using DL. (a) Normal chest. (b) Right chest lung cancer.

angles around the patient's body and a system that produce a 3D image of soft tissues and bones. This technique provides more detailed information as compared to X-rays. In this technique, X-Rays are used which produce ionizing radiation which is harmful and sometimes may damage DNA and often lead to cancer. A patient diagnosed with fever for six days is recommended for RT-PCR and CT as shown in Figure 7.5. Early diagnosis of Covid-19 is crucial for treatment and control as compared to the RT-PCR test. CT technique provides 98% sensitivity for Covid-19 infection as compared to RT-PCR with 71% sensitivity. It is considered a routine tool to diagnose pneumonia at an early stage in a faster manner [10,29,35].

7.3.3 Magnetic Resonance Imaging (MRI)

This is an MI technique used to form images of the anatomy and physiological process of the human body. It comprises magnetic fields, magnetic field gradients, and system-generated

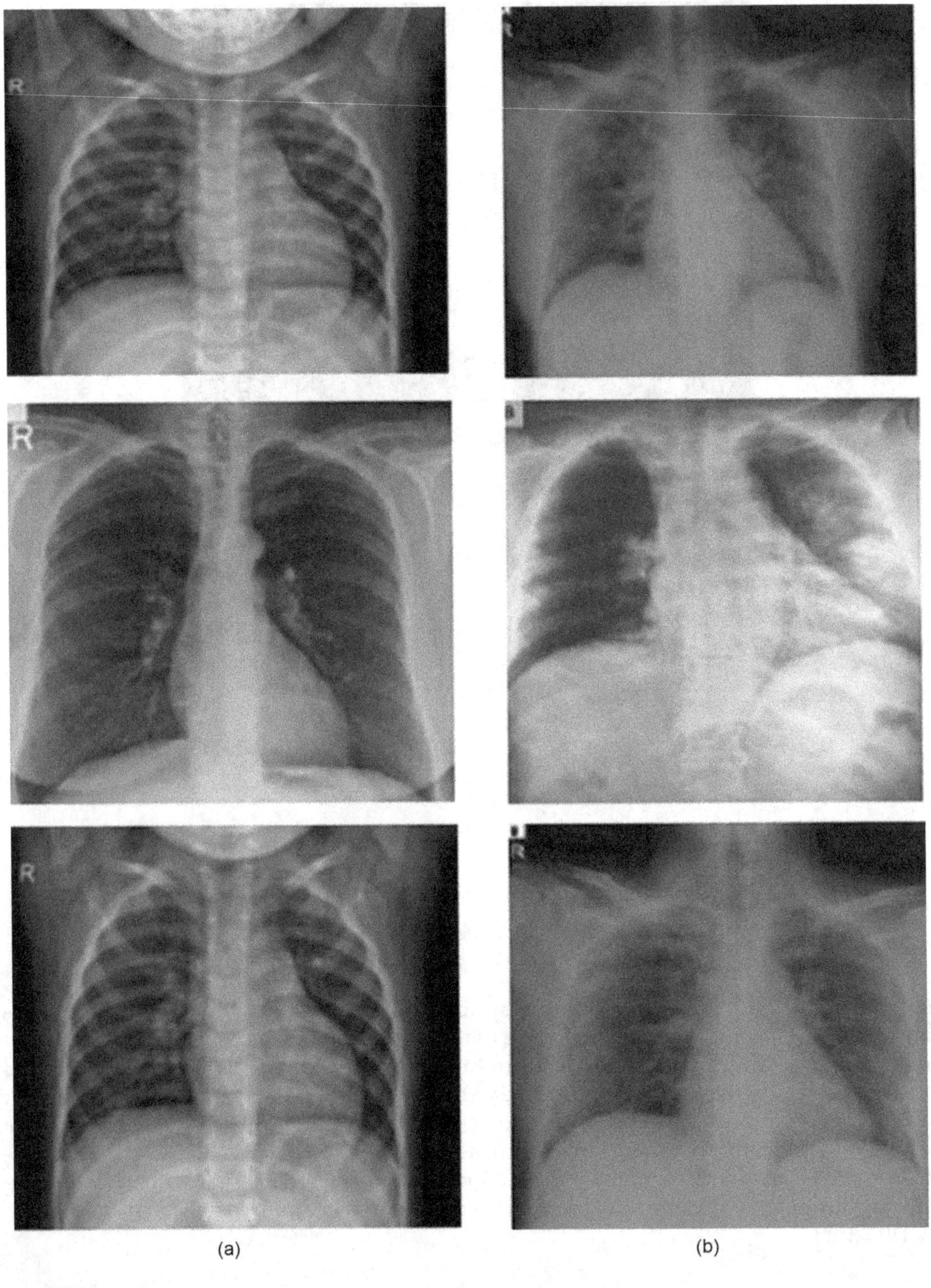

FIGURE 7.4
MI using X-Ray. (a) Normal lungs. (b) Covid-19 infected lungs.

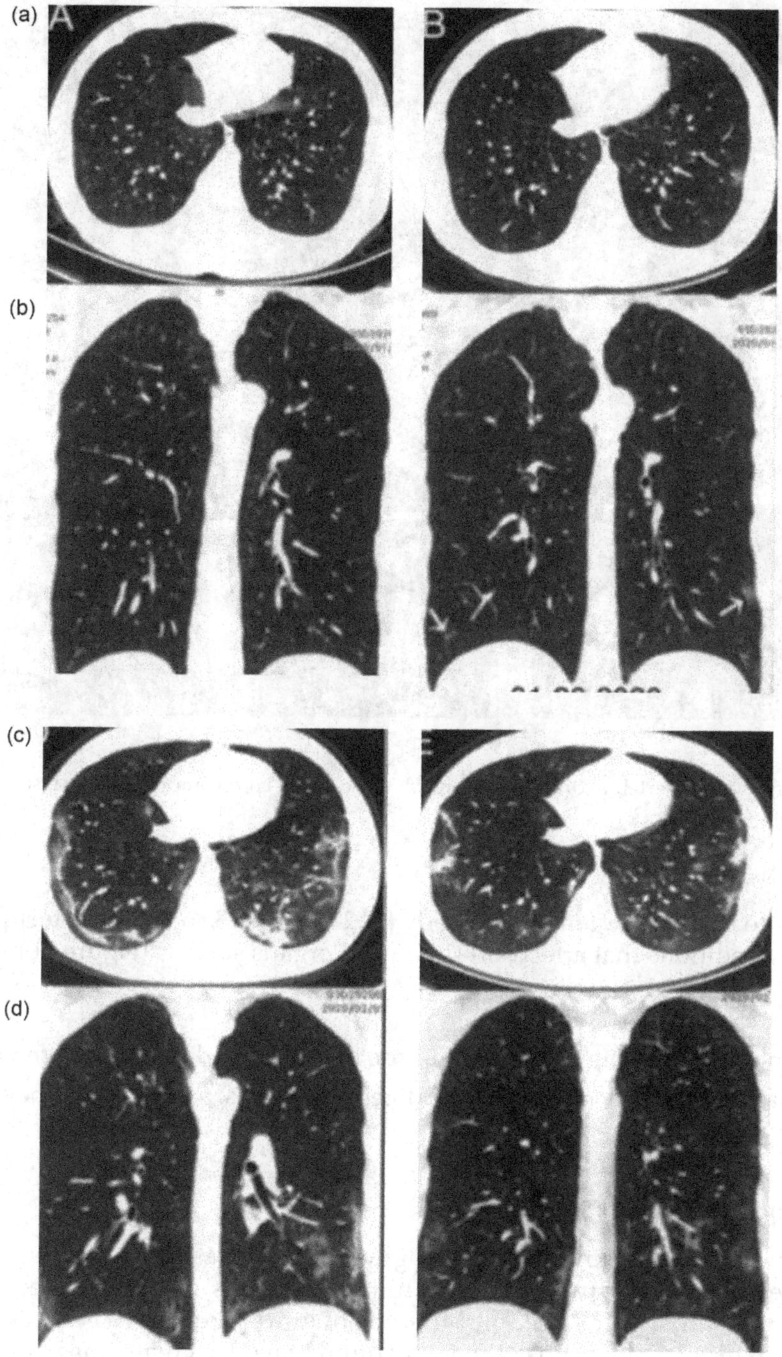

FIGURE 7.5
MI using CT. (a) Normal chest CT with axial and coronal planes was obtained at the onset. (b) Chest CT shows minimal ground-glass opacities in the bilateral lower lung lobes. (c) Chest CT shows the progression of pneumonia with mixed ground-glass opacities and linear opacities in the subpleural area. (d) Chest CT shows the absorption of both ground-glass opacities and organizing pneumonia.

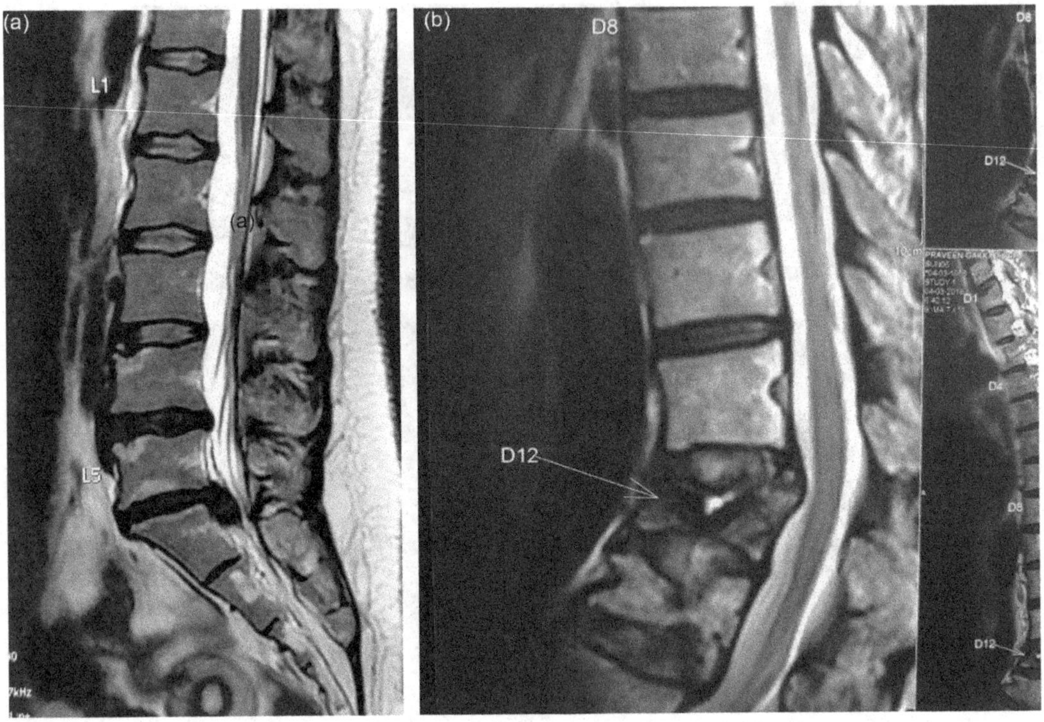

FIGURE 7.6
MRI of a patient. (a) Intervertebral disc herniation at L5-S1 causing cord compression. (b) Osteoporotic fracture at D12 vertebrae.

radio waves to scan the organs in the body [10,21–23,28–33,36–41]. This technique is used to diagnose conditions that affect soft tissue and organs such as (Figure 7.6):

- Tumor or cancer.
- Joint injury, spinal injury, and soft tissue injuries like damaged ligaments.
- Internal organ injury such as brain, digestive organs, and heart, kidney.

7.3.4 Diagnostic Medical Sonography

This technique is also known as sonography or ultrasound that is used to scan high-frequency sound waves. This technique generates images of the internal body of a patient. CV algorithms used in MI are shown in Figure 7.7. This technique is safe during pregnancy because it uses echo or sound waves to create images of the internal body [10,21,29,32,38–41]. Some of the important key features of this technique are as follows:

- Scanning procedure is safe and widely used for diagnosis.
- Widely used to check the progress of a pregnancy.
- No prior preparation is required for this test.

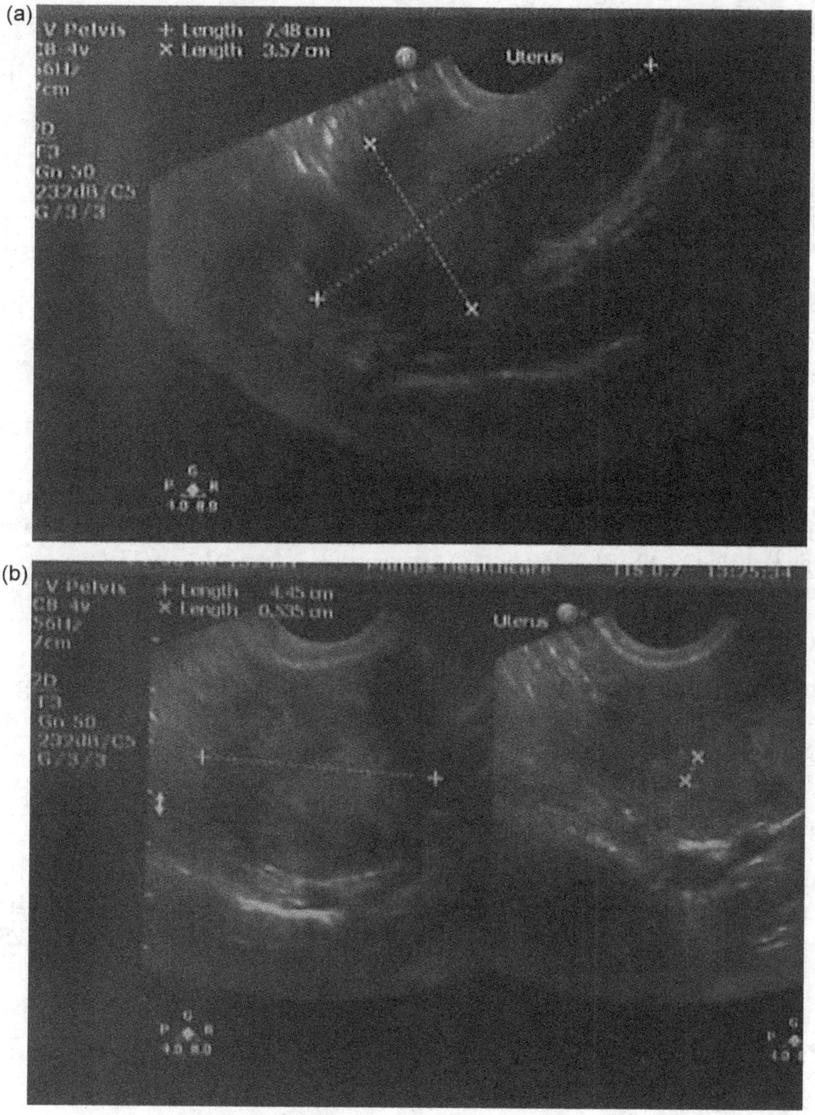

FIGURE 7.7
Diagnostic medical sonography of female uterus. (a) Endometrium thickness. (b) Ovaries of females.

7.4 Digital Imaging Standards in MI

There are two standards in MI, which are listed below.

7.4.1 Digital Imaging and Communication in Medicine

This is used for the loading and exchange of medical images [10,22,29]. Such a typical file comprises a header and an image. The header contains metadata and imaging techniques of a patient.

7.4.2 Picture Archiving and Communication System

It is a system used for loading and exchange of image files. This is used for implementing a workflow of images from taking to analysis to archiving.

7.5 CV Algorithms Used in MI

AI-empowered CV technologies play a vital role in the healthcare industry. In literature, various CV algorithms are present that are widely used in MI. Some of them are mentioned below.

7.5.1 Classification

These algorithms will classify an image into two or more classes. For example, an algorithm trained to detect lung nodules can classify chest X-Ray based on the presence or absence of a nodule [23,42,43].

7.5.2 Localization

In this, algorithms can point to the location of a pattern by creating a bounding box around the problematic region. For example, in the case of mammography, the algorithm can draw a box around a suspected tumor on a breast X-Ray image of a patient [44,45].

7.5.3 Segmentation

It is a technique that can color-code a set of pixels in an image representing regions of detected patterns. This is widely used to determine the size of the anomaly. The size of the change of anomaly can be compared by images captured at a different time [25,46].

7.6 Use Cases AI and CV in MI

AI-empowered CV models are trained on the medical images that can be used to detect, classify, and locate the pattern such as tumors and bone fractures [10–20,22,27–32,38–40]. These techniques have a high accuracy rate in the medical image.

7.6.1 Diagnostic Assistance

AI-empowered CV techniques help a radiologist to diagnose the anomalies in the image or any region by saving time spent in diagnosis [23,43–46].

7.6.2 Screening and Sortation

AI algorithms can analyze and sort the images in picture archiving and communication system that help a radiologist to diagnose the anomalies at an early stage and treatment can be started accordingly [24,36,43].

7.6.3 Monitoring

To analyze response to the treatment in oncology, for example, diseased tissues images are captured and compared at a distinct interval of time. The variation in diseased tissue size can help in understanding the response to the treatment [10,29].

7.6.4 Charting

Post review of findings prepared by AI tool by the medical practitioner provides the essential inputs required for medical charting that otherwise need to be recorded manually [21,40].

7.7 Application of CV in Imaging

Over the past few decades, AI and CV technologies have contributed significantly to the resolution of a large number of biomedical problems, including oncology, cardiovascular analysis, neurology, ophthalmology, MRI CT, and X-Ray [10,21,22,29,47,48]. Some of the important applications are listed below.

7.7.1 Cardiovascular Image Analysis

This analysis has been performed at the Iowa Institute of Biomedical Imaging. Since the 1990s X-Ray coronary angiography, 2D/3D/4D ultrasound (including intravascular), MRI, and CT have led to the established and validated software for numerous medical queries and real-time applications such as coronary artery disease, heart transplant, and arrhythmia [10,29].

7.7.2 Oncology

AI-empowered CV algorithms have shown promising results in cancer diagnostics and treatment, improved response estimation, and predicted medical outcomes [8,46].

7.7.3 Ophthalmology

In ophthalmology, CV technology concentrates on diseases having incidences such as diabetic retinopathy, retinopathy of prematurity, age-related problem, congenital cataract, and retinal vein occlusion [10,21].

7.7.4 Neurology

It allows researchers and neuroscientists to build improved models to simulate the human brain. Neural networks act as "virtual brains" that capture the representation of the human brain. Some of the applications of AI-based neurology are stroke detection, brain tumor assessment, and treatment. Emblem et al. proposed an ML algorithm using MRI-based blood volume distribution data to foresee preoperative glioma survival. Researchers from the Heidelberg University Hospital and the German Cancer Research Centre have trained

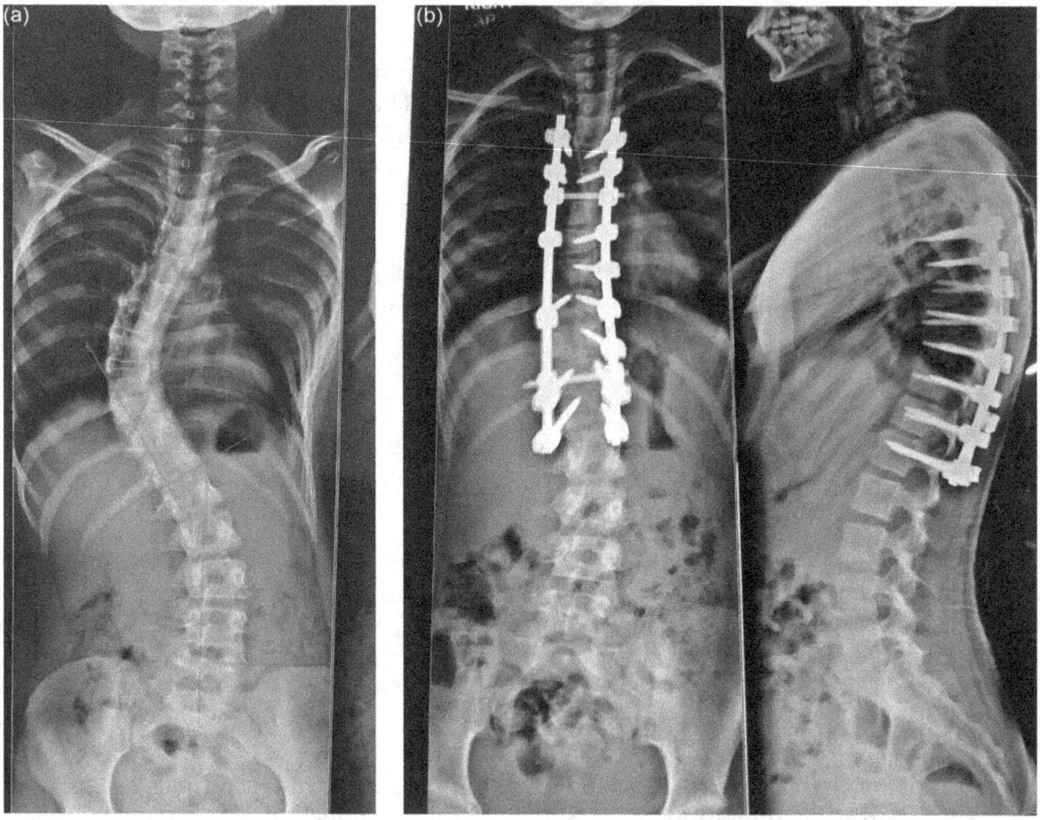

FIGURE 7.8
X-Ray analysis of a patient. (a) Preoperative X-Ray spine of scoliotic deformity. (b) After correction.

an ML algorithm using the app. It consists of 500 MRI scans of patients suffering from brain tumors [10,49].

7.7.5 Orthopedic

This field requires a combination of surgical skills and innovative technology. Over the last few decades, there is a rapid adoption of AI-empowered CV technologies into this field such as spinal deformity detection and surgical outcome calculator (see Figure 7.8). In CV, convolutional neural network (CNN) has proved useful for these purposes. Pretrained CNNs enable the transfer of information to a particular new fracture detection, without the requirement of any computational training that seems to be very time-consuming [29,36].

7.7.6 Emergency Medicine

CV research is rapidly growing in emergency medicine. Various AI interventions help in the improvement of emergency care especially for prediction-based diagnoses and acute radiographic imaging. High-quality evidence is required to further assess the short-term and long-term medical outcomes [29,47,48,50]. In emergency medicine, patients are assessed with a limited amount of information and clinicals often find themselves balancing the probability for risk and decision making. Emergency medicine can surely get benefits due to its potential value in prediction during triage and diagnosis of time-sensitive patients.

7.7.7 MRI Brain Interpretation

CV technologies are used as an adjunct to MRI brain interpretation as the AI-Rad Companion Brain MR for Morphometry Analysis can automatically detect roughly 30 brain segments on MRI images, calculate their volumes, and compare the outcomes to data in a normative reference database for brain morphometry from the Alzheimer's Disease Neuroimaging Initiative [8].

7.7.8 X-Ray Analysis

CV technology presents an important opportunity for radiologists to improve the quality of interpretation and enhance the value of radiology in patient care and health. Lung cancer detection, dentistry interpretation, pneumonia detection, tuberculosis screening, and radiologist performance can get a boost from a CV algorithm that identifies previously undetected.

7.7.9 Surgery

In modern surgical practice, various kinds of surgical cameras are used that provide powerful visual information. This information is used by surgeons to visualize the clinical site and decision making. This signal can be used by CV models to provide support in the detection of instruments, or any activities in real-time during operations in the surgical room and postoperatively for analysis and surgical insight [10,29,32,33,36,43].

7.8 Critical Success Factor

CV-based AI/ML/DL algorithms can be used in MI to enhance the productivity of clinicals. These algorithms can help early diagnosis and treatment of patients. The regulatory agencies like the Food and Drug Administration allow these products to be used in diagnosis. The important critical success factors are discussed as follows.

7.8.1 Accuracy

CV Products are already trained on extensive data. However, all DL/ML/AI products are based on training data. It is important to validate that the product is correctly predicting for every new installation.

7.8.2 Seamless Integration

CV products cannot exist in isolation. It needs to be integrated seamlessly in the application environment so integration should be well designed to save time such as less uploading time of images on the system.

7.8.3 Training

A training session should be arranged for clinicals regarding correct usage of products such as sonography, X-Ray, MRI, and CT.

7.8.4 Productivity Metrics

The business case is normally based to boost productivity. To validate it, a metric should be built in the integration or within the AI Product.

7.8.5 Data Security

CV-based technologies such as AI/ML/DL are well democratized. Throughout the world, various companies are using CV-based technology for product making. So, it is very important to check whether the AI vendor is registered with a particular country or not. Another important factor is to verify where the product is hosted where developers who have access to the data are based at.

7.9 CV in Surgery and MI

With the rapid availability of surgical video sources and the merging of technologies around storage, processing and understanding of video, medical solutions, and product support vision has become an essential component of advanced surgical capabilities (see Figure 7.9). The data availability and computational advancement and AI/ML/DL methodology have led to rapid developments in this domain [10,22]. In the last few years, surgery has gradually moved toward the minimally invasive surgery (MIS). Most of the surgical operating theatres are currently well equipped with digital cameras to visualize the surgical site. The video generated from digital cameras contains valuable information related to motion, shape, size, and functions of the anatomy and instrumentation within it [21,33,50–58].

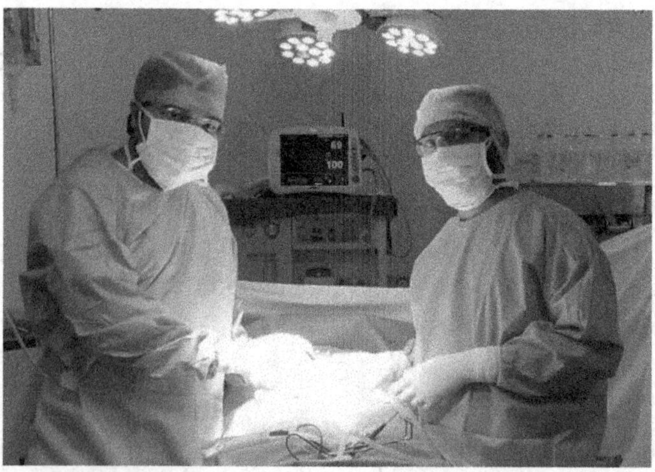

FIGURE 7.9
A surgical theatre with advanced surgical capabilities.

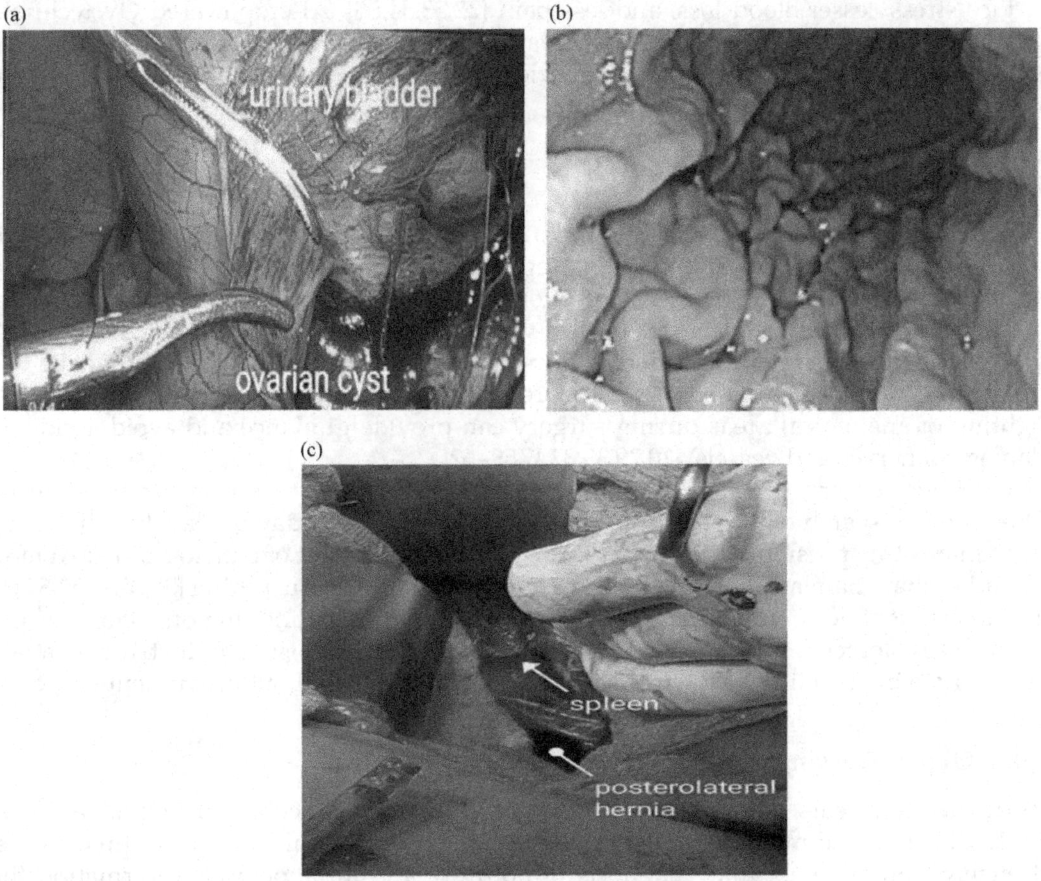

FIGURE 7.10
Minimal access surgery. (a) Ovarian cyst. (b) Endoscopy. (c) Congenital diaphragmatic hernia.

At the time of surgical operations, this video is recorded that contains information about the surgical process, instruments used, complications, steps taken for surgery, and risk. These recorded videos can be used further for educational purposes, clinical meetings, conferences, or case reports for research purposes [10,11,21,33]. To make vision problems manageable, computational methods focus on a subdivision of human vision systems such as object detection and tracking and spatial understanding in surgical images (see Figure 7.10).

Various surgical datasets are publicly available for training and validating new algorithms that result in fast progress toward automatic analysis of surgical data. The use of CV technology to process data from wide-ranging intraoperative imaging modalities or digital cameras can be categorized into three different applications.

7.9.1 Understanding of Surgical Procedure

A surgical procedure can be categorized into various steps such as dissection, anastomosis, and suturing. Detection and localization of these tasks allow for surgical modeling and workflow assessment. This enables the MIS to practice toward the establishment of standard protocols and guidelines for task execution, optimal tool position w.r.t. anatomy, less complex, and cost-effective surgical operations. This surgery results in high security, reduce stress, lesser blood loss, and less pain [22,33,54,55]. AI-empowered CV technologies are well equipped with digital cameras and sensors that make the surgical procedure more effective. Intraoperative videos provide a lot of information about the quality and workflow from these videos.

7.9.2 Object Detection

Object detection in digital images is the main component of CV. Surgical videos aid in visualizing medical goals and sensitive regions to optimize and increase the procedure safety such as surgical instrument detection (Automatic detection and localization helps in accurate positioning and ensure critical structures are not injured), lesion detection (AI vision systems can provide aids highlighting lesions and anomalies that could not be possible without AI systems) as shown in Figure 7.11, anatomy detections (Detection and highlighting of anatomical areas during surgery can provide guidance and avoid accidental damage of nerves and vessels) [10,29,33,37,47,59–62].

In robotic surgery, this information provides active guidance during needle suturing. Due to this research on the surgical instrument is increasing day by day. In robotic surgery, interesting possibilities is there to generate a huge amount of data to trained AI models rather than manual labeling which is expensive and time-consuming [33,37,47,53,59,60]. The instrument detection has also received interesting possibilities in non-robotic surgeries such as colorectal, spine, pelvis, and retinal. In these cases, vision for instrument analysis can help in modeling a system which can report analytics about instrument usage.

7.9.3 Object Tracking or Computer-Assisted Navigation

In the last few years, there is a rapid growth in AI and CV techniques which are used for localization and mapping of the environment using surgical cameras. In this manner, it enables the more accurate diagnosis of both pre and intraoperative information that helps in decision making. In MIS, these systems help in locating the position of endoscopy within the organ and simultaneously detect the shape of the explored organ. Simultaneous localization and mapping is a very challenging problem but the DL algorithm provides better feature detectors and descriptors in laparoscopy, sinus endoscopy, and colonoscopy (see Figure 7.12) [9,26,48,50,51,53].

In robotic surgery, the da Vinci Surgical System uses stereo endoscopes over monocular endoscopes and has better accuracy to capture 3D measurements (Figure 7.13). Vision-based augmented reality in image-guided surgery facilitates the surgeons to access radiological images. Using these images surgeons plan the surgery. Its main goal is to integrate surgical navigation with virtual planning [10–21,33,52,63–66]. Augmented reality-guided surgery is the graphical representation of the 3D complexity of the human anatomy.

7.10 Deployment Issues of Vision-Based Systems

Computer-aided surgery addresses visual, perceptual, and dexterous constraints issued by facilitating presurgical diagnostic imaging and planning, intraoperative visual guidance, and surgical robotic applications. Some of the important open challenges associated with the vision-based system in Operation theatre or MIS surgery are listed below:

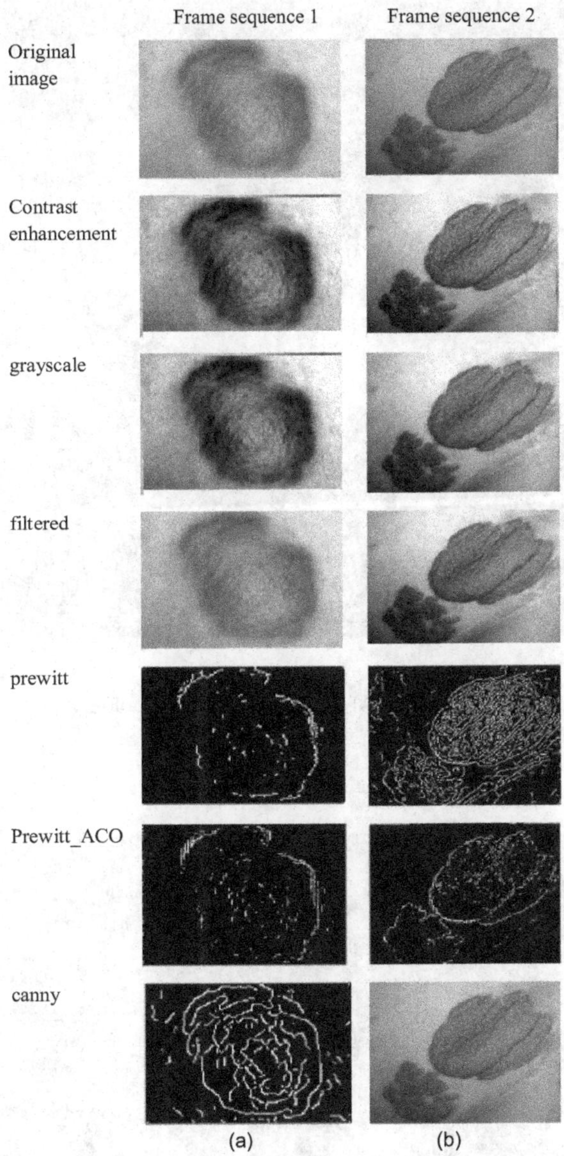

FIGURE 7.11
Skin lesion detection using AI technique. (a) Frame Sequence 1. (b) Frame Sequence 2.

- **Infrastructure**
  To implement the vision-based system, a good bandwidth with latency is required. There should be ease of access to the cloud.
- **User Interfaces Design**
  It is very important to ensure that only appropriate information is provided to the surgical team. There should be a practitioner surgical platform for direct communication in advanced vision-based solutions.
- **Visual Limitations**
  There is no direct visual access to the operation site and anatomical structure, unintuitive hand–eye coordination, and constraints to port openings.

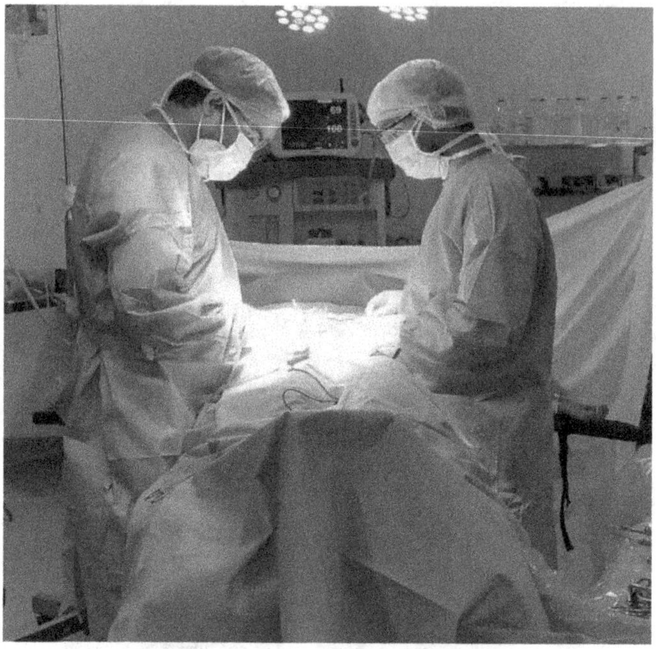

FIGURE 7.12
Surgical operation theatre well equipped with AI-empowered CV technology.

FIGURE 7.13
da Vinci robot system for robotic surgery.

7.11 Conclusion

CV is the processing of digital images or frames sequence from video data for object detection and classifications. Vision-based algorithms can be used in MI to enhance the productivity of medical professionals. These algorithms provide early diagnoses such as cancer or tumor or lesion detection and treatment of patients. In this chapter, we discussed topics such as the basic concept, evolution of CV, MI techniques (X-Ray, CT, MRI, and sonography), CV algorithms, applications, and use cases in MI. Furthermore, the role of vision-based technique in surgical operation theatre includes procedural understanding, detection of an object, and computer-aided navigations. In further sections, deployment issues of vision-based techniques are discussed.

References

1. Computer Vision. Available at: https://newscapeconsulting.com/2021/06/11/ai-computer-vision-in-medical-imaging/ [accessed on 10 June 2021].
2. Computer Vision and Natural Processing Language. Available at: https://www.logikk.com/articles/computer-vision-nlp-healthcare/ [accessed on 10 June 2021].
3. Computer Vision Based Surgery. Available at: https://www.marktechpost.com/2021/02/27/this-surgical-intelligence-platform-combines-artificial-intelligence-and-computer-vision-to-better-the-surgeons-performance/ [accessed on 10 June 2021].
4. Breast Cancer. Available at: https://www.activesilicon.com/news-media/news/eye-tracking-breast-surgery/ [accessed on 10 June 2021].
5. Computer Vision Applications. Available at: https://tel.archives-ouvertes.fr/tel-01557522/document [accessed on 10 June 2021].
6. Computer Vision and Artificial Intelligence in Healthcare. Available at: https://bdtechtalks.com/2020/04/29/ai-computer-vision-health-care/ [accessed on 10 June 2021].
7. Transformation of Healthcare Using Computer Vision. Available at: https://www.allerin.com/blog/how-computer-vision-can-transform-healthcare [accessed on 10 June 2021].
8. Real-Time Applications of Computer Vision. Available at: https://www.appliedradiology.com/communities/Artificial-Intelligence/fda-clears-siemens-ai-based-mri-interpretation-assistants#:~:text=Based%20on%20AI%20algorithms%2C%20the,Disease%20Neuroimaging%20Initiative%20(ADNI) [accessed on 10 June 2021].
9. Liu X, Zheng Y, Killeen B, Ishii M, Hager GD, Taylor RH, Unberath M. Extremely Dense Point Correspondences Using a Learned Feature Descriptor. 2020. https://arxiv.org/abs/2003.00619.
10. Sharma L, Garg P (Eds.). *From Visual Surveillance to Internet of Things*. New York: Chapman and Hall/CRC, 2020. https://doi.org/10.1201/9780429297922.
11. Sharma L, Lohan N. Performance Analysis of Moving Object Detection Using BGS Techniques in Visual Surveillance. *Int J Spatiotemporal Data Sci*. Inderscience. 2019 Jan;1:22–53.
12. Anand A, Jha V, Sharma L. An Improved Local Binary Patterns Histograms Techniques for Face Recognition for Real Time Application. *Int J Recent Technol Eng*. 2019 Jul;8(2S7):524–529.
13. Sharma L, Yadav DK. Histogram Based Adaptive Learning Rate for Background Modelling and Moving Object Detection in Video Surveillance. *Int J Telemed Clin Pract*. Inderscience. 2016 June. ISSN: 2052-8442. https://doi.org/10.1504/IJTMCP.2017.082107.
14. Sharma L, Lohan N. Performance Analysis of Moving Object Detection Using BGS Techniques. *Int J Spatio-Temporal Data Sci*. Inderscience. 2019 Feb;1:22–53.

15. Sharma L, Lohan N. Internet of Things with Object Detection. *Handbook of Research on Big Data and the IoT*, IGI Global, 2019 March, pp. 89–100. ISBN: 9781522574323. https://doi.org/10.4018/978-1-5225-7432-3.ch006.
16. Sharma L. Introduction: From Visual Surveillance to Internet of Things. In: Sharma L. and Garg P. (eds) *From Visual Surveillance to Internet of Things*. Boca Raton, FL: Taylor & Francis, CRC Press, 2019, Vol. 1, p. 14.
17. Sharma L, Garg PK. Block based Adaptive Learning Rate for Moving Person Detection in Video Surveillance. In: Sharma L. and Garg P. (eds) *From Visual Surveillance to Internet of Things*. Boca Raton, FL: Taylor & Francis, CRC Press, 2019, Vol. 1, p. 201.
18. Makkar S, Sharma L. A Face Detection Using Support Vector Machine: Challenging Issues, Recent Trend, Solutions and Proposed Framework. In: Singh M., Gupta P., Tyagi V., Flusser J., Ören T., Kashyap R. (eds) *Advances in Computing and Data Sciences. ICACDS 2019.* Communications in Computer and Information Science, Vol. 1046. Springer, Singapore, 2019. https://doi.org/10.1007/978-981-13-9942-8_1.
19. Sharma L, Garg PK. IoT and Its Applications. In: Sharma L. and Garg P. (eds) *From Visual Surveillance to Internet of Things*. Boca Raton, FL: Taylor & Francis, CRC Press *From Visual Surveillance to Internet of Things*. Taylor & Francis, CRC Press, 2019, Vol. 1, p. 29.
20. Sharma L, Yadav DK, Bharti SK. An Improved Method for Visual Surveillance Using Background Subtraction Technique. *2015 2nd International Conference on Signal Processing and Integrated Networks (SPIN)*, 2015, pp. 421–426. https://doi.org/10.1109/SPIN.2015.7095253.
21. Sharma S, Verma S, Kumar M, Sharma L. Use of Motion Capture in 3D Animation: Motion Capture Systems, Challenges, and Recent Trends. *2019 International Conference on Machine Learning, Big Data, Cloud and Parallel Computing (COMITCon)*, 2019, pp. 289–294. https://doi.org/10.1109/COMITCon.2019.8862448.
22. Artificial Intelligence Technologies, Applications, and Challenges. Available at: https://www.routledge.com/Artificial-Intelligence-Technologies-Applications-and-Challenges/Sharma-Garg/p/book/9780367690809 [accessed on 20 July 2021].
23. Schultheiss M, Schober SA, Lodde M et al. A robust Convolutional Neural Network for Lung Nodule Detection in the Presence of Foreign Bodies. *Sci Rep*. 2020;10,12987. https://doi.org/10.1038/s41598-020-69789-z.
24. Chen YB, Chen OTC. Image Segmentation Method Using Thresholds Automatically Determined from Picture Contents. *EURASIP J Image Video Process*. 2009;2009:1–15.
25. Unnikrishnan R, Pantofaru C, Hebert M. Towards Objective Evaluation of Image Segmentation Algorithms. *IEEE Trans Pattern Anal Mach Intell*. 2007;29:929–944.
26. Rau A, Edwards PJE, Ahmad OF, Riordan P, Janatka M, Lovat LB et al. Implicit Domain Adaptation with Conditional Generative Adversarial Networks for Depth Prediction in Endoscopy. *Int J Comput Assist Radiol Surg*. 2019;14:1167–1176.
27. Yadav DK, Sharma L, Bharti SK. Moving Object Detection in Real-Time Visual Surveillance Using Background Subtraction Technique. *2014 14th International Conference on Hybrid Intelligent Systems*, 2014, pp. 79–84. https://doi.org/10.1109/HIS.2014.7086176.
28. Sharma L, Singh A, Yadav DK. Fisher's Linear Discriminant Ratio Based Threshold for Moving Human Detection in Thermal Video. *Infrared Phys Technol*. Elsevier. 2016 Mar;78:118–128.
29. Sharma L (Ed.). *Towards Smart World*. New York: Chapman and Hall/CRC, 2021. https://doi.org/10.1201/9781003056751.
30. Sharma L. Human Detection and Tracking Using Background Subtraction in Visual Surveillance. In: Sharma L. (ed) *Towards Smart World*. New York: Chapman and Hall/CRC, 2020, pp. 317–328. https://doi.org/10.1201/9781003056751.
31. Sharma L, Yadav DK, Kumar M. A Morphological Approach for Human Skin Detection in Color Images. *2nd National Conference on "Emerging Trends in Intelligent Computing & Communication"*, *GCET*, Gr. Noida, India, 26–27 April 2013.
32. Sharma L, Sengupta S, Kumar B. An Improved Technique for Enhancement of Satellite Images. *J Phys.: Conf Ser*. 2021 Jan; 1714:012051.

33. Carpenter M. An Accidentally Detected Diaphragmatic Hernia with Acute Appendicitis. *Asian J Case Rep Surg.* 2021;9(2):19–24. Retrieved from https://www.journalajcrs.com/index.php/AJCRS/article/view/30260.
34. Computer Vision. Available at: https://arxiv.org/abs/2004.05436v1 [accessed on 10 June 2021].
35. Computer Vision in Diagnosis and Treatment. Available at: https://www.itnonline.com/content/ct-provides-best-diagnosis-novel-coronavirus-covid-19 [accessed on 10 June 2021].
36. Real-Time Applications of Computer Vision. Available at: http://blog.peekmed.com/machine-learning-in-orthopedics [accessed on 10 June 2021].
37. Colleoni E, Edwards P, Stoyanov D. Synthetic and Real Inputs for Tool Segmentation in Robotic Surgery. *Conference on Medical Image Computing and Computer Assisted Intervention*. 2020. Palm Springs, CA.
38. Singh S, Sharma L, Kumar B. A Machine Learning Based Predictive Model for Coronavirus Pandemic Scenario. *J Phys.: Conf Ser.* 2021 Jan; 1714:012023.
39. Jha G, Sharma L, Gupta S. Future of Augmented Reality in Healthcare Department. In: Singh P.K., Wierzchoń S.T., Tanwar S., Ganzha M., Rodrigues J.J.P.C. (eds) *Proceedings of Second International Conference on Computing, Communications, and Cyber-Security.* Lecture Notes in Networks and Systems, Vol. 203. Springer, Singapore. 2021. https://doi.org/10.1007/978-981-16-0733-2_47.
40. Jha G, Sharma L, Gupta S. E-Health in Internet of Things (IoT) in Real-Time Scenario. In: Singh P.K., Wierzchoń S.T., Tanwar S., Ganzha M., Rodrigues J.J.P.C. (eds) *Proceedings of Second International Conference on Computing, Communications, and Cyber-Security.* Lecture Notes in Networks and Systems, Vol. 203, Springer, Singapore. 2021. https://doi.org/10.1007/978-981-16-0733-2_48.
41. Kumar S, Gupta P, Lakra S, Sharma L, Chatterjee R. The Zeitgeist Juncture of "Big Data" and its Future Trends. *2019 International Conference on Machine Learning, Big Data, Cloud and Parallel Computing (COMITCon)*, 2019, pp. 465–469. https://doi.org/10.1109/COMITCon.2019.8862433.
42. Hamberlin J, Kocher MR, Waltz J et al. Automated Detection of Lung Nodules and Coronary Artery Calcium Using Artificial Intelligence on Low-Dose CT Scans for Lung Cancer Screening: Accuracy and Prognostic Value. *BMC Med.* 2021;19,55. https://doi.org/10.1186/s12916-021-01928-3.
43. Yoo H, Kim KH, Singh R, Digumarthy SR, Kalra MK. Validation of a Deep Learning Algorithm for the Detection of Malignant Pulmonary Nodules in Chest Radiographs. *JAMA Netw Open.* 2020;3(9):e2017135. https://doi.org/10.1001/jamanetworkopen.2020.17135.
44. Chan BK, Wiseberg-Firtell JA, Jois RH, Jensen K, Audisio RA. Localization Techniques for Guided Surgical Excision of Non-Palpable Breast Lesions. *Cochrane Database Syst Rev.* 2015 Dec 31;12:CD009206. https://doi.org/10.1002/14651858.CD009206.pub2. PMID: 26718728.
45. Tardioli S, Ballesio L, Gigli S, DI Pastena F, D'Orazi V, Giraldi G, Monti M, Amabile MI, Pasta V. Wire-guided Localization in Non-palpable Breast Cancer: Results from Monocentric Experience. *Anticancer Res.* 2016 May;36(5):2423–2427. PMID: 27127152.
46. Tripathi P, Tyagi S, Nath M. A Comparative Analysis of Segmentation Techniques for Lung Cancer Detection. *Pattern Recognit Image Anal.* 2019;29:167–173. https://doi.org/10.1134/S105466181901019X.
47. Du X, Kurmann T, Chang PL, Allan M, Ourselin S, Sznitman R et al. Articulated Multi-Instrument 2-D Pose Estimation Using Fully Convolutional Networks. *IEEE Trans Med Imag.* 2018;37(5):1276–1287.
48. Bano S, Vasconcelos F, Tella Amo M, Dwyer G, Gruijthuijsen C, Deprest J et al. Deep Sequential Mosaicking of Fetoscopic Videos. *Conference on Medical Image Computing and Computer Assisted Intervention*. 2019. Springer, Cham.
49. Real-Time Applications of Computer Vision. Available at: https://www.quantib.com/blog/artificial-intelligence-neurology-promising-research-and-proven-application [accessed on 10 June 2021].
50. Ma R, Wang R, Pizer S, Rosenman J, McGill SK, Frahm J-M. Real-Time 3D Reconstruction of Colonoscopic Surfaces for Determining Missing Regions. *Conference on Medical Image Computing and Computer Assisted Intervention*. 2019. Springer, Cham.

51. Maier-Hein L, Mountney P, Bartoli A, Elhawary H, Elson D, Groch A et al. Optical Techniques for 3-D Surface Reconstruction in Computer Assisted Laparoscopic Surgery. *Med Image Anal.* 2013;17:974–996.
52. Chen F, Wu D, Liao H. Registration of CT and Ultrasound Images of the Spine with Neural Network and Orientation Code Mutual Information. *Med Imag Augment Reality.* 2016;9805:292–301.
53. Ward TM, Mascagni P, Ban Y, Rosman G, Padoy N, Meireles O, Hashimoto DA. Computer Vision in Surgery. *Surgery.* 2021 May;169(5):1253–1256. Epub 2020 Dec 1. PMID: 33272610. https://doi.org/10.1016/j.surg.2020.10.039.
54. Chaudhary A, Singh SS. Lung Cancer Detection on CT Images by Using Image Processing. *2012 International Conference on Computing Sciences*, 2012, pp. 142–146. https://doi.org/10.1109/ICCS.2012.43.
55. Ginneken BV, Romeny BM, Viergever MA. Computer-Aided Diagnosis in Chest Radiography: A Survey. *IEEE Trans Med Imag.* 2001;20(12):1228–1241.
56. Beucher S, Meyer F. The Morphological Approach of Segmentation: The Watershed Transformation. In: Dougherty E. (ed) *Mathematical Morphology in Image Processing*, pp. 43–481. New York: Marcel Dekker, 1992.
57. Suzuki K et al. False-Positive Reduction in Computer-Aided Diagnostic Scheme for Detecting Nodules in Chest Radiographs by Means of Massive Training Artificial Neural Network. *Acad Radiol.* 2005 Feb;12(2):191–201.
58. Nguyen HT et al. Watersnakes: Energy-Driven Watershed Segmentation. *IEEE Trans Pattern Anal Mach Intell.* 2003 Mar;25(3):330–342.
59. D'Ettorre C, Dwyer G, Du X, Chadebecq F, Vasconcelos F, De Momi E et al. Automated Pick-Up of Suturing Needles for Robotic Surgical Assistance. *IEEE International Conference on Robotics and Automation.* 2018; pp. 1370–1377. Brisbane.
60. Bouget D, Allan M, Stoyanov D, Jannin P. Vision-Based and Marker-less Surgical Tool Detection and Tracking: A Review of the Literature. *Med Image Anal.* 2017;35:633–654.
61. Sengupta S, Mittal N, Modi M. Improved Skin Lesions Detection Using Color Space and Artificial Intelligence Techniques. *J Dermatolog Treat.* 2020 Aug;31(5):511–518. Epub 2020 Jan 3. PMID: 31865822. https://doi.org/10.1080/09546634.2019.1708239.
62. Sengupta S, Mittal N, Modi M. Edge Detection in Dermascopic Images by Linear Structuring Element. *2018 7th International Conference on Reliability, Infocom Technologies and Optimization (Trends and Future Directions) (ICRITO)*, 2018, pp. 419–424. https://doi.org/10.1109/ICRITO.2018.8748610.
63. Peters TM, Linte CA, Yaniv Z, Williams J. *Mixed and Augmented Reality in Medicine.* CRC Press, Boca Raton, FL. 2018.
64. Luo H, Hu Q, Jia F. Details Preserved Unsupervised Depth Estimation by Fusing Traditional Stereo Knowledge from Laparoscopic Images. *IET Healthc Technol Lett.* 2019;6(6):154–158.
65. Colleoni E, Moccia S, Du X, De Momi E, Stoyanov D. Deep Learning Based Robotic Tool Detection and Articulation Estimation with Spatio-Temporal Layers. *IEEE Robot Autom Lett.* 2019;4:2714–2721.
66. Park BJ, Hunt SJ, Martin C, Nadolski GJ, Wood BJ, Gade TP. Augmented and Mixed Reality: Technologies for Enhancing the Future of IR. *J Vasc Interv Radiol.* 2020;31:1074–1082.

Part 3

IoT with Computer Vision for Real-Time Applications

8

Self-Driving Cars: Tools and Technologies

Kavish Gupta, Deepa Gupta, and Lavanya Sharma
Amity Institute of Information Technology, Amity University, Noida, India

CONTENTS

8.1 Introduction

In recent years, self-driving cars have gone from "dreams" to "reality." Companies like Tesla and Google-backed Waymo have started their lineup of self-driving cars and are making millions of it. Uber is also developing technology for self-driving cabs. Food delivery companies like Uber Eats are also working toward automated drones to deliver food to the customers. Luxury car brands like Porsche and Bentley are also putting millions of dollars toward developing self-driving luxury cars. As there are advances made in the self-driving technology, there have been cases of fatal accidents involving self-driving cars which shows there are areas to refine self-driving technology. There have also been reports that in the future self-driving technology will cause many people to lose their jobs [1–4].

Simply speaking, a self-driving car is an intelligent car which obtains information from its sensors and arrives at the required destination. The car uses technologies such as automatic control, artificial intelligence, and computer vision to perform its functions. This chapter discusses these technologies and different levels of automation. Indian companies

DOI: 10.1201/9781003244165-11

like Mahindra and Tata are making progress in developing market viable automated driving technology to work in local conditions. Similarly, Intel is also developing its collision avoidance system [1–3,5–11].

The companies are also developing technology in which the car arrives to its owner itself by providing instruction through a mobile application. This chapter also discusses which technology is better to use in self-driving car: camera or LIDAR system. When compared LIDAR system is much more reliable and gives a more clear view of the surroundings which helps the car to navigate to its location easily, while cameras are less reliable as they cannot give a clear view because of the size of their sensors [7]. Though autonomous or self-driving cars are becoming a trend in foreign countries, they still seem far from reality in India. The main reason is that India does not have the proper infrastructure for self-driving cars to work and the government is also not very supportive of the concept as it will result in huge unemployment. This chapter also discusses this matter deeply [12–16].

8.2 Tools and Technologies

Self-driving cars use navigation systems, path planning, and environment perception technologies to replace the human driver [3,4,12,17–23]. There are five levels of automation as follows:

- **Level 0: No-Automation**
  In this level, there is no automation in the vehicles.
- **Level 1: Function-Specific Automation**
  In this level, specific controls are automated, such as automatic turning or automatic braking.
- **Level 2: Combined Function Automation**
  Two controls can be automated: cruise control and lane guidance.
- **Level 3: Limited Self-Driving Automation**
  The controller can fully take control of all the main functions. The car knows when conditions require the computer to retake control and provides sufficient time for the controller to try and do it.
- **Level 4: Full Self-Driving Automation**
  The vehicle performs all the main safety functions for the entire journey, with the driver not expected to take control of the vehicle at any time. The car will perform all functions from start to end and can also park itself [1,5] (Figure 8.1).

The technologies used in self-driving cars have four parts: environment perception, path planning, car navigation, and car control.

8.2.1 Car Navigation System

In a self-driving car, there are two problems: where is the car now and how to go from its position to the destination. The car should be able to locate itself and the destination of arrival and the path to the destination. To solve this issue the car navigation system was

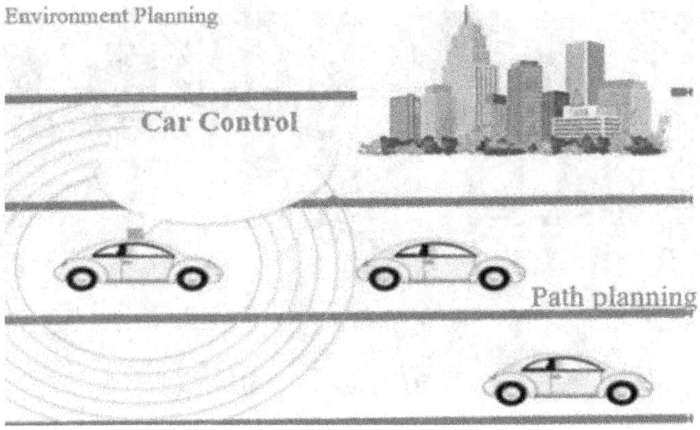

FIGURE 8.1
Different technologies used by self-driving cars include planning of path, controlling the vehicle, and planning of environment.

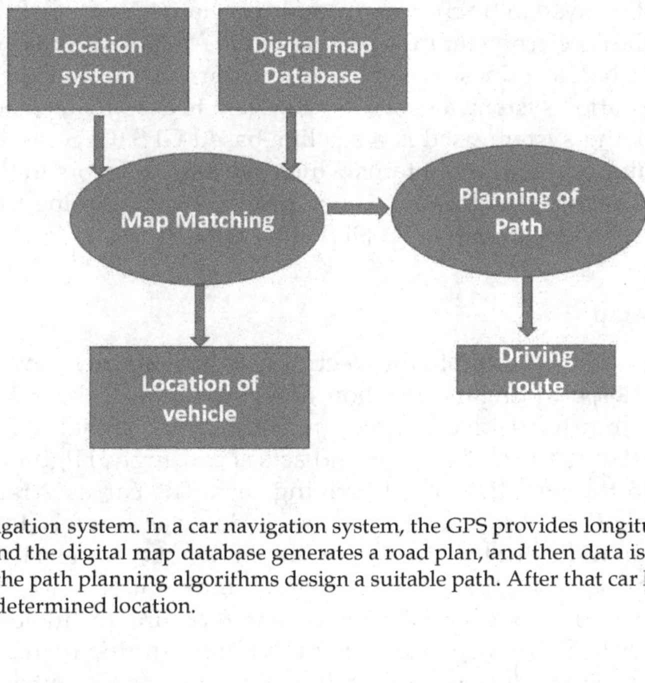

FIGURE 8.2
Working of a car navigation system. In a car navigation system, the GPS provides longitude and latitude. Here, the location system and the digital map database generates a road plan, and then data is fed in the map matching system in which the path planning algorithms design a suitable path. After that car locates itself and starts its journey to the predetermined location.

installed in the self-driving cars. Figure 8.2 shows how the car navigation system works. In a car navigation system, the Global Positioning System (GPS) provides longitude and latitude, the location system and the digital map database generates a road plan, and the data is fed in the map matching system in which the path planning algorithms designs a suitable path, then the car locates itself and starts its journey to get to the predetermined location [9,19].

8.2.2 Location System

The location system determines the location of the vehicle and these locations can be of three types: relative, absolute, and hybrid. In relative location, the moving distance and

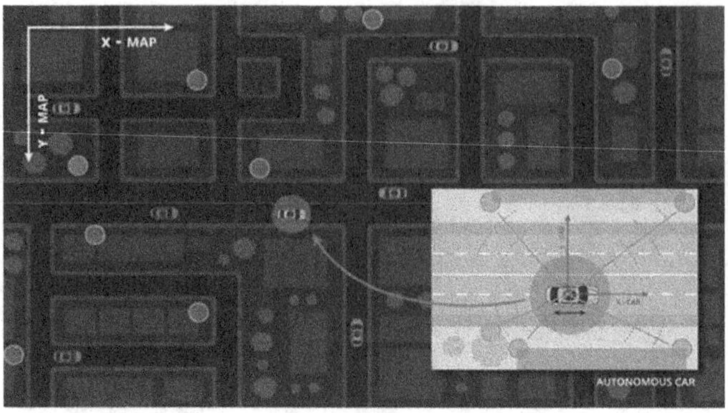

FIGURE 8.3
Using GPS to find location.

the direction to the prior position are added to determine the current position of the car. An inertial navigation system (INS) is an example of relative location. In the INS, the gyroscope sensor and the accelerometer in the car obtain the vehicle angular velocity and accelerated velocity which determines the car's relative course angle and speed.

In the absolute location system, a positioning system is used to determine the location—mainly the positioning system used is a satellite-based GPS. GPS can be interfered with by the weather, buildings, and mountains which can lead to errors in the position [12]. In the hybrid location system, the location is determined by combining the relative location system and the absolute location system [9] (Figure 8.3).

8.2.3 Electronic Map

Digital map information is stored in the electronic map; it mainly consists of traffic information, road facilities, building information, and geographical characteristics. Electronic maps (Ems) are currently designed to recognize the traffic signals and the traffic signs, and the car understands the traffic signs and acts accordingly. High-definition maps are now being used in the new line of self-driving cars. The car uses radars and cameras to get the location of the car and can narrow it down accurately to centimeters. The high-definition map indicates the location of the car. It not only tells the location but also builds a map of the surrounding of the car which makes it easy for the car to navigate properly to its destination. But with the increase in mapping technology the legal issues are also increasing, the mapping system always keeps scanning roads and private driveways through the high-resolution cameras in the car, in some countries the government heavily regulates their geospatial data and even forbid the exporting geospatial data (Figure 8.4).

8.2.4 Map Matching

The geographical information and the location from the EM are used in map matching. Map matching uses the fusing technique to combine the latitude and the longitude to find a path which is accurate and time-saving. The map matching algorithm also analyzes the traffic in the route to its destination and suggests the most efficient path [24] (Figure 8.5).

FIGURE 8.4
Working of electronic map.

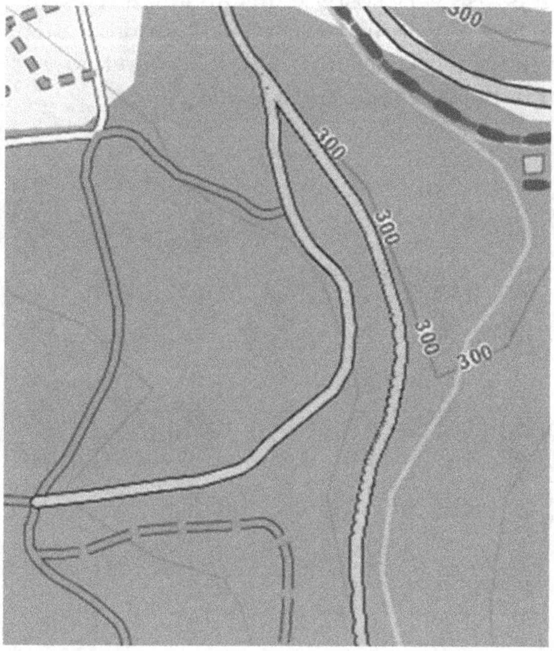

FIGURE 8.5
Working of map matching algorithm.

8.2.5 Global Path Planning

It is used to determine the best path to the destination. The key algorithms that are used in path planning are Dijkstra, Bellman-Ford, and Floyd. The global path planning system is fully developed and is being deployed in commercial cars [3].

8.2.6 Environment Perception

Environment perception consists of components such as laser and camera. The car senses the surrounding environment and changes its control accordingly to avoid any obstacle in its way. The car uses a combination of cameras, lasers, radar systems, and multi-sensors to determine the surroundings. Environment perception also eliminates the problem of navigation failure due to the absence of network connectivity. The laser sensor differentiates between the real and data world, the radar sensor is used for distance measurement, and the visual sensor reads the traffic lights and the road signs. The car also uses the LIDAR system to percept the environment [6,8,10]. The radar in combination with the camera determines the obstacles at a distance, and so the car changes its speed, direction, and other control settings to avoid the same (Figure 8.6).

8.2.7 Laser Perception

In laser perception, the car emits lasers continuously and the transmitter receives the lasers and by measuring the time taken by the laser to return and the strength at which the laser is returning, the system determines the position, size, speed, and also the obstacle type [6–8]. Laser perception may be the best and the most accurate way for the car to determine the surroundings, but it is also very costly to install it in a self-driving car which makes it less affordable to the buyers. The three-dimensional omnidirectional laser radar installed in the Google self-driving car costs around $80,000; therefore, reducing the price of the laser systems is also an important issue (Figure 8.7).

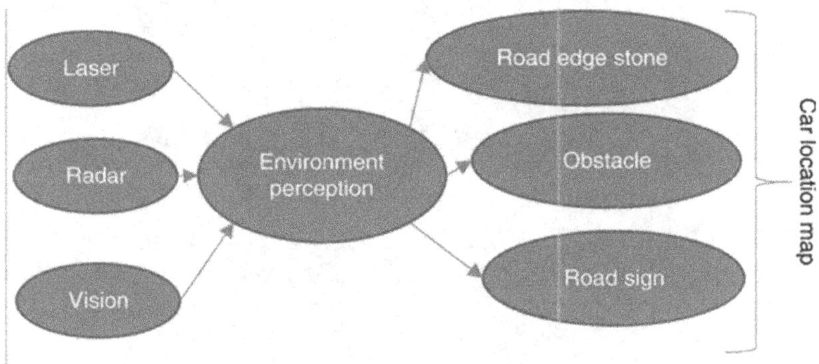

FIGURE 8.6
Components of environment perception.

FIGURE 8.7
Working of laser perception.

FIGURE 8.8
Radar determining the surroundings.

8.2.8 Radar Perception

Radar perception detects the distance between the car and the objects around it, the radar system calculates the distance by measuring the time it takes the millimeter wave to come back after colliding with the objects around the car. The main producers of radar systems are Northrop Grumman Corporation, Lockheed Martin Corporation, Israel Aerospace Industries, BAE systems [7] (Figure 8.8).

8.2.9 Visual Perception

Visual perception helps the car to detect the traffic lights and the traffic sign which are necessary for the car. Deep learning is currently being used in visual perception. Algorithms like LaneNet are used to make the car understand when there is a need for changing the lane [7].

Algorithms like YOLO and SSD are used for object detection and are constantly updated. But there are also some problems with the deep learning, if two people are standing close to each other the algorithm faces difficulty to determine there are two people or not. So LIDAR detection using deep neural networks is currently trending.

8.3 Vehicle Control

Vehicle direction and speed come under vehicle control. All the information on vehicle status, environment perception, traffic regulations, driving target is fed to the computer in the car, then the computer performs calculations and sends the instructions to the vehicle control system and then the vehicle control system performs according to the

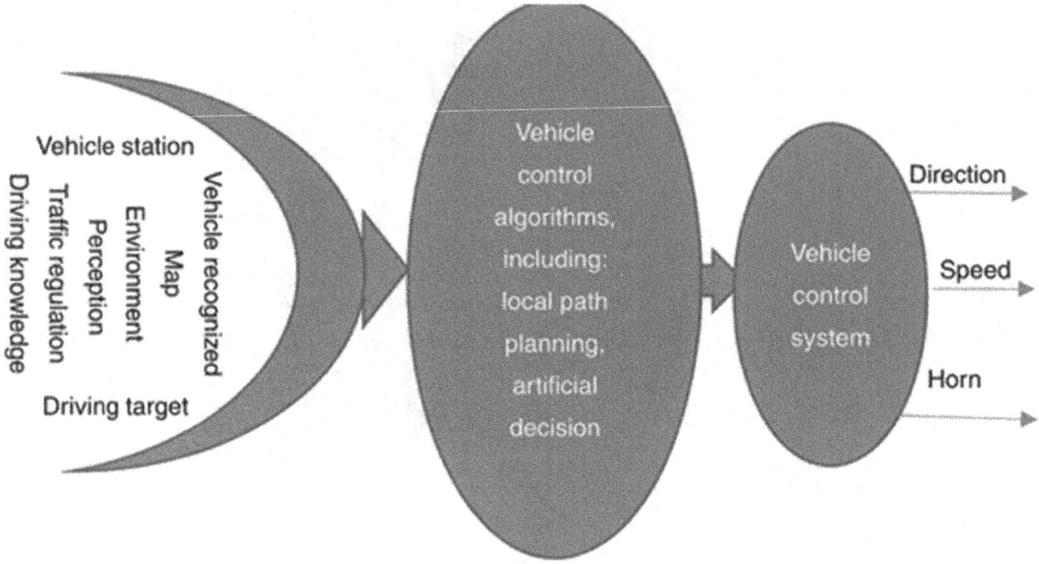

FIGURE 8.9
Components of vehicle control system.

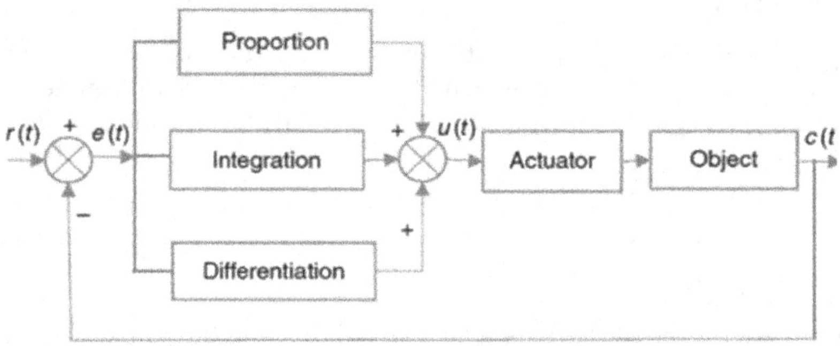

FIGURE 8.10
Working of vehicle control method.

given instructions. The vehicle control system controls all the major components such as braking, steering, locking doors, automatic transmission, and cruise control system (Figure 8.9).

8.4 Vehicle Control Method

The vehicle control method uses the Proportion Integration Differentiation (PID) algorithm. In Figure 8.10 $r(t)$ is an input expired signal, a feedback error signal is $e(f)$, the control signal calculated by the PID algorithm is $u(t)$, and the current actual output signal is $c(t)$. The PID algorithm uses three parameters: proportion, integration, and differentiation. There are some issues with the PID algorithm such as low adaptability and complex parameter adjustment. But the improved PID algorithm resolves all the issues with the classic PID algorithm [11].

8.5 Comparison between Camera and LIDAR

The two key sensors used in a self-driving car are cameras and LIDAR. LIDAR is able to sense the surrounding of the car more clearly; however, it is very expensive. In comparison, cameras are the cheaper option but it does not provide an accurate and clear vision of the surroundings. In self-driving cars, Google uses the LIDAR system, while Tesla uses cameras. The reports say that in the coming days the LIDAR will replace the cameras [22] (Figure 8.11).

8.6 Disadvantages of Self Driving Cars

Though self-driving cars or automated vehicles have many advantages and are continuously getting updated it also brings some disadvantages with them. The main challenges could be the user resistance to giving up control, loss of driving skills, increased risk of hacking, liability for damage issues, need to frame a new law, loss of driving-related jobs, loss of situational awareness, and so on [2].

There is no law in place to decide if an accident happens from a self-driving car, who will be liable for it—the company or the owner? These situations require an entirely new legislative framework to decide who is responsible for those accidents. The automated vehicles have cameras and capture their surroundings, which raises concern about the privacy of the people as it stores the locations and captures the places where the person is going. Moreover, the information can be leaked if a hacker hacks the car. The loss of driving-oriented jobs is the main reason behind the Indian government not supporting the

FIGURE 8.11
Lidar determining the surroundings.

introduction of automated vehicles in the Indian market as there are more than 3 million people who are employed in driving-related jobs [2].

8.7 Legal Issue

In India, the automobile industry makes 7.1% of the GDP, but the government is reluctant to introduce autonomous vehicles to the market because of the job losses, big automobile manufacturers are putting millions of dollars toward developing self-driving vehicles but are not able to enter India. There had been talks with Tesla to bring their self-driving cars to India. Replying to a question of bringing their cars to India, Elon Musk, CEO of Tesla, said on Twitter there is no proper infrastructure in India to make their electric self-driving cars work [2].

India's Minister of Road Transport and Highways Nitin Gadkari said the government is not against technology, but the government cannot afford to provide jobs to 4 million drivers who will lose their jobs because of the autonomous cars. Because of it, many investors don't see India as a market for self-driving vehicles. But the lack of interest of the government is not the only issue with self-driving vehicles in India, there is a need for policy transformation to introduce autonomous vehicles. The automobiles in India are regulated by the Motor Vehicle Act, 1988 which does not permit the use of self-driving vehicles in India. Another issue is that in case the AV hits a person or another car there is no law about who will be liable for that, the owner, or the insurance company [23].

8.8 Conclusion

This chapter discusses technologies such as car navigation systems, location systems, LIDAR, and environment perception used by self-driving cars or automated vehicles.

FIGURE 8.12
Algorithm determining the status of surrounding vehicles.

Automated vehicles or self-driving cars use a car navigation system to navigate their way, while the LIDAR system uses lasers to determine its surroundings and identify the obstacles in its way.

The computer in the car uses a map matching algorithm in combination with a location system and GPS to determine the best route to its destination. The car uses all the information vehicle status, environment perception, traffic regulations, driving target is fed to the computer in the car, then the computer performs calculations and sends the instructions to the vehicle control system and then the vehicle control system performs according to the given instructions.

This chapter also discusses the disadvantages of self-driving or automated vehicles which include liability issues in case of an accident, privacy issues as the car keeps track of the location and captures the images of the places it is being driven to, loss of driving-related jobs, need for an entirely new legislative framework as the Motor Vehicle Act, 1988 does not permit the use of self-driving or automated vehicles use in India (Figure 8.12).

References

1. Zhao, J., Liang, B. and Chen, Q., 2018. The key technology toward the self-driving car. *International Journal of Intelligent Unmanned Systems*, 6(1), pp. 2–20.
2. Bagloee, S.A., Tavana, M., Asadi, M. et al., 2016. Autonomous vehicles: challenges, opportunities, and future implications for transportation policies. *Journal of Modern Transportation*, 24, pp. 284–303. https://doi.org/10.1007/s40534-016-0117-3.
3. Wang, J., Steiber, J. and Surampudi, B., 2009. Autonomous ground vehicle control system for high-speed and safe operation. *International Journal of Vehicle Autonomous Systems*, 7(1–2), pp. 18–35.
4. LIDAR System, available at: https://ieeexplore.ieee.org/abstract/document/8067701 [accessed on 21 July 2021].
5. Stilgoe, J., 2018. Machine learning, social learning and the governance of self-driving cars. *Social Studies of Science*, 48(1), pp. 25–56.
6. Hecht, J., 2018. LIDAR for self-driving cars. *Optics and Photonics News*, 29(1), pp. 26–33.
7. Vellinga, N.E., 2017. From the testing to the deployment of self-driving cars: legal challenges to policymakers on the road ahead. *Computer Law & Security Review*, 33(6), pp. 847–863.
8. Hirz, M. and Walzel, B., 2018. Sensor and object recognition technologies for self-driving cars. *Computer-Aided Design and Applications*, 15(4), pp. 501–508.
9. Wallace, L., Lucieer, A., Watson, C. and Turner, D., 2012. Development of a UAV-LiDAR system with application to forest inventory. *Remote Sensing*, 4(6), pp. 1519–1543.
10. El-Rabbany, A., 2002. *Introduction to GPS: The Global Positioning System*. Artech House, Boston, MA.
11. Zhu, H., Yuen, K.V., Mihaylova, L. and Leung, H., 2017. Overview of environment perception for intelligent vehicles. *IEEE Transactions on Intelligent Transportation Systems*, 18(10), pp. 2584–2601.
12. Introduction to Self-Driving Cars, available at: https://ieeexplore.ieee.org/abstract/document/8220479 [accessed on 22 July 2021].
13. Electronic Maps, available at: https://www.tandfonline.com/doi/abs/10.1207/s15327108ijap0603_3 [accessed on 27 July 2021].
14. Sharma, L. (Ed.). 2021. *Towards Smart World*. New York: Chapman and Hall/CRC, https://doi.org/10.1201/9781003056751.
15. Sharma, L. and Garg, P.K. (Eds.). 2020. *From Visual Surveillance to Internet of Things*. New York: Chapman and Hall/CRC, https://doi.org/10.1201/9780429297922.

16. Sharma, L. and Garg, P.K. "IoT and its applications", *From Visual Surveillance to Internet of Things*, USA: Taylor & Francis, CRC Press, Vol. 1, p. 29.
17. Giesbrecht, J., 2004. Global path planning for unmanned ground vehicles (No. DRDC-TM-2004-272). Defence Research and Development Suffield (Alberta).
18. Vehicle Control System, available at: https://www.inderscienceonline.com/doi/abs/10.1504/IJVAS.2009.027965 [accessed on 18 July 2021].
19. Environment Perception, available at: https://ieeexplore.ieee.org/abstract/document/7857073 [accessed on 20 July 2021].
20. Global Positioning System, available at: https://books.google.co.in/books?hl=en&lr=&id=U2JmghrrB8cC&oi=fnd&pg=PR13&dq=GPS+system%5D&ots=9Ny_mVTFAX&sig=mJL9a4UV24x_DPbo_3PVuLXdjhg&redir_esc=y#v=onepage&q=GPS%20system%5D&f=false [accessed on 21 July 2021].
21. Self-Driving Car and the Law, available at: https://ieeexplore.ieee.org/abstract/document/8220479 [accessed on 24 July 2021]
22. Ethical and Social Aspects of Self-Driving Cars, available at: https://arxiv.org/abs/1802.04103 [accessed on 24 July 2021].
23. LIDAR in Self-Driving Cars, available at: https://www.osapublishing.org/DirectPDFAccess/75474D39-9457-4F8D-85495D43F2016D8D_380434/opn-29-1-26.pdf?da=1&id=380434&seq=0&mobile=no [accessed on 24 July 2021].
24. Challenges in India, available at: https://www.dqindia.com/self-driving-cars-still-distant-dream-india/ [accessed on 26 July 2021].

9

IoT and Remote Sensing

Yaman Hooda

Faculty of Engineering and Technology, Manav Rachna International Institute of Research and Studies, Haryana, India

CONTENTS

9.1 Background of Internet of Things

Internet of Things (IoT) is a network which is made up of physical elements or "things" that are rooted with various software, sensor and emerging technologies, for the only purpose of connecting various devices or systems with each other to exchange information in the form of data among them by using the power of Internet. The basic technologies that form the background of formation of "things" consist of junction of various phases of different technologies such as machine learning, commodity sensors, real-time analytics, and embedded systems. Conventional areas of networking of wireless sensors, embedded systems, building automation contribute to the formation of IoT in the real world [1–8].

DOI: 10.1201/9781003244165-12

In September 1985, the term and concept of "IoT" first made its appearance in the speech given by Peter Lewis at the 15th Annual Congressional Black Caucus Foundation in Washington, DC. Lewis stated that "The Internet of Things, or IoT, is the integration of people, processes and technology with connectable devices and sensors to enable remote monitoring, status, manipulation and evaluation of trends of such devices." However, the actual revolution in the field of IoT came in 2010, when it was reported that the things-to-people ratio grew from 0.08 to 1.85 in 2003, as estimated by the CISCO Systems, by considering their concept of IoT as the fact in time when there are more "things" than people are connected with the Internet. Sometimes, the background of IoT also refers to installing short-range sensors (such as mobile transceivers) into various gadgets for forming the mode of communication between the things and people while performing the work of daily necessities [9–13].

One of the main characteristics of the IoT devices is that they can upgrade their software without or negligible involvement of the user. Also, for communication between the devices, the process of the setup of new devices with the existing device is also simple and doesn't consume much time. With the advancement in the technology and the development of Interoperable Communication, IoT enables devices to connect with each other, having the same or different architectures [1–3,9–22]. The "Functional Block" of the IoT systems offers different skills of identification, sensing, inclination, process of communication, and data transfer and management. The various parts of this functional block are shown in Figure 9.1.

Currently, IoT finds its applications in nearly all phases of life: from connecting people via IoT-enabled mobile devices to IoT Edge and cloud facilities. Data management becomes more effective and efficient with the formation of IoT-enabled data centers. With the help of IoT in automation, devices in the residential and commercial buildings can easily be

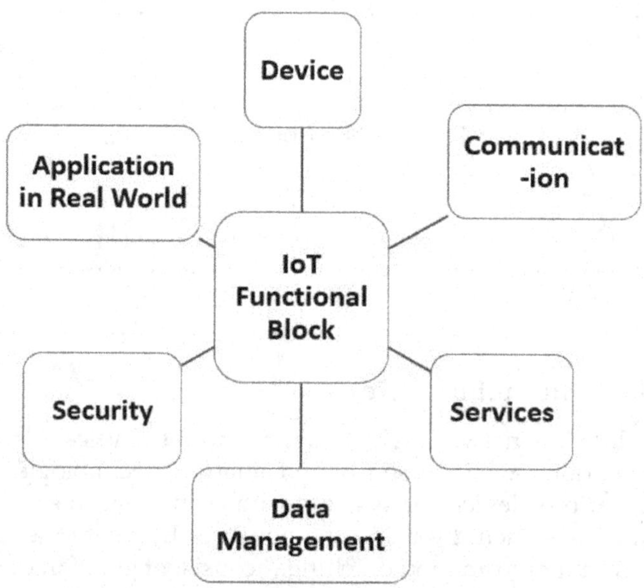

FIGURE 9.1
Parts of IoT functional blocks.

managed. Also, the concept of smart cities has come into existence because of the application of IoT in factories, transportation industries, and construction companies.

9.2 Background of Remote Sensing

Remote sensing can be considered as an art or science for gathering information about an object, an area, or any phenomenon with the help of the data fetched by devices or "sensors," provided that these devices or "sensors" are not in direct contact with the object, the area, or the phenomenon taken into consideration. Thus, it may be defined as the process of gathering, processing, interpreting, and investigating the data obtained by the devices, without any physical contact with the element under investigation [23–37]. Therefore, it may be considered that the methodology of remote sensing is studying the physical as well as chemical characteristics of the considered element from a particular distance.

Remote sensing is originated from the concept of aerial photogrammetry. Aerial photogrammetry is the technique of collecting photographs that are under consideration from a considerable height with the help of cameras installed or mounted on the aircrafts. The photographs thus obtained are known as "Aerial Photographs." The basic advantage of the remote sensing is the synoptic view of the area it provides, which enables a larger area to be captured or analyzed within a small duration of time, and then the data collected can be interpreted and processed to generate the outputs in the form of a map. In the process of remote sensing, the data are collected with the help of high-power, high-range sensors, which can be used in the form of camera, radar, scanner, etc. installed on platforms such as satellites and aircrafts. These sensors then manipulate and modify the information gathered and sent it back to storage units. These stored data are then formatted and processed to generate photographs and other forms of digital media [1,2,5,9,14–19,30,37]. The digital data or photographs can be understood digitally and/or visually to generate maps. Therefore, remote sensing may be defined as a technological process that consists of all the different methods to obtain data from a considerable distance, in the form of pictures or other relevant electromagnetic (EM) records of the surface of the earth, by processing and interpreting the same. Infrared bands (reflected and thermal) or the microwave region of the EM spectrum is used in the technology of remote sensing.

There are two basic processes of remote sensing which can be used to gather the information about the resources of the earth:

1. Data acquisition
2. Data analysis

The process of data acquisition consists of six different phenomena: (i) energy sources, (ii) energy propagation through atmospheric layers, (iii) energy layers of earth interaction, (iv) retransmission of energy through atmospheric layers, (v) sensors, and (vi) the data obtained by the sensors in the digital or pictorial form. On the other hand, the process of data analysis consists of three major steps. The first step involves arrangement of devices that can help view and interpret the pictorial as well as digital data; the second is formation of maps, both in hard copy and digitalized format, with the help of data being gathered; and the last step is to present the information to the users in the most understandable way so that they can use the same in their decision-making process [23–38].

Thus, *remote sensing can be considered as a methodology for "sensing" the earth's surface with the help of the properties of electromagnetic spectrum, the waves emitted, refracted, or reflected by the sensors, for improving the land-use resource management and thus creating a sustainable environment.*

9.3 Process of Remote Sensing

To understand the process of remote sensing, one should know the basic elements which constitutes a remote sensing imaging system, given as follows [23–38]:

a. **Energy Sources**
   An energy source is considered to be the element that provides illumination to the object under consideration. This illumination sometimes refers to an electromagnetic energy.
b. **Interaction between Radiation and Atmospheric Layers**
   After the propagation of energy from the source to the object under consideration, it comes into direct contact with the atmosphere and interact with the different atmospheric layers. This interaction between the energy and the atmospheric layers occurs again after the energy ray is reflected from the object to the sensors. This interaction involves scattering and absorption of energy.
c. **Interaction with Earth's Surface**
   Depending upon the characteristics of the radiation and the object, interaction between the energy and the object takes place. The interaction between the earth's surface and the incident energy causes the energy to reflect, transmit, scatter, absorb, and emit.
d. **Recording of Energy by Sensors**
   The energy that is scattered or emitted from the object is then received, collected, and recorded by the sensors. These sensors are installed on the platforms at a considerable distance from the surface of the earth. The sensors used in remote sensing can be of two types: active sensors and passive sensors, *which will be explained in the next section.* And with the advancement in the technology, these sensors are IoT-enabled, making the process of gathering and interpreting the data simpler than the conventional methods.
e. **Processing**
   The signals recorded by the IoT sensors is then communicated in an electronic system so as to process the data obtained in the form of an image. This image obtained so can be of a photographic film or magnetic tape or digital data.
f. **Interpretation and Analysis**
   The image obtained and processed so is then interpreted by visual or digital or electronic means for the purpose of extracting information about the illuminating object. This information extracted from the various means of images makes us understand the object under consideration in a better way, by revealing some new information or assisting in solving complex problems.

IoT plays an important role in the form of sensors that collect the data and then reflect them back to the final station for its processing and final output. Now, let us study about the two types of IoT sensors and their characteristics, which are used in the remote sensing process.

9.4 IoT-Based Remote Sensing Sensor Systems

The technological advancements in the field of remote sensing have experienced a massive development in the past decades, including the usage of the IoT in different elements of the remote sensing. IoT-based sensors in remote sensing are generally used for gathering data in the form of electromagnetic waves. These sensors then transmit the gathered data to the final station, and the data are interpreted with IoT-enabled devices. These IoT-based sensor systems are mounted on an appropriate platform, which can be lift to a considerable height to perform its function. They can be of either stationary type or mobile type. The former includes a tripod stand for making the observations in the field, while the latter includes the collection of the data from satellites and aircrafts. Aircrafts are used for covering a smaller area, as compared to that for retrieving information from the satellites to get a synoptic view.

IoT-based remote sensing sensor systems can be classified into two broad categories on the basis of the source of energy: manmade and naturally occurring. The IoT sensors that have the capabilities of sensing the natural radiations, which are either emitted or reflected from the surface of the earth, are known as *passive IoT sensors*, whereas the IoT sensors that carries an electromagnetic radiation of a considerable wavelength or a band of wavelength for the purpose of illuminating the surface of the earth are called *active IoT sensors*. Both of these sensors can be either imaging or non-imaging. The imaging sensor is used to measure the radiation received from a specific point of a particular area, sensed with image-making devices including camera, computer screen, and television [3,9–14,16–22]. These types of sensors yield a 2D spatial distribution of emitted or reflected magnitude of the electromagnetic radiation. On the contrary, a non-imaging sensor is used to measure all the radiations collected from every point under the area of consideration. Also, these sensors integrate the collected radiations and generate the final output in the form of electrical signals with different strengths. The non-imaging sensors focus upon the wavelength resolution, whereas the imaging sensors focus on the spatial resolution. With the help of IoT in the sensor system of remote sensing, now the process of measuring the properties of the signals, both point by point and instantaneously, has become simpler and easier (Figure 9.2).

9.4.1 IoT-Enabled Passive Sensors

Passive sensors is a system that depends on the exiting energy source for performing their function. In IoT-enabled passive systems of remote sensing, the sample is considered to be the reflected rays emitted from the surface of the ground; here, the source of the energy is not dependent on various instruments used for the recording purpose. Some of the examples for IoT-embedded passive systems are television cameras, electro-optical scanners, photographic cameras, imaging spectrometers, and Return-Beam Vidicon (commonly known as RBV). In the case of television cameras, the sensors installed in the cameras formed the pattern of the images collected in the form of electrical charges, which is further scanned with the help of an electron beam and transformed to electrical signals. With the help of IoT, these signals can be transferred to a very long distance, and can also be

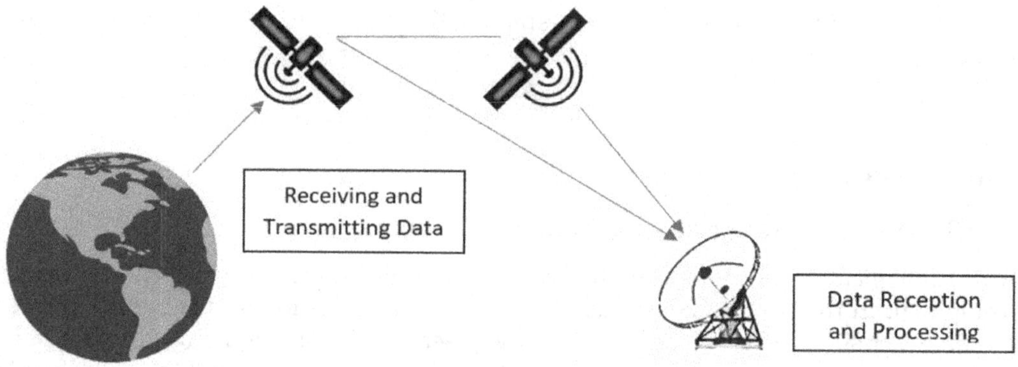

FIGURE 9.2
Simple schematic route of remote sensing.

stored on the cloud for future references. In photographic camera system, the image is recorded on the film base with the help of a photographic emulsion. With the sensors enabled in this camera system, the recording is much faster and efficient, and also the final image produced is more understandable and easier to recognize the data printed on the final output. RBV is an advanced version of TV camera system, which basically works on the information gathered by the IoT-enabled sensors in a black-and-white configuration. IFOV or Instantaneous Field of View is the technology used for the determination of ground – cell resolution. With this technology, scan lines can easily be measured in electro-optical scanners. These scanners have been proved to be more efficient than camera-based systems as firstly, the former is able to record a wide range of energy waves, and secondly, the scanner systems are able to record the data from the target area directly from the spectral information (Figure 9.3).

IoT technology is also used in the reception system of the scanner. The detectors are used in the scanner systems to convert the oscillating recorded waves to the desired electrical signals, which will then be converted into numerical data and stored on a magnetic tape. The hyperspectral sensor, commonly known as an image spectrometer, is a system of sensors which are able to attain the images from numerous very narrow infrared bands of the EM spectrum. These sensors are able to retrieve the data in a continuous spectral manner, covering the entire range or bandwidth; as a result, when these sensors record and transfer the data at the final station, the data obtained so will have a very high spectral resolution, thus avoiding the problem of breaking of the pixels when the final image is concentrated at a particular point for observation and gathering information simultaneously.

Thus, with the help of IoT in the sensors, the data can easily be captured, transferred, and then converted into a desirable format so that it can be easily understood by the users. Also, the biggest advantage of enabling IoT in these systems is that there is no loss of any information or data. Once collected, the data are securely stored and transferred to the desirable location. And with the help of the cloud technology, these data are stored firmly and can be used as and when required.

9.4.2 IoT-Enabled Active Sensors

The system of IoT-enabled sensors that uses the manmade energy sources for the purpose of data collection is called an active sensor; for example, using a flash bulb while taking

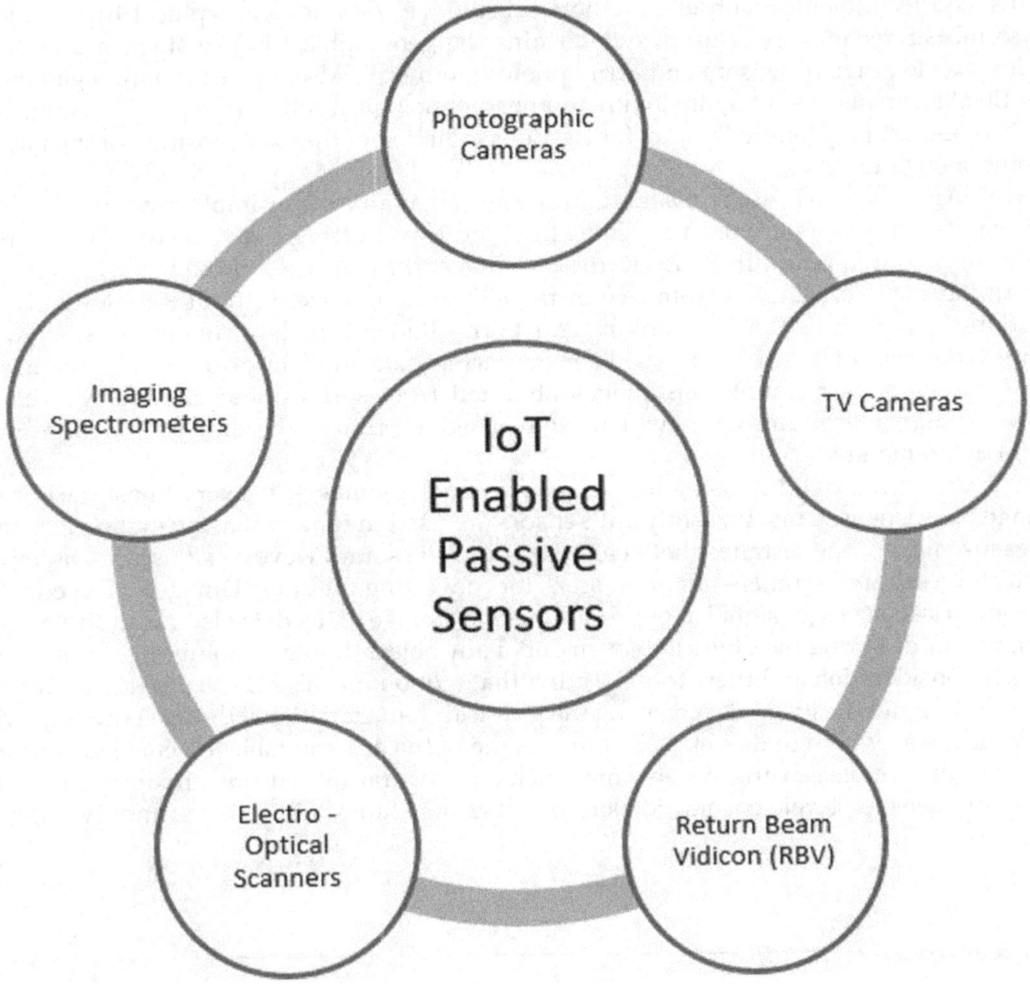

FIGURE 9.3
IoT-enabled passive sensors.

photographs of an object in a dark environment. The two basic examples that fall under the category of active sensor systems are laser scanners and RADAR.

In the case of laser scanners, the pulses are generated and emitted mostly in the infrared space of EM spectrum. The pulses are communicated to the target station, and then, the reflected energy waves or signals are recorded on a receiver. With the enhancement of IoT in the remote sensing technology, these scanners are not only able to measure the intensity and amplitude of the signals, but also help in generating a Digital Surface Model (DSM). When these scanners are collaborated with other advanced and sophisticated technologies, a Digital Elevation Model (DEM) can be generated. An example is the technology of LIDAR (LIght Detection And Ranging). The combination of LIDAR and remote sensing proves to be most beneficial in the field of surveying for the determination of leveling and elevation. With the capability of recording multiple signals from the same exact pulse, the elevation of the surface of the earth as well as on the surface of the earth can be easily determined. The data received from the LIDAR systems are directly unified with one more advanced technological subset of remote sensing, i.e., GIS or Geographical Information System. GIS technology is used to determine the geographical information of a system with a wide range of sensors and aerial photogrammetry. Also, the information gathered by LIDAR can also be integrated with another technological subset of remote sensing, i.e., GPS or Global Positioning System, for the determination of the exact position of the target point or station.

RADAR or RAdio Detection And Ranging system is another example of active systems in which the sensors are used for collecting and transmitting the data from the source point to the final point. In RADAR, the pictures and the images of the topography and landscapes can be derived from two of the either radar systems: (i) SLR (Side-Looking Radars) and (ii) SLAR (Side-Looking Air-borne Radars). In both of these radar systems, the technology of IoT is used for gathering data in form of images or pictures, which accurately resemble the images obtained from aerial photogrammetry, with a low-angular solar illumination which can be used to produce the shadow effects, as per the requirements.

The increasing trend of using IoT in every technology makes it cover almost the entire range of sensor systems. Presently, IoT sensors are used in remote sensing systems for the measurement of the distances between two points via sound waves – *ultrasonic sensors*; to detect the infrared signals – *infrared sensors*; for converting the optical images received into the electrical waves (or signals) for presenting and storage of the data electronically – *image sensors*; for detecting the physical movement of any object (living or nonliving) in the area under consideration and then, transforming that motion into signal – *motion detection sensors*; and for measuring and converting the physical characteristics of the light rays into the electrical waves or signals – *optical sensors.* Some of the other notable IoT-enabled sensors used in the remote sensing systems may include temperature sensors, proximity sensors, pressure sensors, level sensors, accelerometer sensors, humidity sensors, and gyroscope sensors.

9.5 Remote Sensing and Its Types

Remote sensing can be broadly classified into two categories:

a. Ideal Remote Sensing Systems
b. Real Remote Sensing Systems

As the name suggests, an ideal remote sensing system consists of an ideal source, which is used to produce electromagnetic waves of known uniform intensities and wavelengths. With an ideal system of remote sensing, it has been predicted that there is no loss occurred during the whole process of gathering the data and then, transferring the same data into electrical signals for proper understanding of the information collected. Also, the relationship between the propagation of the returning signal and energy of the incident waves (of all wavelength of any intensity) is linear (Figure 9.4).

In reality, the response of the wavelength vs intensity is recorded and then processed into an explainable arrangement. This further is being recognizable as a unique entity to a specific target in its specific state.

An ideal remote sensing system consists of the following main components:

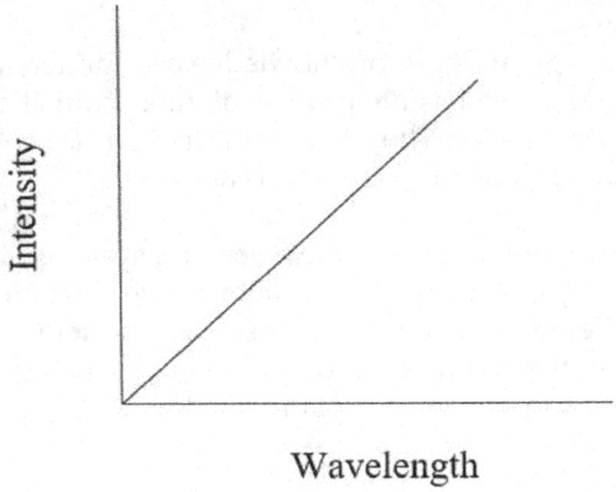

FIGURE 9.4
Intensity versus wavelength curve for ideal remote sensing system.

a. **Energy Source**
The source of energy is used to provide the energy at a known and constant high level of input, over the entire range of wavelengths regardless of place and time.

b. **Atmosphere and Atmospheric Layers**
In an ideal system of remote sensing, the atmosphere and atmospheric layers are considered to be in a "noninterfering" nature, with the consideration of no modification in the energy from the source point, regardless of the fact that the wave is hitting onto the surface of the earth and after being reflected back. This consideration can be fulfilled irrespective of place, time, and wavelengths, with sensing altitude involved.

c. **Series of Interactions at Earth's Surface**
Ideally, when the wave is generated from the source and comes in contact with the surface of the earth, the interaction happened so, causing the generation of reflected and/or emitted signals, would be selective in terms of wavelengths. Also, the nature of the interaction would be invariant, known, and exclusive to every feature type of earth's surface under consideration.

d. **Super Sensor**
This type of sensor would be reliable, simple, and sensitive to the entire range of wavelengths. The sensors yield detailed data spatially on the radiance or absolute brightness, as a derivative relation of wavelength. The other advantages of such type of sensors include accuracy, economical, simple in operation, and doesn't require any space or power virtually.

e. **Real-Time Data-Handling System**
In real-time data-handling system, the moment when the response of a radiance versus wavelength is generated over a terrain element, it will be processed in an interpretable format. The recognition of this response is unique for a specific terrain element. By providing information timely, the procession is performed instantaneously or in "real time." Due to the consistency among the matter–energy interactions, no reference data had been used while doing the analysis. The final data will give an insight into the physical–biological–chemical state of every single feature of the element under consideration.

f. **Multiple Data Users**
These users have a great depth of knowledge and understanding of their concerned fields of study as well as the process of data acquisition and data analysis in remote sensing technology. The "data" collected so can be used and proved to the "information" in various forms to different users.

In reality, there is no concept of an ideal remote sensing system. The various components of an ideal remote sensing system discussed above have certain disadvantages, which formed the concept of the other type of remote sensing system, i.e., real remote sensing systems. The difference in the components of real remote sensing systems in comparison to that of ideal remote sensing system is explained below:

a. **Energy Source**
The nature of the spectral distribution of the emitted and/or reflected waves from the various features of surface of the earth is far from uniform. Also, the levels of the solar energy varies in accordance with location and time, and with different materials present in the surface of the earth; the emitted energy showcased varying levels of efficiency. All the sources of energy used in the real systems are nonuniform in accordance with wavelengths. The properties of such systems also vary with respect to location and time.

b. **Atmosphere and Atmospheric Layers**
In real system of remote sensing, the various layers of the atmosphere interfere with the properties of the rays and thus, modify the characteristics of the energy waves. The atmospheric layers tend to modify the strength as well as the spatial distribution of the energy.

c. **Series of Interactions at Earth's Surface**
The process of remote sensing would be simpler only when the emitted or reflected energy is having a unique characteristic in a known way. Even though the signatures or patterns of the spectral response play a vital role in the determination, identification, and analysis of materials on the earth's surface, the spectral world is full of ambiguity.

d. **Sensor**
In reality, there is no any single sensor that is sensitive to the entire range of wavelengths. All the sensors have a fixed boundary of spectral sensitivity. Each sensor is having different characteristics, and the images produced by the sensors are different from one another.

e. **Data-Handling System**
The data obtained so by the data acquisition process are converted into an understandable format, which requires proper instrumentation, well-defined experience, considerable thoughts, time efficiency, and reference data.

f. **Multiple Data Users**
The successful application of the data obtained via remote sensing system is made by the users. The "data" obtained by the remote sensing process can only be considered as an "information" to the users if and only if the users can understand that information. This process of understanding the "information" includes the successful interpretation of the data and using the same in the optimum application areas. Also, there is no single methodology by which the data can be interpreted and

used. With the different types of sensors available in the market and progression in technology, every set of sensors require a different approach toward the interpretation of the data and using it as an "information" in their respective application areas. The basic differences between the components of an ideal remote sensing system and real remote sensing system can be summarized as follows (Table 9.1).

9.6 Data Acquisition and Data Interpretation

9.6.1 Data Acquisition

IoT plays a significant role in the process of recording the images from the remote sensing process into a digital format. These images are then processed with the help of computer software to generate images for the next operation of data interpretation (Figure 9.5).

TABLE 9.1
Ideal Remote Sensing System *versus* Real Remote Sensing System

Components	Ideal Remote Sensing System	Real Remote Sensing System
Energy Source	Energy with constant wavelength.	Energy with varying wavelength.
Atmospheric Layers	Does not interfere or modify the properties of the energy.	Modify the properties of the energy as it passes through the different layers of atmosphere.
Interactions at Earth's Surface	The nature of interaction is known and invariant, without any ambiguity.	The spectral responses do not have unique characteristics.
Sensor	"Super Sensors" are used.	Different types of sensors are used depending upon the requirement.
Data-Handling System	No reference data are used, providing the final data to give an insight into the physical–biological–chemical states of every single feature of the element under consideration.	The data obtained by the data acquisition process is converted into an understandable format, with the help of reference data and considerable and experienced thoughts.
Multiple Data Users	The "data" collected can be used and proved to the "information" in various forms to different users.	The "data" obtained by the remote sensing process can only be considered as an "information" to the users if and only if the users can understand that information. This process of understanding the "information" includes the successful interpretation of the data and using the same in the optimum application areas.

FIGURE 9.5
Role of IoT in data acquisition.

The technology of IoT not only helps in the process of data acquisition, but also helps in the further steps to get the final data which are easily interpreted by the users, including the steps of data processing and data storage. The images produced so can be in the form of digital or photographic film, and the former is delivered to the users in computer-formatted readable tapes. But, in the case of photographic film, the variation in the brightness is used to explain the variation in the scenic characteristics; thus, the part on the film which reflects high energy is the brightest portion, and the darkest portion represents the less energy areas.

The digital data acquisition is the procedure of generation of images by considering the illuminating source and phenomenon of absorption and reflection of energy. IoT-enabled imaging sensors are used in the transmission of illuminating energy into the digital format. Electromagnetic (EM) spectrum is considered to be the prime source for the images. The spectral bands vary from the highest energy (gamma rays) to the lowest energy (radio waves). This electromagnetic energy is detected either electronically (for scanners) or photographically (for cameras). The advantages of using photographic systems include (i) economical, (ii) relatively simple, and (iii) higher degree of geometric integrity. On the contrary, electronic sensors are expensive and complex in nature, but offer several advantages including (i) improved calibration potential, (ii) broader spectral range of sensitivity, and (iii) ability to transmit data electronically. Also, the records of the detected signals can easily be obtained from a developed photograph where the film can act as both a medium for detecting and recording. The signals from the electronic sensors are recorded on a magnetic medium, which are converted to a photographic film or an image, in which the former serves only the purpose of the recording medium, whereas the latter can be used as both the detecting and recording media.

The images obtained by satellites from the operation of remote sensing are recorded digitally, and then, they are processed by computers so as to produce the final images for interpretation. A digital image can be expressed as an array of numerical characterized by a series of bits of finite numbers. The images obtained from satellites are used in several ways including (i) tracking earth's resources, (ii) flood control, (iii) fire control, (iv) monitoring crops' growth, and (v) environmental studies. The image obtained so by the IoT sensors can be represented as a 2D function of any two variables such as x and y, $f(x, y)$, where 'x' and 'y' are considered to be spatial coordinates. The amplitude of the function considered at any specific level is the intensity of the image at that specific level. An image is considered to be a discrete image if the value of the spatial coordinates and the amplitude are finite and discrete. These images are made up of a finite number of known components, each components having its own value and particular location. These particular components are pixels, pels, image elements, and picture elements. One of the basic advantages of IoT-enabled image sensors is that the images obtained are of good quality and can easily be understandable as compared to the traditional sensors. "Pixels can be defined as smallest individual element in an image holding quantized values which are used to represent the brightness of a specific color at any specific location. These pixels can be stored in the memory of a computer system as a raster map or raster image (a two-dimensional array of small integers). These values are often stored or transmitted in a compressed form." [21–25]

Each of these pixels is having the coordinates in the form of (x, y) representing the continuous selection of the surface of the earth in the discrete space. In this pixel coordinate system, the origin is taken the left upper corner of the image. Every pixel of the image is linked with a specific number, which is represented as DN or Digital Number. DN is considered to be the digitally brightness-quantized value. Thus, a number of arrays of DNs are used to make a digital image, representing the distribution of field parameters (including topographic elevations and EM radiations, etc.). DN is generally used to depict the mean brightness or radiance of a smaller area relatively within a specific scene. Normally, the values of DN that generate a digital image, are documented over numeral ranges as follows:

0 to 1023	$2^{10}=1024$
0 to 511	$2^9=512$
0 to 255	$2^8=256$
0 to 127	$2^7=128$
0 to 63	$2^6=64$

The above-mentioned ranges represent a set of integers which are recorded by using 10-, 9-, 8-, 7-, and 6-bit binary coding computer scales, respectively. "The size of an area of a particular consideration affects the reproduction of the details within the scene," implying that the pixel size is inversely proportional to the details of the scene [20–29]. Thus, in digital representation, the smaller the size of the pixels, the more the detailing of the scene. In two-dimensional space of images, the address of pixel can be specified in distances, measures along the rows and columns. Image intensity, gray value, and image brightness are some of the terms used for pixel magnitude.

Another way of obtaining a digital image is by the process of quantization or sampling. The process of quantization or sampling involves conversion of an analog or a continuous image into a digital format. Considering a continuous image function $f(x, y)$, sampling is referred to the process of digitizing the coordinate values of 'x' and 'y', whereas the process of quantization involves the digitizing of intensity or amplitude values.

9.6.2 Data Interpretation

In remote sensing, the process of data interpretation includes the analysis of digital data and images or photographs. This process can be accomplished either with the help of a computer system well equipped with the latest analysis methodologies and techniques, or with the help of visual interpretation. The analysis of images with visual interpretation is considered to be the first-ever analysis approach toward the remote sensing imagery. In visual interpretation, the position, shape, and size of the object under consideration can easily be analyzed, along with the color saturation and contrast. If the height of the considered object has to be determined, indirect visual analysis can be performed (Figure 9.6).

The computer-assisted software techniques are considered to be better than the visual interpretation as the former takes less time in the data interpretation. Also, the rate of successful interpretation in visual interpretation process is only high when the interpreter is having a good experience. But, these two analysis techniques are complementary to each other, and if both of these techniques are used for the process of the data interpretation, the results would be better when considered either of the two (Figure 9.7).

The primitive form of remote sensing is photo interpretation. This involves identification and recognition of objects in the photographs obtained from aerial photogrammetry and then judging their significance. As stated earlier, the success of the photo interpretation depends upon the ability to interpret the images by the concerned engineers accurately. The accurate interpretation also involves the knowledge and understanding of concerned area, ingenuity, and skill.

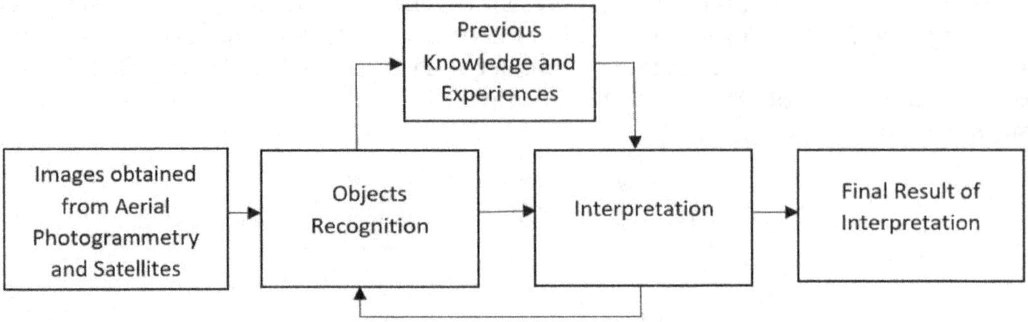

FIGURE 9.6
Flow chart of data acquisition and data interpretation in IoT-enabled remote sensing.

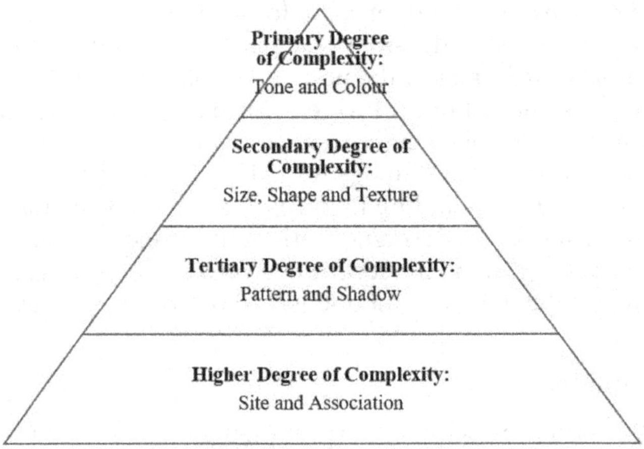

FIGURE 9.7
Degree of complexity in image interpretation.

In image interpretation, there are four different degrees of complexity, namely:

1. Primary Degree of Complexity
2. Secondary Degree of Complexity
3. Tertiary Degree of Complexity
4. Higher Degree of Complexity.

The first degree of complexity in image interpretation is known as Primary Degree of Complexity. This degree of complexity deals with the complexities in two major characteristic domains of image interpretation, i.e., tone and color. The *tone* of an image may be considered as the varying range of brightness in an image, from the lightest part to the darkest part. This is considered to be one of the most important elements in the process of image interpretation and has to be done by an experienced person. With upcoming invention of IoT in sensor technology, these image sensors installed are now able to capture the tonality of the image in a perfect range. The *color* of an image includes the information about the color in the form of pixels. A colored image has three channels or values per pixel. The color of an image is used to measure the chrominance and intensity of light. The Second Degree of Complexity in the image interpretation involves the complexities that may involve in the size, shape, and texture of an image. The *size* of an image can be determined by describing the dimensions of the image in width and height in pixels. The maximum size of an image can be determined by megapixels of the camera used for taking the image. The *shape* of an image defines an enclosed, flat area of a space. Though the shape of an image can be formed by geometrical curves and colors, it is limited to a two-dimensional entity, described by its width and length. The *texture* of an image in image processing is a calculated set of metrics, which is designed to compute the apparent texture of image. The texture of an image gives the finding about the spatial arrangement of the whole image or a specific part of an image in the form of its color or intensity.

The complexities that arise in the *pattern and shadow of an image* fall under the Tertiary Degree of Complexity in image interpretation. The *pattern* of an image can be defined as a set or an array of images that have something common in between them. For example, if the images are collected for a particular theme, such as topography of a region, they are known as thematic images. Other examples may include a *pattern* of smooth images or frightening images or sharp images. A *shadow* of an image can be defined as a real dark area of an image when the source of the energy or light is blocked by an opaque element. The cross section of a shadow of an image is a 2D profile, or may be considered as an inverse projection of an element. It surrounds every point of the 3D volume space behind the element, with the source of light facing the element. Higher Degree of Complexity in image interpretation includes two parts: site and association. The complexities in the *site and association* of an images are not that much of magnitude. After ruling out the various complexities at the lower degrees in the process of image interpretation, the siting of an image is possible with great accuracy and no error. These four degrees of complexities have to be taken into consideration while performing the process of image interpretation. The final output received after considering all these complexities listed, will be then transferred to users for their applications in various areas.

A *reference data* can be defined as an accompanying data which is used in the observations or measurements regarding any element, area, or phenomenon, available in any arrangement and derived from any point of source. The considered element, area, or phenomenon has to be remotely sensed. The main usage of these reference data is in the process of verification of the information which is extracted from the remote sensing data. It also helps as a support not only in the analysis of the data received, but also in the process of image interpretation. The procedure of acquiring the reference data is time consuming and expensive, but for the purpose of the accuracy and reliability, it is considered to be perfect aid for remote sensing operations. Also, these reference data can be used for the calibration purpose of the IoT sensors for gathering and transmission of the information.

The most important property of any remote sensing system is *spatial resolution.* Spatial resolution is used in the determination of the capability of any system of remote sensing to record the details of the data spatially. In analog photography, the extent of the sharpness of an image is considered to be the spatial resolution of an image. There are many factors on which the spatial resolution depends such as the image motion during exposure, resolving power of the camera lens or the film, and the atmospheric environments during image exposure. The power of the lens as well as that of a film can be quantifiable. "The Resolving Power of the camera lens can be determined by means of a resolution test pattern, which is made up of a number of sets of varying thickness parallel black lines, which are being separated by the white spaces of same thickness, known as line pair" [10–15].

The system of the spatial resolution for an imaging system is demarcated through different criteria such as:

1. The geometrical characteristics of the imaging system
2. An ability for finding out the difference between the point targets
3. An ability in measurement of spectral properties of tiny targets
4. An ability in measurement of the periodicity of targets, showing a repetitive manner.

The spatial resolution can be further classified into two categories: (i) high spatial resolution and (ii) low spatial resolution. The basic difference between the high spatial resolution and low spatial resolution is that finer details of a particular area can be easily visible in the former as compared to the latter. As a result of this, high spatial resolution images will contain more detailed data, while low spatial resolution images will be able to show the coarse features only. However, both the categories have their own application areas, but for more efficient, reliable and accurate image interpretation, high spatial resolution images are used more as compared to low spatial resolution images (Figure 9.8).

9.7 Application Areas of IoT in Remote Sensing

With the inclusion of IoT technology in different parts of the remote sensing process, the remote sensing offers a wide range of application areas. IoT-enabled remote sensing process does not only provide an affordable and practical way for a continuous and accurate

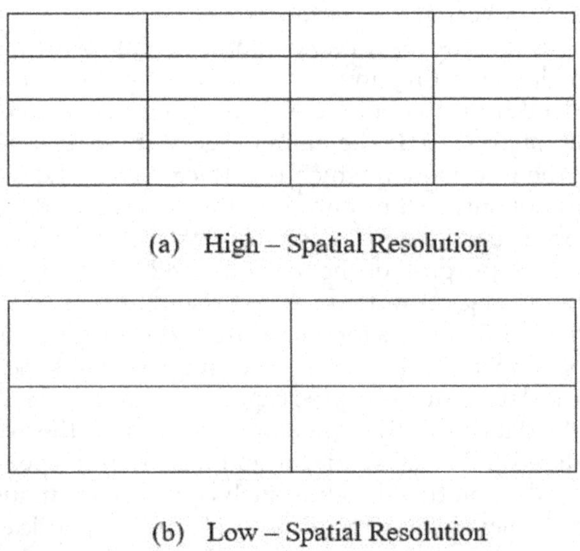

FIGURE 9.8
(a) High and (b) low spatial image resolutions.

observation and monitoring of all the resources in the surface of the earth, but also helpful in the determination of the impact and effect of the activities of human beings on land, water, and air, i.e., whole ecological balance. The data obtained from IoT-enabled sensors are used to provide the necessary information for making accurate decisions and formulation of the acts and policies with respect to resource development and land use. With the help of IoT in remote sensing, the forest fires can be detected before the appearance of smoke in the sky; by sensing the existence of pollutants in the environment, it helps in detection of any diseased plant before the identification of speckles by naked eyes; and the exact location of the icebergs can be determined. The practice of remote sensing started in India in the early 1900s, and after the evolution of its usage in monochromatic aerial photography, it was used for geological exploration work and surveying extensively.

9.7.1 Mineral Exploration

IoT-enabled remote sensing system is used to carry out the reconnaissance as well as in-depth exploration of resources such as fossil fuels and minerals. With the advancement of the remote sensing operations, this technique is used to carry out the following functions:

1. To provide the data used in the applications of geological interest.
2. In controlling the ore deposits individually so as to map the local patterns of folds, faults, and fractures.
3. In soil exploration.
4. In determination of the mining areas by mapping the regional contours.
5. In detection of the hydrothermally reformed rocks and stones.

9.7.2 Disaster Management

Naturally occurring disasters such as earthquakes, volcanic eruptions, floods, forest fires, avalanches, landslides, and tsunamis would have a great impact on the society, including both the loss of life and property. There is no technology developed that can totally contradict the effects of these disasters or may stop the occurrence of these disasters. However, in remote sensing technology, signals can be used for detecting the occurrence of these disasters, thus minimizing the effects of the same on the society. With IoT being implemented at every stage, the disastrous effects of these natural hazards can be minimized in the following ways:

1. By accessing the occurrence of the disaster, the damage caused by it and then, thereby helping in the public rescue operations. Along with this, the operations for providing medical aid can also be run simultaneously.
2. Identification of risk-prone areas around the world. This can only be possible by analyzing the geological information accurately, effectively, and efficiently.
3. Warnings and precautions can be issued on the basis of the successful prediction of ground movements in a specific area, which is useful in minimizing the impact of the disaster on that specific area.

9.7.3 History and Archeology

History and archaeology are application areas which have the widespread usage of remote sensing techniques. The images obtained at the final point of remote sensing system are used in determining the historical and archaeological outlines of the usage of the land in prehistoric time. In the findings of remote sensing, it was observed that some of the present agricultural fields and forest areas are the towns or small districts of yesteryears era. With the modification in the water content in the layers of the soil, accompanied by the change in the characteristics of the materials present in different layers of the soil, IoT sensors are now able to identify the important archaeological buried sites and also give the significance and evidence of the existence of various important civilizations around the world. The growth of both IoT and remote sensing opened up new means in investigation of those remote regions or areas, which was not possible before because of the limitations of conventional form of surveying.

9.7.4 Environmental Observations

The technology of remote sensing is considered to be the chief source of providing information regarding environment and its balance in nature, as is it the prime way of gathering the fundamental information of the environment, specifically on either a repetitive scale or a regional scale. For this very purpose, satellites had been launched that gather and record all the information regarding the motions that occur in the environment. For example, most of the satellites being launched are for weather forecasting. With the help of IoT, the weather forecasting of even a remote area is possible. The study of oceans has become easier with IoT and remote sensing, and as a result, very minute details and processes that occur the oceans are now being studied resourcefully. The serious environmental problems like ozone layer depletion and global warming can also be tracked using artificial intelligence and remote sensing with full-time monitoring, data gathering, storage, and transmission. Also, the increased levels of harmful pollutants and reducing air quality index can also be supervised and scrutinized using this technology.

9.7.5 Land Cover Analysis

Land-cover analysis or land-use analysis is another important application area of remote sensing. The land-use pattern at a particular area is directly linked with the population of that area. As the population around the world is rapidly increasing, the pattern of the land use and land cover is also changing rapidly. It has been observed that the forest areas are decreased for increasing the areas for cultivation as the area dedicated to the latter is increasingly urbanized. These changes are not natural, and thus most of these changes are not good for the interest of human beings. If the same continues, it will prove to be disastrous to human beings as well as total ecological balance. Thus, for preventing this damage to the environment, a proper planning has to be done to control these changes. The planning phase is the most important phase as it requires gathering of the data from the existing land-use patterns, which can be done precisely with the help of remote sensing and IoT [28–37].

But there is a difference between land use and land pattern. *Land use* is a method for describing the usage of a plot of land for specific purposes such as agricultural land, land dedicated for residential buildings, commercial buildings, or even for industrial setups.

On the contrary, the *land pattern* includes the types or *patterns* of materials that exist on a specific piece of land such as land showing the patterns of buildings, forest areas, vegetation, hilly areas, and many more [37–41]. An evergreen forest can be considered to be an area under land pattern, and the various activities on any specific piece of land on that evergreen forest such as extraction of oil and lumbering are considered to fall under the category of land use. Therefore, accurate and up-to-date information about the land under consideration is essential for many aspects of gathering data from that particular piece of land.

The technique of remote sensing and IoT has become an intense part of mapping of land for defining both land use and land pattern because of the following advantages:

1. Images obtained are unbiased, resulting in the formation of a permanent data set which can be interpreted for a vast variety of land use or land cover including agricultural fields, forest areas, and rapid urbanization.
2. The process of image interpretation is less expensive and less time consuming in comparison to the different branches of traditional way of surveying which includes direct ground contact.
3. The remotely sensed images provide a detailed "perspective" that is not possible in the case of ground surveying.
4. One of the main advantages of using remote sensing technique is that the images obtained so eliminate the issue of surface access, that every so often hinders the ground surveying.
5. The degree of accuracy is very high in the case of images obtained from IoT sensors, as compared to the conventional way of surveying because of the spatial resolution which precisely ties the level of detailed required for data gathering.
6. The land with large areas can be mapped easily and rapidly.

But, with all these advantages, the usage of remote sensing has several disadvantages too, which are as follows:

1. To carry out the surveying of smaller areas, mobilization of remote sensing operations is found to be uneconomical.
2. In most of the images obtained by aerial photogrammetry and IoT technology, the lack of perspective in the horizontal direction is observed, which proves to be essential in identification of various classes of land use.
3. Sometimes, the problem of distinguishing the land on images has occurred.

The System of Classification in Land Use and Land Pattern not only recognizes the resources for land coverage, but also its related activities under land usage. When this system of classification utilizes the concept of gathering the data from IoT-assembled remote sensing technology, this system must meet the following measures (Figure 9.9):

1. The level of the accuracy of image interpretation must be equal approximately for every category.
2. This system must be valid for widespread areas.

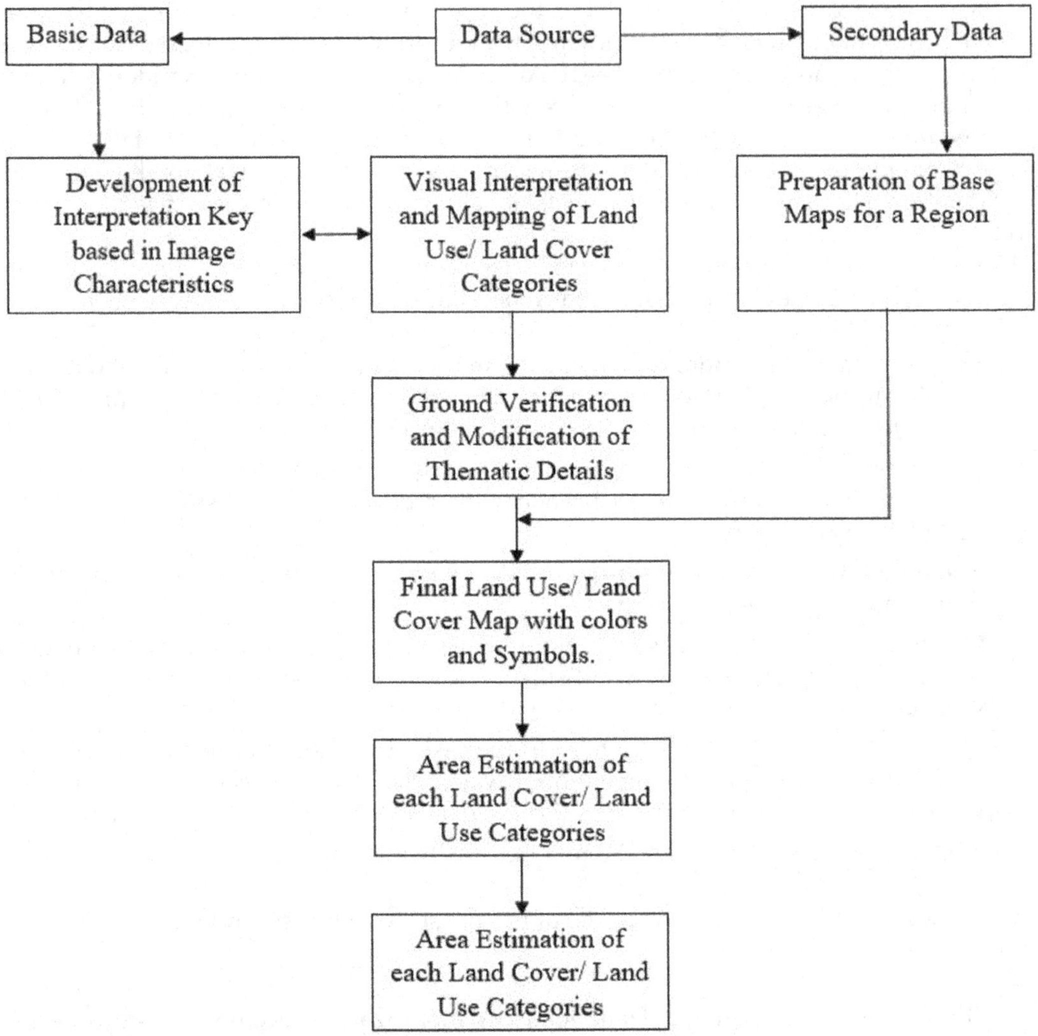

FIGURE 9.9
IoT remotely sensed image interpretation process for land use/land cover.

3. The system should be such that it can use the subcategories which are derived from either surveying on ground or by using remote sensing technique.
4. As population and urbanization are increasing rapidly, it must consider the future land in comparison with existing land data.
5. In identifying the categories of land cover or land usage from the remotely sensed data, the level of precision must not exceed 85 percent.
6. The results that require to be repeated after a certain interval of time, must be obtained via interpreter–interpreter communication.
7. The system must use the IoT–remote sensing technique to gather the data or information throughout the year.
8. The recognition of the multiple land use should be done.

9.8 IoT and GIS

Geographical Information System or GIS is a system that allows us to better understand the world and thus enables the "development of spatial intelligence for logical reasoning." The system of GIS uses the data and the information gathered from the remote sensing operations and finds application in various areas. It is considered to be a computer-based system that includes gathering, storing, and manipulating data and then analyzing and displaying the data for the purpose of solving complex problems in the fields of management, planning, and research. The main advantage of GIS is that it can handle both spatial and nonspatial data types. This system is considered to be a specialized database system for the preservation of its recorded information.

The combination of IoT and Geographical Information System (GIS) assures to provide improved situational awareness by analytical indicators, faster understandings and perceptions, and the duration of the time for making important decisions. As, in the present world, every worker, asset, or even a sensor has its specific geographical position, GIS plays a vital role in network operations and IoT. An IoT system consists of four basic components: (i) devices (or commonly known as *sensors*), (ii) connectivity or communication link, (iii) data processing system, and (iv) user interface. GIS also has five basic subsystems which completely blend with the components of IoT:

1. Data input
2. Data management
3. Data processing
4. Data display
5. Data output.

In GIS, data input deals with the creation of an image-based GIS system with the help of multiple geo datasets. The purpose of data management is efficient storage of data and also its easy retrieval. The data processing includes the phenomena such as data classification, data feature enhancement, and data manipulation. The data collected so are displayed, and the final output is generated to provide different products in the form of images and maps (basically thematic maps) for their applications (Figure 9.10).

9.8.1 Real-Time GIS and IoT

The real-time GIS refers to the capability of the GIS system to use the real-time data in everyday decision-making processes. This will enhance transformation of the data collected and the information being utilized at any situation. The technological linkup of GIS with real-time condition is fed by IoT, providing views to the everyday procedures revolving around the organization of data and enhancing the decision-making process with the futuristic strategies and ideas [28–41]. With IoT enabled in GIS operation, the user while operating the latter application can find the answers of the following questions:

1. What is happening at any moment?
2. Where the incident is taking place?
3. Who all are going to be affected by that incident?

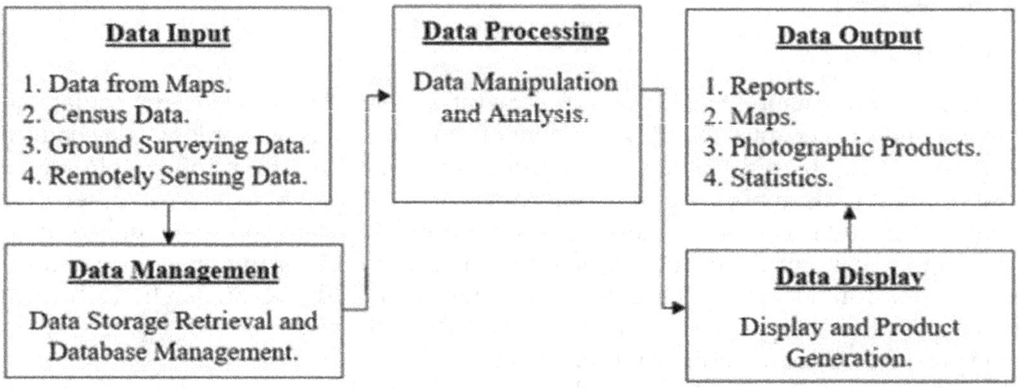

FIGURE 9.10
Subsystems of GIS.

FIGURE 9.11
Procedure of real-time GIS and IoT.

4. What are the various assets available?
5. Where are the beloved people of the users located?

Some of the application areas of this real-time IoT-enabled GIS are as follows (Figure 9.11):

1. To monitor the public utilities that serve for the public such as electricity, wastewater, and clean water.
2. To monitor and track the air traffic worldwide by aviation organizations and airport authorities.
3. To monitor the crime as it happens by law enforcement agencies. The new function of "SOS" can also be implemented efficiently.
4. To issue initial warnings in case a place is going to experience any natural disaster by monitoring weather conditions, floods, earthquakes, droughts, and even forest fires.
5. People can use IoT-enabled electronic gadgets such as smartwatches, smartphones, fitness bands, and many more to capture every single information about their own activity anytime.
6. The local and state government authorities can use this application in managing the operations related to cleanliness of an area by proper tracking of trash truck and snow poles.
7. The transportation departments can monitor the traffic flows and road blockages on highways and expressways, and can also monitor the rail traffic.

8. Many organizations use real-time social media feeds to monitor the emotional state of employees on any particular issue.
9. To monitor the safety of the people during large gatherings by emergency management agencies such as Olympic Games and World Cup Championship.

The system of real-time GIS is characterized by an unceasing stream of activities forwarded from IoT sensors. Each activity in this process measures the latest or current magnitude of any state recorded and transmitted by a sensor such as location, distance, speed, altitude, temperature, level of water, and amount of current. The best way for viewing, monitoring, or responding to any data in the real world is to use maps.

With the advancement in the technology, IoT and Augmented Intelligence (AI), as well as its subdisciplines such as machine learning, deep learning, and fuzzy logics, are finding their application in all the sectors surrounding us. In lieu of the same, many organizations and industries are investing in huge funds in IoT technology-based concepts such as electric and autonomous vehicles, smart cities, telecommunication infrastructure, public utilities, and safety measures. The sensors used for the statics purpose are digitizing the planet so efficiently and effectively to a level which none of us have ever seen or thought before. For example, in the case of automatic vehicles, the early systems are able to detect their location with the help of mobile phones, with less accuracy and supporting data. This technology has been continued to enhance the features by installing the new categories of sensors, which monitor the location with more precision than the former one and are also economical comparatively [3–5,9–13,20–22].

In present scenarios, the system of real-time GIS is complementing strongly with the solutions provided by IoT with the expansion of the possibilities to integrate continuous space–time analysis. A perfect example is autonomous vehicles which not only account their locations, but also make observations regarding the different pavement conditions. The observations recorded by these vehicles are used in the detailed analysis of the pavement conditions and will be helpful in providing alternative routes or generating warnings in case of any problematic situation, as and when required. Thus, IoT systems gather information from multiple sensors of different types, process it, and produce the results with optimum accuracy and is also able to solve and manage critical operations in any area or field [8,23–27,42–45].

Incorporation of various networking of sensors, which are intelligently combined in a geospatial background for the purpose of optimization of different operations, is one of the major contributions of IoT in GIS. Even the information that is recorded previously and stored in database system can be taken altogether to observe all different ways of a problem for making smarter conclusions, thereby optimizing services, improving the level of accuracy and efficiency, and reducing cost.

But one has to also focus on the various sources through which real-time data can be collected or gathered by the sensors. Real-time data can be of several various forms and thus, are used in a wide range of applications. Maps play an important role as a source of real-time data in the following scenarios:

1. To predict the path of hurricane or cyclone.
2. To determine the hourly wind condition pattern for a specific location.
3. To predict the damage caused by natural hazards such earthquakes, avalanches, and landslides.
4. To monitor the positioning of metro transportation systems.

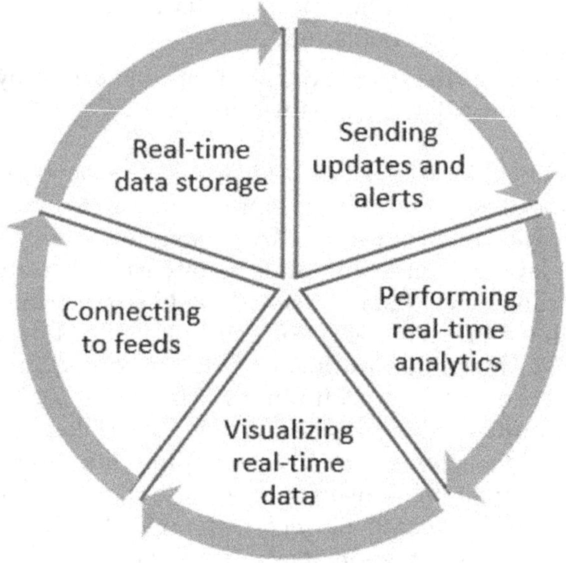

FIGURE 9.12
Five components.

5. Used in steam gauges for determination of water levels to monitor the risk of drought and floods.
6. To monitor the severe weather conditions worldwide.
7. To monitor the geolocated live tweets from Twitter and Instagram posts.

9.8.2 Capabilities of Real-Time GIS Platform

IoT servers and real-time GIS applications are responsible for bringing the real-time data on just a click from your finger, allowing the users to connect virtually, processing and analyzing any type of data at any time, with the features of updates installing automatically [23–41]. With this application, it allows the users to make complex decisions faster and respond to any situation with an increased level of mindfulness when they experienced any changes. There are five basic components of the capabilities of real-time GIS system (Figure 9.12):

1. **Connecting to Feeds**
   IoT-enabled servers are able to receive and interpret the real-time data input from any source via virtual mode. The system is not only able to know the receiving process of the real-time data, but also the formatting of the data. Various types of input connectors are used to capture this real-time data from different sources.
2. **Sending Updates and Alerts**
   The connectors, commonly known as *output connectors*, prepare and send the treated data to the users in an anticipated arrangement. These output connectors are used for translating the events into an arrangement, which is capable of being transferred through a specific communication network.

3. **Performing Real-Time Analytics**
   These services allow the user to play a significant role in the definition of the stream of data event and also to perform any filtration or processing on the data while it streams through output connector. With this component, users are able to focus and discover on those events, places, and operations that are important to them.
4. **Visualizing Real-Time Data**
   With various operations, the application allows the users to visualize and display the key information regarding the specified operation. The views obtained so can be easily stored on IoT cloud service, available on various GIS-based software applications, and thus can be shared to different members of the same organization, different groups in the same organization, and also with public when required.
5. **Real-Time Data Storage**
   In most of the scenarios, geodatabase system is used for capturing the data streamed in real-time environment. The finest practice to reinforce the historic archiving of actions is using the temporal or historic class feature for storage of all the actions perceived from the information. Thereby, the nature of every element can be gathered and indefinitely stockpiled, covering all the information from the very initial action until the present time. But, if the data are going to be stored from the beginning, the size of the statistics will grow with time. The frequency and size of arriving information are the two basic factors on which the rate of growth of data or information depends. Thus, a retention policy has to be defined and enforced to determine the amount of history which has to be maintained actively in a particular geodatabase (Figure 9.13).

9.9 IoT and GPS

As an important sub-application area of remote sensing, GPS or Global Positioning System is a weather radio navigation system, which is completely space-based. The basic advantage of GPS is it can provide quick and accurate magnitude of position, velocity, and time of any object, located at any location worldwide at any specific time. This particular system depends on three basic pillars: (i) satellite signals, (ii) exact time, and (iii) sophisticated algorithms. The working of these three pillars is dependent on each other, and thus helps in the generation of distances in order to triangulate places anywhere worldwide. The GPS is made up of three sections:

1. Space section
2. Operational control section
3. User equipment section.

Space section, also known as satellite constellation, consists of satellites that revolve around the earth in a fixed orbit. The primary function of the space section is to provide a proper range or frequency of signals and messages in the form of data, to the customer's equipment. The operational control section is also known as a ground monitoring/control

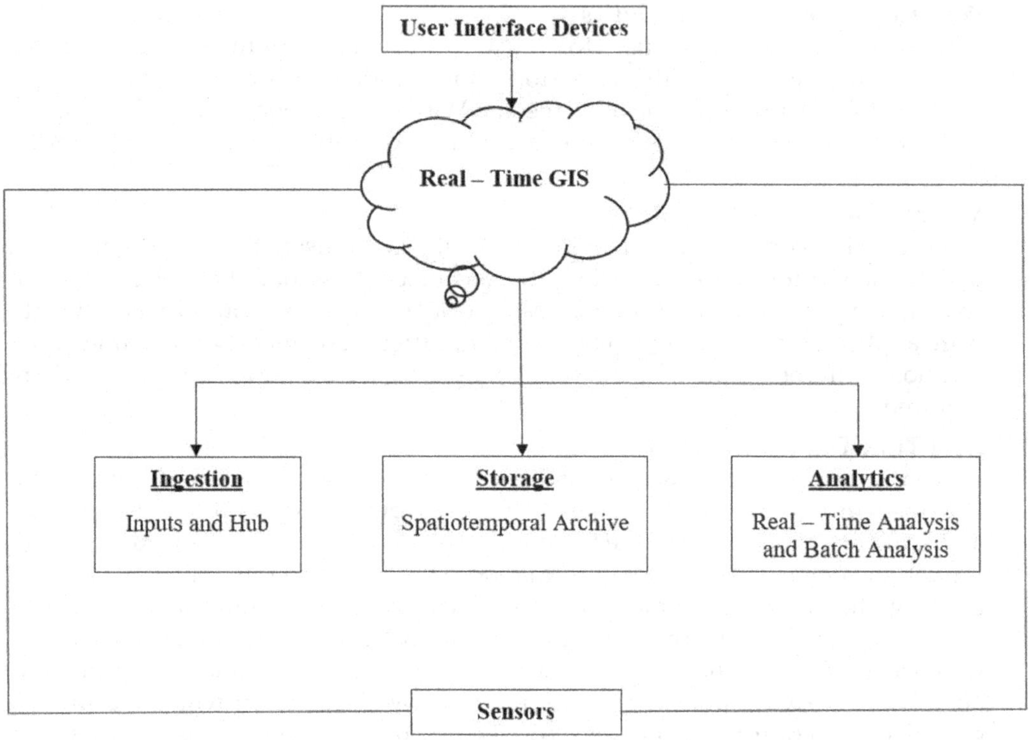

FIGURE 9.13
Real-time analytics with IoT.

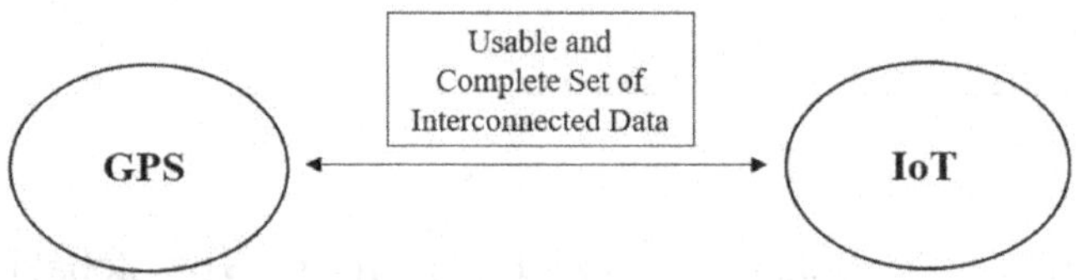

FIGURE 9.14
GPS and IoT communication link.

network. The primary function of OCS is tracking the signals and also maintaining the position of satellites in space. It maintains the satellite configuration in the orbit and helps in monitoring the health of satellites and signal integrity. This section is also responsible for updating the satellite-clock corrections and various other parameters which are necessary in determining the time, velocity, and position of the user. The user equipment section performs functions of timing and navigation, such as in surveying.

GPS is irreplaceable to an IoT system as it helps in computing and recording speed, location, direction, and time. Since GPS is considered to be the initial approach in tracking and cataloging the data of the physical world in digital format, it had a crucial impact on the IoT-related technological approaches. IoT technology is used for collecting and quantifying huge amounts of information in the form of data, ranging from healthcare to vehicles, whereas GPS is used to track the information about the location of the above stative objects (Figure 9.14).

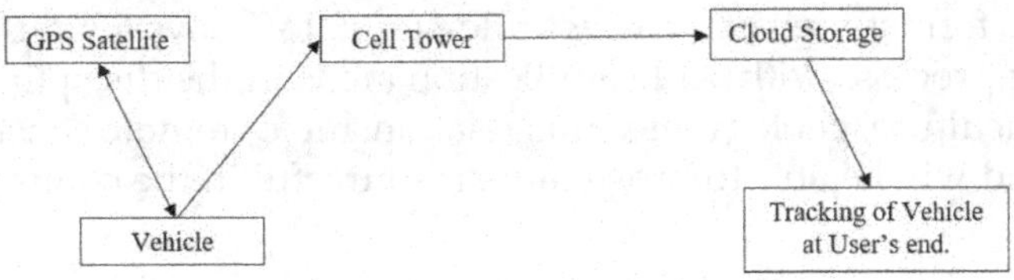

FIGURE 9.15
Tracking of vehicles with GPS and IoT.

The technology of IoT helps in monitoring the objects and hardware systems, so as to provide the information regarding the real-time scenario and all the related data regarding the various operations of the devices, whereas GPS provides the actual physical longitudes and latitudes of the object under consideration.

Let us take an example of how it works. The wide application of GPS and IoT can easily be experienced in tracking the location of vehicles. For tracking the location of any vehicle, GPS tracking devices are installed on the vehicles. GPS satellite monitors the position of the vehicle every time and transfers the information via signals to the cloud storage hub. With the availability of Internet services, the owner of that particular vehicle is able to monitor every single movement of that vehicle (Figure 9.15).

9.10 Future Scope

IoT and remote sensing play important roles in each other's disciplines. With IoT performing various operations at different levels not only in the field of remote sensing, but also in the subdisciplines of remote sensing, i.e., GIS and GPS, one can easily solve any kind of problem in the complex environment. With all these technological structures working together in tandem, these systems are actually laying foundations for even bigger scientific and commercial approaches such as smart cities, electric or self-driven vehicles, and automation. Not only in engineering applications, this system could also work in healthcare industries and an interconnected, gigantic ecosystem, allowing the smart devices to interact with complex positioning abilities for achieving the goals that are not possible with traditional or conventional methods and approaches.

9.11 Conclusion

This chapter deals with the interaction of two emerging technologies: remote sensing and IoT. The art of remote sensing has emerged as one of the most captivating themes in the past four decades. The observation of the earth from the space through remote sensing system have proved to be advantageous in the fields of dynamics of the surface of earth, competent management of natural resources, and the whole environment inclusively. With the interference of IoT in the remote sensing system, the system becomes more effective and efficient. From the whole process of gathering the information via sensors to interpret to the users in the format they can easily understand, IoT plays a vital role in every stage of remote sensing process. With wide application areas in the disciplines of engineering, management, agriculture, geology, etc., this relationship of remote sensing and IoT proves to be beneficial and will be able to prove more resourceful in the coming future.

References

1. Drone usage in Covid-19. Available at: https://www.unicef.org/supply/media/5286/file/%20 Rapid-guidance-how-can-drones-help-in-COVID-19-response.pdf [accessed on 10 July 2021].
2. L. Sharma, P. K. Garg (Eds.) (2020) *From Visual Surveillance to Internet of Things*, Chapman and Hall/CRC, New York. DOI: 10.1201/9780429297922.
3. L. Sharma, S. Sengupta, B. Kumar. (January 2021) An improved technique for enhancement of satellite images. *Journal of Physics: Conference Series* 1714:012051.
4. S. Singh, L. Sharma, B. Kumar. (January 2021) A machine learning based predictive model for coronavirus pandemic scenario. *Journal of Physics: Conference Series* 1714:012023.
5. G. Jha, L. Sharma, S. Gupta, Future of augmented reality in healthcare department. In: Singh P.K., Wierzchoń S.T., Tanwar S., Ganzha M., Rodrigues J.J.P.C. (eds) *Proceedings of Second International Conference on Computing, Communications, and Cyber-Security*. Lecture Notes in Networks and Systems, vol. 203. Springer, Singapore, 2021. DOI: 10.1007/978-981-16-0733-2_47.
6. G. Jha, L. Sharma, S. Gupta, E-health in Internet of Things (IoT) in real-time scenario. In: Singh P.K., Wierzchoń S.T., Tanwar S., Ganzha M., Rodrigues J.J.P.C. (eds) *Proceedings of Second International Conference on Computing, Communications, and Cyber-Security*. Lecture Notes in Networks and Systems, vol. 203. Springer, Singapore, 2021. DOI: 10.1007/978-981-16-0733-2_48.
7. S. Kumar, P. Gupta, S. Lakra, L. Sharma, R. Chatterjee, "The zeitgeist juncture of "Big Data" and its future trends", *2019 International Conference on Machine Learning, Big Data, Cloud and Parallel Computing (COMITCon)*, 2019, pp. 465–469. DOI: 10.1109/COMITCon.2019.8862433.
8. S. Sharma, S. Verma, M. Kumar and L. Sharma, "Use of motion capture in 3D animation: Motion capture systems, challenges, and recent trends", *2019 International Conference on Machine Learning, Big Data, Cloud and Parallel Computing (COMITCon)*, 2019, pp. 289–294. DOI: 10.1109/COMITCon.2019.8862448.
9. S. Makkar, L. Sharma. (2019) A face detection using support vector machine: Challenging issues, recent trend, solutions and proposed framework. In: Singh M., Gupta P., Tyagi V., Flusser J., Ören T., Kashyap R. (eds) *Advances in Computing and Data Sciences*. ICACDS 2019. Communications in Computer and Information Science, vol. 1046, Springer, Singapore. DOI: 10.1007/978-981-13-9942-8_1.
10. L. Sharma, P. K. Garg. IoT and its applications. In: Sharma, L., GArg, P. K. (eds) *From Visual Surveillance to Internet of Things*, Taylor & Francis, CRC Press, Boca Raton, FL, Vol. 1, pp. 29.
11. L. Sharma, D. K. Yadav, S. K. Bharti, "An improved method for visual surveillance using background subtraction technique," *2015 2nd International Conference on Signal Processing and Integrated Networks (SPIN)*, Amity University, Noida, India, Feb. 19–20, 2015, pp. 421–426. DOI: 10.1109/SPIN.2015.7095253.
12. D. K. Yadav, L. Sharma, S. K. Bharti, "Moving object detection in real-time visual surveillance using background subtraction technique", *2014 14th International Conference on Hybrid Intelligent Systems*, Gulf University for Science and Technology, Kuwait, December 14–16, 2014, pp. 79–84. DOI: 10.1109/HIS.2014.7086176.
13. L. Sharma, A. Singh, D. K. Yadav. (March, 2016) Fisher's linear discriminant ratio based threshold for moving human detection in thermal video. *Infrared Physics and Technology* 78:118–128, Elsevier.

14. L. Sharma, N. Lohan. (January 2019) Performance analysis of moving object detection using BGS techniques in visual surveillance. *International Journal of Spatiotemporal Data Science* 1:22–53.
15. A. Anand, V. Jha, L. Sharma. (July 2019) An improved local binary patterns histograms techniques for face recognition for real time application. *International Journal of Recent Technology and Engineering* 8(2S7):524–529.
16. L. Sharma, D. K. Yadav. (June, 2016) Histogram based adaptive learning rate for background modelling and moving object detection in video surveillance. *International Journal of Telemedicine and Clinical Practices*. ISSN: 2052-8442, DOI: 10.1504/IJTMCP.2017.082107.
17. L. Sharma, N. Lohan. (March, 2019) Internet of things with object detection. In: *Handbook of Research on Big Data and the IoT*, IGI Global, pp. 89–100. ISBN: 9781522574323, DOI: 10.4018/978-1-5225-7432-3.ch006.
18. L. Sharma. Introduction. In: *From Visual Surveillance to Internet of Things*, Taylor & Francis, CRC Press, Boca Raton, FL, Vol.1, pp.14
19. L. Sharma, P. K. Garg. Block based adaptive learning rate for moving person detection in video surveillance. In: *From Visual Surveillance to Internet of Things*, Taylor & Francis, CRC Press, Boca Raton, FL, Vol. 1, pp. 201.
20. L. Sharma (Ed.) (2021) *Towards Smart World*, Chapman and Hall/CRC, New York. DOI: 10.1201/9781003056751.
21. L. Sharma. (December 2020) Human detection and tracking using background subtraction in visual surveillance. In: *Towards Smart World*, Chapman and Hall/CRC, New York, pp. 317–328. DOI: 10.1201/9781003056751.
22. L. Sharma, D. K. Yadav, M. Kumar, "A morphological approach for human skin detection in color images", *2nd national conference on "Emerging Trends in Intelligent Computing & Communication"*, GCET, Gr. Noida, India, 26–27 April 2013.
23. D. Zachary, S. Dobson. (2021) Urban development and complexity: Shannon entropy as a measure of diversity. *Planning Practice and Research* 36(2):157–173.
24. A. Saxena, M. K. Jat. (2020) Analysing performance of SLEUTH model calibration using brute force and genetic algorithm–based methods. *Geocarto International* 35(3):256–279.
25. G. V. Jain, S. A. Sharma. (2019) Spatio-temporal analysis of urban growth in selected small, medium and large Indian cities. *Geocarto International* 34(8): 887–908.
26. E. Woźniak, W. Kofman, S. Lewinski, P. Wajer, M. Rybicki, S. Aleksandrowicz, A. Włodarkiewicz. (2018) Multi-temporal polarimetry in land-cover classification. *International Journal of Remote Sensing* 39(22):8182–8199.
27. S. K. Mithun, S. Chattopadhyay, B. Bhatta. (2016) Analyzing urban dynamics of metropolitan Kolkata, India by using landscape metrics. *Papers in Applied Geography* 2(3):284–297.
28. G. M. Tsarouchi, A. Mijic, S. Moulds, W. Buytaert. (2014) Historical and future land-cover changes in the Upper Ganges basin of India. *International Journal of Remote Sensing* 35(9):3150–3176.
29. X. Yu, A. Zhang, X. Hou, M. Li, Y. Xia. (2013) Multi-temporal remote sensing of land cover change and urban sprawl in the coastal city of Yantai, China. *International Journal of Digital Earth* 6(sup. 2):137–154.
30. G. Sandhya Kiran, U. B. Joshi. (2013) Estimation of variables explaining urbanization concomitant with land-use change: a spatial approach. *International Journal of Remote Sensing* 34(3):824–847.
31. Md. J. Alam, M. M. Ahmad. (2010) Analysing the lacunae in planning and implementation: spatial development of Dhaka city and its impacts upon the built environment. *International Journal of Urban Sustainable Development* 2(1–2):85–106.
32. A. A. Bindajam, J. Mallick, S. Talukdar, A. R. Md. Towfiqul Islam, S. Alqadhi. (2021) Integration of artificial intelligence–based LULC mapping and prediction for estimating ecosystem services for urban sustainability: past to future perspective. *Arabian Journal of Geosciences* 14:18.
33. V. Chettry, M. Surawar. (2021) Urban sprawl assessment in eight mid-sized Indian cities using RS and GIS. *Journal of the Indian Society of Remote Sensing* 69:101–109.

34. A. Shukla, K. Jain, R. Ramsankaran, E. Rajasekaran. (2021) Understanding the macro-micro dynamics of urban densification: A case study of different sized Indian cities. *Land Use Policy* 107:105469.
35. J. Mallick, S. AlQadhi, S. Talukdar, B. Pradhan, A. A. Bindajam, A. R. Md. Towfiqul Islam, A. S. Dajam. (2021) A novel technique for modeling ecosystem health condition: A case study in Saudi Arabia. *Remote Sensing* 13(13): 2632.
36. S. Talukdar, K. U. Eibek, S. Akhter, Sk. Ziaul, A. R. Md. Towfiqul Islam, J. Mallick. (2021) Modeling fragmentation probability of land-use and land-cover using the bagging, random forest and random subspace in the Teesta River Basin, Bangladesh. *Ecological Indicators* 126:107612.
37. K. Kushwaha, M.M. Singh, S. K. Singh, A. Patel. (2021) Urban growth modeling using earth observation datasets, Cellular Automata-Markov Chain model and urban metrics to measure urban footprints. *Remote Sensing Applications: Society and Environment* 22:100479.
38. C. Shikary, S. Rudra. (2021) Measuring urban land use change and sprawl using geospatial techniques: A study on Purulia Municipality, West Bengal, India. *Journal of the Indian Society of Remote Sensing* 49(2):433–448.
39. P. K. Sethi, S. Sankalp, S. N. Sahoo. (2021) Quantifying the dynamics of urban growth modes in Bengaluru, India. *Proceedings of the Institution of Civil Engineers - Urban Design and Planning* 174(1):1–14.
40. R. M. Hasnine. (2021) Population pressure and urban sprawl in Kolkata Metropolitan Area. In: *Habitat, Ecology and Ekistics*, Springer International Publishing, pp. 163–178.
41. S. Chakraborty. (2021) Remote sensing and GIS in environmental management. In: *Environmental Management: Issues and Concerns in Developing Countries*, pp. 185–220.
42. M. K. Jat, D. Khare, P. K. Garg. (2009) Urbanization and its impact on groundwater: a remote sensing and GIS-based assessment approach. *Environmentalist* 29:17. DOI:10.1007/s10669-008-9176-2.
43. C. Barber, C. J. Otto, L. E. Bates. (1996) Evaluation of the relationship between landuse change and groundwater quality in a water supply catchment using GIS technology: the Gwelup wellfield Western Australia. *Journal of Environmental Geology* 4(1):6–19.
44. F. De Smedt, O. Batelaan. (2003) Investigation of the human impact on regional groundwater systems. *Transactions on Ecology and the Environment* 64, WIT Press. www.witpress.com, ISSN 1743-3541.
45. S. Jamal, W. S. Ahmad. (2020) Assessing land use land cover dynamics of wetland ecosystems using Landsat satellite data. *SN Applied Sciences* 2:11.

10

Synthetic Biology and Artificial Intelligence

Vukoman Jokanović
ALBOS doo, Innovative Company, Belgrade, Serbia, Vinča Institute of Nuclear Sciences, Belgrade, Serbia

CONTENTS

10.1 Introduction

CRISPR-Cas9 genome-editing technique allows the manipulation of target genes and genomic regions. CRISPR is an acronym for clustered regularly interspaced short palindromic repeats, while Cas9 is a dual RNA-guided DNA endonuclease enzyme associated with CRISPR. The efficiency of a particular single guided RNA (sgRNA) is not uniquely defined by the identity of the sequence with the target site (TS). CRISTA (CRISPR Target Assessment) presents a new software of the machine learning (ML) network, which defines the tendency of a genomic TS to be separate from a corresponding sgRNA, describing the 3D structure and stickiness inside of the entire TS [1–3].

CRISPR with protein 9 (Cas9) is the basis of the microbiological adaptive immune system, which is recently used for modulation of DNA sequences inside the endogenous genome not only in "in vitro" but also in the whole organisms "in vivo" [4,5].

DOI: 10.1201/9781003244165-13

Besides, it is known that Cas9 endonuclease activates programmable sgRNA to activate double strands at specific TS [6]. This leads to adequate sgRNA, as a suitable guide, with a neglected deviation from TS. As a result, the number of non-compliances is the predictable range, while the discrepancies at the protospacer adjacent motif (PAM) is negligible. Proximal positions are more active than PAM-distal positions (PAM presents a 2–6-base pair of DNA sequences), including DNA sequence targeted by the Cas9 nuclease in the CRISPR bacterial adaptive immune system [4,5].

Generally speaking, an autonomous system, software or robotic, is a system that goes beyond the behavior programmed with an initial algorithm. Autonomous systems can learn from data and the environment, so their behavior is unpredictable. Their contextual behavior is unpredictable because it depends on an indefinite number of variables. Similarly, an SB entity or a biological entity modified by SB techniques operates at different levels, from the cellular to the real world, interacting with all components of such a context [5,7].

The "programmed" nature of these entities, together with their complexity, makes it difficult to predict the behavior of the actual system. SB entities are autonomous like autonomous AI systems, although they are generated differently. The problem of the AI prediction algorithm is present not only in genetic editing but also in the cases of simulation of evolutionary processes and self-replicating systems. The problem of prediction related to the artificial construction of super-intelligence has an analogous problem in the general goal of building complex SB systems. AI emphasizes the unpredictable aspects of algorithms because its main goals include creating autonomous systems and communicating with its products in an open context. Previous SB researches include designing and constructing new biological systems, and redesigning existing biological systems for suitable application [8–13]].

Some authors direct great attention to the information theory and algorithmic programming in explaining biological phenomena. This attention is especially directed to the possible relationship between the probability of disease and the state of the immune system, particularly from the point of cell reprogramming. Such approaches to software engineering are based on a computer interpretation of cell behavior. Gene therapy is just one possible example of such a methodology. The main aim of the application of the algorithms is to control the behavior of the cell systems, despite their possible strong perturbations, during the cell replications [14,15].

Cell reprogramming is achieved by natural or artificial methods, which are used in SB. The CRISPR-Cas9 papers show that genome engineering is defined as the process of introducing desired modifications into the genome. Therefore, recently many programmable technologies for nuclease-based genome editing are discovered [15,16].

Cas9 is an enzyme that mediates the cleavage of the targeted DNAs. Analogous to the search function in modern word processors, Cas9 leads us to specific locations within complex genomes by a short RNA search sequence. Using this system, DNA sequences and their functional outputs are now easily edited or modulated in virtually any chosen organism, with an obvious analogy with word processors, suggesting a wide potential application of Cas9 [17,18].

If we consider the parallelism of the Cas9 behavior with AI and computer algorithms, in both cases, a more general risk is related to the uncertainty of the prediction either of the AI or Cas9 process. Particularly, the outcomes of Cas9 are not controlled in the long run due to too many factors involved in gene expression and cell reproduction. Therefore, the SB, whose subfield is genome engineering, has the same long-term prediction problem as AI because it aims to produce autonomous systems, requiring spontaneous communication among the SB products. SB can have complex systems as an outcome or goal, like (for example) synthetic multicellular systems, while AI uses computer simulation and models, specially adapted to the needs of SB, to study their processes and behaviors over resulting outcomes [17,18].

Genetic engineering, including genome-editing techniques, can lead to reaching biological super-intelligence through recognition of biological mating features or through some other controlled modes of evolutionary action, such as implantation of edited DNA into embryos and embryo selection through numerous generations. This can lead to such a form of super-intelligence, which would create smarter and smarter human persons by speeding up evolution. This process leads to a large number of increasingly intelligent people, who then contribute to the accelerated development of artificial super-intelligence, repeating this cycle over and over again. Therefore, by SB techniques and methods, it seems possible to gain control over human beings [18,19].

In return, from the aspect of AI, evolutionary computing is the shortest path to achieve an artificial super-intelligence if human intelligence can be observed as an evolution product, as Moravec believes [20]. The prediction of the process outcome and its control afterward is the most important goal of such an AI development strategy. In turn, the results of evolutionary computation help in the faster and safer development of SB. In particular, the regulatory role of such predictions inevitably leads to the development of complex autonomous systems that could have a profound impact on human life and society, which is why it is necessary to redefine the ethical, legal, and political problems of AI and SB regulation [21].

10.2 CRISTA Method of Machine Learning

The CRISTA method is based on an ML paradigm for predicting the affinity to cleave a TS using a given sgRNA. This approach considers a wide range of characteristics that include those specific to genomic composition, characteristics that thermodynamically describe sgRNA, and traits related to the comparative resemblance of sgRNA and genomic target. This method achieves higher prediction accuracy than other widely used methods. Cross-validation determines the degree of predictability of the model, which follows the basic features of genome cleavage combined with various detection methods [22].

Previous studies were not conducted at the genome level because they analyzed non-targets pre-selected based on sequence resemblance. Certain experimental techniques for acceptable profiling of the whole genome were observed, particularly including the method of the integrated oligonucleotides in double-stranded DNA strands detected by sequencing (GUIDE-Sec) or method of the genome-wide translocation sequencing detected by hypertext transfer protocol (HTGTS) [21], or by direct labeling by BLESS (software which use genome-wide approach to map DSBs at the level of the nucleotide resolution, capable to detect telomere ends and nuclease-induced DSBs). These studies have shown that CRISPR non-targets can be found in unexpected locations, like the places with alternative PAM sequences (sites that contain a high number of mismatches and splits out of targets at higher probabilities than previously predicted). This evidence clearly follows that a complex set of attributes has crucial importance in the CRISPR-Cas9 function [23–26].

The CRISTA ML algorithm for predicting the probability of splitting is an MK method based on a random noise regression model [23–26]. CRISTA was studied through certain experimental genome researches to combine a big set of traits to calculate the probability to cleave targeted DNA using sgRNA. The resulting regression function of CRISTA consists of numerous interactions between built-in patterns described by a series of decisions. The performance of CRISTA prediction in cross-validation of the effects of a single sgRNA was evaluated and compared with alternative tools. First, the square of the Pearson correlation coefficient (r^2) between the experimentally observed and predicted splitting was calculated. The obtained results in cross-validation correspond for r^2 was 0.65. Accordingly, OptCD gives an r^2 equal to 0.13, while for the CCTop, an r^2 was equal to 0.23. The Cutting Frequency Determination (CFD) score correlates best with the r^2 equal to 0.52. A similar trend was obtained when calculating Spearman's rank correlation scores (Spearman's rho coefficients for CRISTA, OptCD (Optimized CRISPR Design), CCTop (Consensus Constrained Topology) prediction, and CFD [27–29].

The receiver performance curve (ROC) is used to distinguish experimentally divided and undivided sites (therefore, for this performance assessment, it is necessary to introduce positive or negative sets). The prediction is used as a threshold for distinguishing between positive and negative results during ROC calculation. CRISTA showed the highest AUC (AUC is the area under the curve for the precision call curve) score of 0.96, while the values of CFD score was 0.91, OptCD 0.85, and CCTop 0.85 AUC values. CCTop determines and ranks all potential sgRNA TS according to their non-target values. It is noticeable that all procedures show high AUC scores, influenced probably by a high number of unbroken locations contained in the dataset. Therefore, the detection and ranking of the positive samples is further compared. Like the ROC (receiver operating characteristic) curve, PRC-AUC closer to 1.0 illustrates extremely good predictions [27–29].

Recently, several types of CRISPR mechanisms are designed to compare experimental and simulated data. One of them is alignment-based models, in which CRISPR conductor sequences are determined on the base of PAM sequences mapping in the genome. The second is a hypothetical model in which gRNA activity is predicted based on particular characteristics such as guanine-cytosine (GC) value. Finally, one group of methods are combined ML and DL, known as machine deep learning (MDL) methods, in which gRNA activity estimation was obtained by algorithms learned on large datasets of CRISPR experiments performed on different cell types [30,31].

As a result of the increase in CRISPR gene editing, the database of the world CRISPR society, MDL-based approaches, are growing, making MDL the most significant method for predicting CRISPR gRNA activity. Furthermore, still, all current forecasting models suffer from a lack of data. ML approaches are more efficient than the other approaches thanks to more efficient data management, but they cannot efficiently predict processes if they do not possess enough data to isolate their key characteristics. In addition, ambiguities in the essential behavior of the CRISPR-Cas9 system have not yet been fully examined by current state-of-the-art algorithms [30,31].

It is indicative that some key properties are still missing, such as the local state of chromatin. Although deep neural networks (DNNs) can automatically extract these characteristics, they cannot efficiently predict processes if there is not enough data to isolate their basic characteristics, like ML. In addition, the ambiguities of the mechanism of the CRISPR-Cas9 system have not yet been examined enough, limiting their ability to be used effectively in recent state-of-the-art algorithms. It is indicative that some basic properties are still missing, such as the local state of chromatin. Furthermore, DNNs can automatically extract these characteristics, but they cannot efficiently predict processes if there is not enough data to extract their essential characteristics. These DNN characteristics and their significance for CRISPR functions still need to be functionally validated. The main reason for such a procedure is the heterogeneity of data generated from different platforms and cell types and the data inconsistencies caused by the number of detected non-target sites, although they are usually significantly lower than the number of predicting targeted sites (TS) [31,32].

The CRISTA method is used for evaluation of the cleavage efficiency of a particular genomic part using a particular sgRNA. Unlike many computer tools developed for these purposes, CRISTA also considers broader genomic attributes. It predicts the sequence which will be targeted by sgRNAs, like DNA protrusion sites (taking in mind that such protrusions are present in approximately 20% targets), while some of these targets are present with a significant frequency of occurrence. These findings are opposite to the Haussler conclusions [22,33] who told that such protrusions are rare and happen only on the split targets with small frequencies. Such forms of mismatch and bulges can be associated with Cas9 activity. Besides the importance of basic genomic attributes, in Cas9 action, significant roles belong to attributes that describe basic pair similarity nucleotide composition. The results of some researchers emphasize the importance of specific genome characteristics, like DNA geometry, its stiffness, double helix width, and DNA enthalpy. These properties are commonly applied to predict genomic entities, like nucleosome conformity and transcription factor binding sites, or to optimally adjust empirical procedures (e.g., PCR) [22,33].

Such characteristics are important for predicting the effectiveness of CRISPR. Some evidence indicates that unifying local DNA geometry and geometry of other genomic traits can improve TS prediction and rankings. The inclusion of genomic characteristics data in the analysis of such systems improves the efficiency of its target ranking. Also, the characteristics concerning RNA thermodynamics conduce greatly to the exact prediction. It should be noted that the variance of these characteristics is still small because they are fairly consistent for all specimens of the same sgRNA. Their significance is much more important when examining the effectiveness of a large number of targeted goals. CRISTA model, which is trained as a regression model, adjusts the splitting efficiency obtained using appropriate experimental studies, where the biggest problem is the need to combine the results of different experimental platforms into one scale, which can introduce a bias effect into the model [34,35].

Alternatively, it is possible to analyze the data within a given classification, wherein the data obtained by profiling CRISPR-Cas9 at the level of the whole genome can be considered as a single, where all truncated places are considered as a set of positive and uncut places as negative. The usually used models for such researches are regression and classification models. A classification approach that largely depends on the quality of experimental data uses the Random Forest classification algorithm. Therefore, the results found using this approach are very close and do not have the same efficiency as results obtained by using the regression model. Better prediction results of the regression model are the consequence of the more precious explained probability of cleavage between cleavage sites. The strictly defined boundary between places prone to cleavage, assigned as positive, and places not prone to cleavage, assigned as negative, represents the noise. Therefore, the main problem is to set a given limit according to experimental data. CRISTA uses currently available data, which includes profiling non-targets across the genome using CRISPR-Cas9 appropriate predictive tools to extract specific genome characteristics [35–39].

10.2.1 Data Labeling

In order to train the controllable guided ML model, one of the beginning steps is to determine the appropriate "labels" for the applied data. In the CRISPR study, such labels include cleavage efficiency, gene knockdown efficiency, and fluorescence-expressed expression intensity. The given data is associated with either attributes "discrete" (high/low) or "continuous" (0–1). The values attributed to them depend on different factors, such as the used algorithm or the desired outcome. Classification algorithms are used for discrete variables, while "regression" algorithms are used for continuous variables [40–43].

Accordingly, the efficiency of sgRNA cleavage is continuous, as the efficiency ranges from 0% to 100%. Recent regression approaches for predicting CRISPR efficacy are not accurate as empirical approaches because they are not well correlated with predictions. Often a target for which 100% efficiency is predicted will be no more effective than a target for which 80% efficiency is predicted due to the complexity of modeling of biological systems, which are usually less predictable. Labeling of sgRNA as "high" or "low" efficiency requires less data than regression model on a continuous scale from 0% to 100% efficiency. The capability of classification programs to distinguish the most active sites from others is practically very important. Nevertheless, sgRNA efficiency is more accurate if it is modeled using regression algorithms. Data discrepancies occur when researchers publish only positive results for CRISPR experiments. One way to avoid data imbalances is to select the optimal threshold when transforming efficiency values from continuous model to binary (high/low). A potential solution is to improve the existing targets sampling method instead of the initially used randomization method [44–46].

10.2.2 Prediction Targeted Activity

Many of MDL approaches for predicting targeted CRISPR activity are based on ML, like sgRNA Designer, sgRNA Scorer, CRISPR-Scan, SSC, and CRISPRater methods [47–51]. These methods cannot be intuitively explained. Other methods are computational. They are based on a complete understanding of the algorithm ("algorithmic transparency"). Such methods are CRISPRater, CRISPR-Scan, and SSC. They represent simple linear models and logistic digression models. Such models are the easiest to interpret the obtained results. These linear models can be easily trained. Therefore, they are very suitable for application at large-scale sgRNA predictions, but they have limitations for processing nonlinear bond characteristics. TUSCAN is a model that is trained by using a Forest of random data. It explains them and does not require normalization of data or parameters adjustment, but it is unreliable when the data form is abruptly changed. Other models, like SVM (Support Vector Machine), are unacceptable for big amounts of data, while another, like GBRT (Gradient Boost Regression Tree), are easily adapted to different data characteristics, are not applicable for precise data interpretation [47–51].

Recently, investigators have successfully used DL techniques in CRISPR research. DeepCas9, DeepCRISPR, DeepCpf1, and CRISPRCpf1 [52–59] use CNNs (convolutional neural networks) to explain sRNA activity by automatic recognition of sequence characteristics. The major power of DL is the sophisticated structure of CNNs, which enables automatic recognition of data characteristics, and subsequently their separation and functional confirmation. For a more efficient application of DL, it is necessary to significantly increase the number of data, using the methods of their artificial duplication. Therefore, publicly available target data is still not enough to build a powerful model of DL. Due to the large variability between species, species-specific algorithms are designed, like friCRISPR for drosophila, CRISPR-P for plants, CRISPR-Scan for zebrafish, and EuPaGDT for pathogens [51,52,60,61].

10.2.3 Prediction Non-Targeted Activity

Previous researches have shown that non-targeted sites (NTS) in the CRISPR-Cas model are not chaotic [52]. It was found that they possess a higher probability occupation in the case of an active mutation at the end of 5′ gRNA, while an active non-target mutation A in C at the 8th location is more likely. This evidence can be partly explained by decreasing the non-target activity when the 5′ end of gRNA is shortened [52]. Moreover, if we cut off part of gRNA, particularly at the 5′ end, the target activity will also decrease. Therefore, a number of algorithms for searching NTS are developed, like conventional alignment algorithms: bovtie, bovtie2, bva, TagScan, and GPGPU-enabled with CUSHAV [52–56].

All of the listed software have two constraints: a limited number of nonconformities and a rigid PAM. Therefore, the new algorithms adapted for CRISPR-Cas non-target location prediction systems have been developed such as CasOFFinder, FlashFri, dsNickFuri, and CRISPR [52–56]. In addition to methods based on harmonization, a part of methods based on learning has been developed. Thus, the MIT server [53,54] manually evaluates the target result based on an estimation of the numerous mismatched nucleotides and the spacing between them. This result can be used to explain whether the gRNA non-target score achieves the boundary value of 66. DL methods can also be applied for CRISPR non-target scoring. Two algorithms called CNNstd and DeepCRISPR use the CNNs approach to foresee the assessment of gRNA relevance for the CRISPR-Cas9 system [57–59,62].

The various methods applied to quantify CRISPR activity may influence heterogeneity among various experiments. The first stage of quantification is measuring gRNA-mediated CRISPR-Cas9 activity to determine phenotypic impacts. Genetic functional knockout (KO) enables the estimation of the quantified gRNA activity by quantifying green fluorescent protein (GFP) intensity [58,59,62]. The GFP-based approach assumes a fluorescence-activated cell sorting analysis (FACS). Therefore, some researchers used tests, which are not sensitive to drugs to measure the effectiveness of gRNA [57,59].

These tests often underestimate the real activity of CRISPR gRNA, showing false-negative results. The second most commonly applied approach of measuring CRISPR gRNA activity is based on serious sequencing of nucleotide sequences at the TS [62,63], directly indicating the existence of indels generated by CRISPR-Cas9. Endogenous DNA repair machines can induce CRISPR activity. Therefore, many CRISPR activity measurements can be performed using a suitable detection method to minimize data heterogeneity [62,63].

10.2.4 Data Scattering

Data scattering is a usual problem in non-target forecasting. Most of the used data on non-target gRNA is obtained following high- and low-throughput sequencing methods such as targeted PCR. For each TS, corresponding target sequences with splitting activity can be registered inside of the genome. These gRNA sets registered by various procedures are labeled as positive, while the gRNA sets with unregistered splitting are more frequent than detected. This causes data discrepancies in the negative group because positive sRNA non-target sites are present only in a highly low partition comparing to the total present sites. The PRC curve sampling estimation method could help to solve this problem. Several characteristics are essential for the targeted activity of gRNAs, such as the composition of the sequences, as well as their thermodynamic properties, specifically their secondary structure properties, and physicochemical properties [26,64].

The composition of the sequence can be classified into the conventional characteristics used in the algorithm of the gRNA designer, which was considered the most modern tool before 2017, for interpreting nucleotide content of the gRNA spacer sequences. Some researchers show that gRNAs with low or high GC partition are usually less active [26,64].

Knowing that the most active gRNAs contain approximately 50% GC (accepting that the percentage of GC in the spacer sequences is particularly important). Biochemical and conformal researches suggest that the thermodynamics of gRNA may also affect the linkage of gRNA of targeted DNA [26,64].

The location where the amino acids are cut is particularly important. In addition, several assumptions have been included to simplify model design, such as secondary spacer conformity and Gibbs free energy. They are essential for the self-bending activity of gRNA spacer sequences, although these are not too large to reach high gRNA activity. This limitation prevents the linkage of gRNA to the TS. Additionally, the length of the gRNA graft is also important. Experiments have shown that 67R and 85nt gRNA grafts can show higher efficiency in comparison with the real graft size [3,65–67].

Epigenetic characteristics, such as the availability of chromosomes, which affect the combination of gRNA and target sites, should also be included in the model [65].

Extraction of functions based on DL methods, such as DeepCpf1, DeepCRISPR, and DeepCas9, which implies automatic extraction of gRNA characteristics, is highly popular because DL methods do not request precious characterization of gRNA sequences. Based on the automatic extraction of essential characteristics of gRNA, DL algorithms can determine very deeply sequences [3,65,66].

10.2.5 Selecting Data

Besides labeling, each sample must contain a set of specific characteristics, consisting of the basic data (genetic, epigenetic, or experimental) in an arrangement proper for ML model. The main provocation is to provide enough data so that the ML software can generate reliable results. Without the inclusion of the additional data, it is very difficult to obtain predicted enough precious results, for certain specific test [66–68].

Therefore, the goal is to create a generalized approach, which gives more accurate predictions. Genome-related models include the sgRNA sequence and PAM and/or adjacent nucleotides. Knowing that the efficacy of sgRNA depends on the presented nucleotide types, the additional advantage of this model is the universality of the obtained sequence information by it. The sgRNA sequence is known to be crucial for guiding CRISPR/Cas9 to the TS. Since all of the applied programs take only sequence information as input, they can successfully foresee the sgRNA efficiency for each cell kind [66–68].

SVM programs based on classification or regression that incorporates sgRNA length are sgRNA design, ge-CRISPR, sgRNA scorer 2.0, CRISPRpred, VU-CRISPR, TSAM, and CRISPR-DT [68–72]. Since the SVM algorithms model are nonlinear, the data should be transformed in a suitable format for a multidimensional presentation if they can be linearly separated. Azimuth [3] and CRISPRpred algorithms are trying to improve prediction precision by incorporating in the model location data, such as targeting exons positions and/or targeting genes position [3].

Although these increase the model performances [3,73,74], such programs result in a reduction in generalization compared to the sequence-only models. Therefore, Azimuth uses the sgRNA design algorithm only if the position of the target cleavage site is not available. In this case, the inclusion of epigenetic data may improve the model precision, enabling specific prediction not only for the species but also for the cell type [30]. Looking to find patterns that give more data and enhance precision, attention should be given to the reduction of the volume of data, regardless of their relevance, because the inclusion of irrelevant characteristics (i.e., experiment ID in the tracking system) only increases the noise and requests additional memory space. Although FORECast program was trained on>40,000 sgRNAs, it requests a limited sample size, which enables a limiting processing time. Somewhere, by applying multiple models, each of them is responsible for a specific type of genome change. Such characteristics show the SPROUT program, which is specialized in deleting irrelevant data, and other unimportant data, for example, for tracking the type of mononucleotide change [3,68,73–75].

10.2.6 Setting Data into Machine-Readable Form

Identified data for incorporation in the learning database should be refined, following specific criteria. This is particularly important for the sequences data because the majority of ML algorithms cannot identify the specificity of certain data, which belongs to the generic data format, presented as a series of numbers. Thus, all nucleotides in the alphabet of DNA (or RNA) should be presented as a number from 0 to 3 (A=0, C=1, G=2, T=3) because today, many ML algorithms can identify a position which is changed. Regardless, this presentation is not appropriate for algorithms working with continuous variables because T (3) is more distinct either to A (0) or to G (2). Nevertheless, we can process them by using 'one-hot encoding' procedure [76,77].

This allows nucleotides to be presented by 0s and 1s, using a special column for each location in the sequence and each nucleotide. These processes can be enlarged to create a new pattern representing, for instance, nucleotide pairs. This simplifies entails generating an extra pattern column for each permutation and arrangement of two nucleotides at each location along the sequence. Moreover, it is possible to process strings using coding using the so-called "hot protocol," which allows nucleotides to be displayed as 0 and 1, applying a separate column for each location in the sequence and each observed nucleotide. It is clear that the given processes can be further extended by characteristics, such as nucleotide pairs [76,77].

10.2.7 Algorithm Selection

The train of chosen model is possible with a well-chosen set of features and carefully selected labels. Currently, many ML algorithms are used to predict CRISPR, each of which has some advantages and disadvantages. The two most common algorithms used in ML are linear regression and logistic regression. Linear regression is suitable for the continuous notation in CRISPR processes, while logistic regression is used in sgRNA design [78]. The last model is suitable for discretely labeled data. To avoid any manual transformation, algorithms should support nonlinear separation, like an algorithm that supports nonlinear separation or vector machines (SVMs). Trained SVM models, supporting classification or regression, use sgRNA scores, sgRNA design, ge-CRISPR or sgRNA 2.0 scorer, CRISPRpred, VU-CRISPR, TSAM, and CRISPR-DT [75,78–81].

Modeling nonlinear data by converting their characteristics into multidimensional presentation requests linear fractionalization data [78,80]. The lack of transparency of data or behavior similar to the "black box" should be balanced by clear models, which refine the modeled data. A second significant characteristic of the CRISPR algorithm is its ability to capture higher-order interactions, like interactive features. In the context of the efficacy of sgRNA, the interactive characteristics may be nucleic, epigenetic, or some other, which can together correlate with satisfying efficiency [75,78–81].

Tree-based software is a set of algorithms that incorporate higher-order interactivities. Their goal is to generate so-called "pure" groups containing exclusive high- or low-efficiency goals. The two of these algorithms used to predict sgRNA productivity, as arbitrary foreseen methods, are given by CRISPRpred, CRISTA, TUSCAN, CRISPR-DT, and CUNE algorithms [81], and advanced algorithms such as Azimuth and SPROUT [75,-78–82]. These algorithms produce programs consisting of multiple decision trees, using more search space. Therefore, they are superior to the single tree method because they can improve generalization or reduce the error range. Today, algorithms consisting of multiple nonlinear sheets, like CNNs, are increasingly used for this purpose [75,78–82].

Such algorithms are successful in analyzing not only images but also the objects in them. Today, the sgRNA foresee area uses DL based on the development of algorithms such as DeepCpf1, DeepCRISPR. DL is just one of the tools in finding the right algorithm, as it is shown by the example of the CRISPR-GNL algorithm, or the Bayesian ridge regression solution, which is often more reliable than DeepCas9 [83]. Algorithms such as CNN can analyze figures containing objects in arbitrary positions, sizes, and angles, without scaling and rotating procedures, not only for previously trained datasets but also for certain new samples [75,81–83].

The ability to detect basic patterns in arbitrary data often requires perfectly defined sets of characteristics, which is particularly important in the case of processing CRISPR data. If an image consists of an extremely high number of pixels, with objects present in unpredictable sizes, positions, and directions, a usual CRISPR TS contains 20–30 bases, with defined coordinates for entities, like the sgRNA TS and PAM [75,81–83].

Therefore, the application of CRISPR target processing by usual ML methods is relatively simple, showing that DL algorithms are redundant for such purposes. Therefore, it is increasingly applied for various Cas enzymes and future uses where the whole genome can be used as an input to find the optimal targets for the entire genome, with improved accuracy. It is obvious that the choice of the optimal approach depends on the given task and its linearity or nonlinearity, and characteristics interactivity which should be included, as from the possibility of recognition of the influential characteristics to the applied model [83,84].

10.2.8 Predicting CRISPR Target Activity: GNL Scorer

The CRISPR/Cas9 gene-editing system constrains two essential constituents: small guided RNA (gRNA) and Cas9 endonuclease. gRNA is a chimerical RNA molecule constituted from tracrRNA (trans-activating CRISPR RNA) and crRNA, where crRNA is a part of RNA that contains the gRNA locating in the specific part of host DNA linking to tracrRNA. It manages the Cas9 protein to the genome TS. Therefore, selecting a target location with pronounced targeted activation and low non-target locations is essential for gene editing. The activation of CRISPR-Cas9 during gene editing in mammalian cells is induced by numerous factors, like its secondary structure and availability of chromatin-guided sequence [85–88].

The results of the research so far have exhibited that the activities of CRISPR gRNA are very fluctuating. Therefore, many web tools and programs are established for silicon gRNA illustration that promotes CRISPR method and its uses. Recently, such tools are based on the ML. However, there are limitations in their application to predict gRNA activation on different types of data [85–88].

Accordingly, a program with cross-species generalization was developed. This targeted gene-editing software called GNL Scorer (GNL and GNL-Human) integrates optimal datasets and problem-solving characteristics between species. The applicability of a possible data extension is tested firstly by GNL software. In particular, both normalized and non-normalized datasets are trained by the Gradient Boosted Regression Tree, the most modern approach used in gRNA Designer (a series of rules, defined by using a series of characteristics). Compared to Doench's (V1+V2) data series, this dataset cannot enhance the Spearman correlation coefficient (SCC) by only increasing the data amount without considering the group effects of the SCC for the normalized dataset [89,90].

Compared to non-normalized datasets, there are no important variations between them and non-normalized Doench datasets. However, when we use coupled normalized data group, SCC is considerably lower ($P=0.044$) than the reliable value when Doench's dataset is used. It can be noticed that strong cumulative influences are present between different datasets. Therefore, these algorithms were developed with a certain set of data. Furthermore, this model, if applied to any other datasets, shows a tendency to include significant uncertainties. Accordingly, the performances of each model in each dataset should be carefully evaluated and selected as the model with the highest SCC [89,90].

Appropriate datasets applied for learning are the most acceptable datasets. Hela, Hct116, and Doench_V1 are the best datasets from the aspect of gRNA activation and its prediction. gRNA activation in Hela and Hct116 datasets is determined using the quantized sequencing detection method (QSDM). Data of sequences of gRNA activation show the best generalization characteristics, i.e., they have a high generalization potential. Like the FC and RES datasets, the Hela and Hct116 datasets possess a clearer indication control during ML. Subsequently, the BRR model was shown to be most suitable after filtering the model over the following datasets: Hela, Hela+Doench_V1, Hela+Hct116, and Hela+Hct116+Doench_V1 [89–92].

These results show that the generalized foreseen efficiency of the integrated Hct116+Hela dataset is equivalent to one Hela dataset and that these two datasets are more acceptable than other dataset combinations. After assessing the significance of epigenetics for model characteristics, a set of the most appropriate traits is chosen. A set of features without epigenetics can be included to reveal the optimal combination of features for predicting gRNA activation. For comparison, a combined set of 2,701 traits is generated, including those associated with the epigenetic. The two most acceptable datasets, normalized with the Hela and Hela+Hct116 program, are used for learning. The program with the greatest SCC is selected. To avoid over-adjustment, a subset of Doench_V1 data is taken to assess the correct choice of properties [89–92].

Finally, two generalized evaluations (GNL and GNL-Human) are obtained, and their performances are evaluated and compared with other state-of-the-art prediction models of the efficiency of sgRNA splitting over SCC. The enhanced CRISPR gRNA activation foreseen algorithms include GNL and GNL-Human model evaluation constructed by using the BBR model. Both of them are based on the two best-performed datasets (Hela and Hct116). Essentially, GNL was learned using a normalized Hela dataset, while GNL-Human uses a Hela+Hct116 dataset. For a general assessment, GNL and GNL-Human models are compared with other dataset sources. It was shown that GNL and GNL-Human programs are the best for most datasets in all performed tests. GNL fits almost all other data types, while GNL-Human is at the top list among human datasets (HEL), clearly showing its application possibilities when used in human cell experiments [89–92].

10.2.9 Insight in CRISPR ML Models and Way of Minimization Errors

Learning models on various subsets of numerous characteristics are unsuitable and require a long processing time. Logical approaches, like regression and tree-based models, are more suitable for identifying influential characteristics [93–95].

Such models are suitable for learning various available features and subsequently ranking them according to their partition in the model or the significance of their impact. Such models can be extended to "generate appropriate hypotheses." Traits such as the number of independent nucleotides, sgRNA positions within proteins, and melting temperature contribute to models in determining DNA cleavage efficiency [93–95].

Understanding how to interpret model performance is crucial in assessing the reliability with which a model presents basic data. Usually, a decrease in bias leads to an increase in variance, which is why balance needs to be found when both types of errors are minimized. A too complex model usually has a large variance, and although it perfectly represents the data on which it is trained, it suffers from the impossibility of self-improvement in determining the efficiency of sgRNA editing. Therefore, it shows low foreseen accuracy on samples on which it is not learned. The complexity of the program is also related to noise or deviations in data during learning. Noise is caused by the presence of specific nucleotides that affect the efficiency of sgRNA when it comes to training-related data [93].

Sometimes, a negative goal (characterized by low efficiency) has sequences that are very dissimilar to other negative goals to resemble the properties of a positive goal sequence with significant efficiency. This may be caused by an experimental error, particularly if the sample size is tiny or characteristics that would identify as a bad target (like epigenetic information) are not contained in the dataset. Training of models on the Forest may influence large deviations of the model, while an oversimplified model induces a high bias. Dissimilar to high-variance approaches, high-bias approaches have a high inadaptability. Such models do not prove to be reliable enough even on the data on which they were trained. In order to modify the way of learning algorithms from data and diminishing bias and variance, it is necessary to adjust their so-called hyper-parameters, which the researchers themselves introduce into the model [93–95].

Every algorithm possesses its own set of hyper-parameters, which change a particular approach to the learning process. If by increasing these hyper-parameters, the complexity of the model grows, by decreasing them, it is possible to reduce the complexity of the model. By trying different hyper-parameters, a balance can be struck between bias and variance, achieving by this way, the lowest foreseen error [93–99].

10.3 Possible Consequences of CRISPR Technology

Unlike our ancestors, we will soon be able to solve many hitherto unsolvable genetic problems. According to the calculations of Nick Bostrom and Karl Schulman, related to research on the impact of IQ growth on the world around us, choosing the 'most intelligent embryo' during in vitro fertilization will increase the IQ of the average baby by about 11.5% in each new generation. This means that after ten generations, according to Shulman's calculations, such a successor would have a coefficient that is 115% higher than his ancestor who lived ten generations before, from which it follows that the average person subjected to such genetic selection would have the intelligence of a genius. And if embryonic stem cells, transformed into male and female gametes, were used, similar results could be achieved in just 6 months. And who then wants to wait two centuries to be a successor of a race of geniuses? [100–102]

It's like that now, and what will happen after ten generations, when it will probably be possible to create the ability to surpass even what you only dream about now by computer programs, which is why it is impossible to predict even in the hottest imagination since it exceeds even the most incredible expectations. The genetic basis of intelligence is very complex because it presupposes several ingredients, which include certain forms of computing capability, spatial consciousness, analytical thinking, and compassion, as well as the inevitable influence of environmental factors. Stephen Hsu, one of the founders of the Cognitive Genomic Laboratory, estimates that about 10,000 genetic variants are likely to affect intelligence. Therefore, he is frightening that despite all the potential dangers, many variants will appear realistically "in the next ten years, in which all available genes will be included in the choice of the smartest embryo." The basis of this world, which threatens to completely break away from control, is a technology called CRISPR-C 99, based on the process of cutting off a part of the DNA sequence present in a gene and gluing another part of the DNA sequence in its place, in a surprisingly fast and precise way. In addition, application of this extremely powerful technology paves the way for manipulating the human genome in a shorter period of time than that was required for researchers previously [103].

It is already a tool that enables a predictable result of reproduction, which is too simple and imperfect without this type of intervention. There is always the possibility that if a man and a woman lack the best genetic heritage, their offspring would be a complete failure. Application of CRISPR technology eliminates such risks, and what makes it even more exciting is the ability to insert a desired genetic trait directly into a male or female gamete, which undoubtedly easily produces a unique child with Bernard Shaw's intelligence and Isadora Duncan's beauty, who would unmistakably carry these genes to their future generations as well. If people with mutations that cause Huntington's or Tay-Sachs disease use these powerful tools, they can get rid of the deadly effects of these embryonic diseases and thus can avoid the many troubles caused to their children by these diseases [100,103].

We are beings who have no boundary for our unreal desires, because some may want to create their own offspring with distinctive traits ranging from the color of eyes or hair to having supernatural powers such as the strength of an elephant, the flexibility of a leopard, have supernatural intellectual abilities, and live almost forever. The seed of an offspring would help reach the dream of all insane people on the scale of evolution, who dreams of enslaving the entire world and thus becoming its eternal masters, throwing a glove of challenge at the face of God himself, until the heart of the Earth and the heart of the cosmos split into half and burst with great sorrow. Do new, unimagined possibilities open before it, or is it a path that leads to the impassable path from way out, because such progress could open the door wide to the winds of evil, who always follow their intentions, which are most often stimulated in their embryo by the most humane desires?

According to Chris Impey, such rapid evolutionary changes, which took thousands of years on the Earth, are now possible in a very short time. He said that it was no longer a dream to make it possible for a man to stay on Mars through such evolutionary changes in man that he will be able to easily adapt to the drastic changes in temperature and climate expected on Mars. It follows that it is quite certain that the establishment of small colonies of people would soon be possible even in such harsh conditions. In return, a stay on Mars will cause human bodies to become significantly taller and thinner than they are on the Earth due to about 40% less gravity of Mars. And just what to say about the many predictions. Natural evolution has demonstrated that many directions lead to the equivalent aim. The power given to us by the extraordinary skills of CRISPR makes real the centuries-old aspirations of people and helps realize many of their dreams by perfecting their genome, which allows them to make their lives much more beautiful and richer, but also much more intriguing, leading them to many paths of temptation. Instead of changing nature and adapting it to ourselves, as we have done so far, this technology for the first time in the human history allows us to change and improve ourselves, and thus overcome many obstacles that seemed insurmountable to us until now [104].

Medicine has always been and remains the main guide in such changes because it seeks to put every technology in the service of what is good for man; therefore, although they are essential, many complex moral issues are left aside. Possibilities in curing, for example, Parkinson's disease with brain pacemakers, which would control the symptoms of this disease, or Alzheimer's disease, adding that our memory is already stored in Google, all suggests that our lives are already only partially biological. People have always been skilled in fitting into their world by applying tools and aids present in their everyday life. But if a double connection is established in which devices and aids actively and automatically constantly adapt to us, just as we adapt to them, then the line between such tools and users begins to disappear completely. Such technologies are becoming less and fewer tools and more and more part of our mental composition. They remain tools only in a fine and ultimately paradoxical sense, while interacting with our unconscious nerve structures, so the work of our brain comes down to answering the question of who I am, because such machines, being much more powerful of us, carry within us all our collective knowledge, which has been accumulated throughout the history of the human race, speaking of us, of our meaning and the nature of the human mind [100,105,106].

10.3.1 Recent Investigations on Unnatural Nuclear Pairs and Exciting Near Future

Ray Kurzweil, a well-known futurist, believes that we will soon be able to even literally transfer the content of our consciousness to the computer. He believes in this hypothetical moment when AI will reach a critical point after which it will be completely superior to human intelligence, radically changing civilization, and perhaps human nature is very close. Accordingly, Dmitry Itskov believes that technology will allow man to live forever in hologram of his body by making them possess superhuman abilities and therefore, he will be able to withstand extreme external conditions, such as high temperature, pressure, radiation, or lack of oxygen. Of particular interest are the neuro-interfaces that allow an electrical signal to be sent to neurons or brain tissue, with a basic computer output to the brain provided by an external computer to control the program. If signals are sent via such an interface to a neuron to turn it on or off by a synaptic 'grafting' mechanism, then it seems obvious that in this way, the extended neuro-interface will have the ability to issue commands to parts of the brain, suggesting when need to be active or inaudible, by successfully coordinating numerous brain regions simultaneously [106].

Hans Moravec [20], one of the leading experts in robotics, predicts that human consciousness will soon be taken over by computer networks, and asks the question: "How will we think without the body?" "Such a step inevitably leads to a radical redefinition of the body and our identity." as well as the scientific understanding of biology in general at the same time, many other changes are taking place that completely change the perception of our world. Research by Floyd Romesberg and his colleagues in a fascinating way managed to expand the natural genetic code, made up of four to six letters by adding two new molecules, which they marked as X and Y. This was a miraculous discovery showing that from the beginning of life on the Earth, the life of all beings has been written in the DNA code with only four letters, G, T, C, and A, denoting the molecules guanine, thymine, cytosine, and arginine, which are paired in a DNA helix. Metaphorically speaking, a suitable combination of these letters created all living organisms with their genetic code, which is now, thanks to the discovery of Romesberg, expanded by two new molecules, thus opening the way for the creation of completely new life forms [107–110].

Romesberg's modification of *Escherichia coli* makes it possible to determine and precisely direct the work and behavior of the organism based on its genetic material so that the microbe can be programmed to produce new types of proteins that can be used as drugs for the treatment of various diseases. Such modified microorganisms represent only the initial phase of research aimed at "creating organisms with completely unnatural attributes and properties," in which cells represent a stable form of semi-synthetic life and the basis for reaching the crucial aim of SB, which tray to design new life forms and functions. While every living being on the Earth carries a record of DNA code, constituted of four base pairs, characterized by the letters G, T, C, and A, modified *E. coli* organisms contain a completely new type of DNA code, with two, added DNA bases, X and Y, which were created by successfully inserting new nucleotide units into the genetic code of *E. coli*. In that way, the first semi-synthetic organism in the world was obtained, with a genetic code consisting of two natural base pairs and two supplementary alien base pairs [107–115].

Thus, such research, which began in 2008, after the selection of 60 potential candidates, who could pair 3,600 combinations, selected base pairs named d5SICS and dNaM, whose molecules proved to be compatible with the "enzyme machine" for DNA copy and translation; however, according to Romesberg's own admission, they had no idea that a synthetic organism could be produced in such a way with such base pairs. The hope that this could happen was given to him by reactions that took place in simple test vessels, thanks to the fact that scientists were able to produce unnatural base pairs for the first time, which could copy themselves and transcribe into RNA, which request these pairs to be recognized by enzymes that evolved using A, T, C, and G. For that, it was only necessary for the cells to accept the foreign base pairs needed to maintain the DNA molecules by repeating the cell division cycle, during which the DNA is copied. Romesberg's team selected *E. coli* to ensure that the altered single-celled algae gene, which encodes the corresponding protein, passes through its bacterial membrane, using a short DNA loop known as plasmid, which contained a couple of foreign bases. With such a diatomic protein, which contained foreign nucleotides, the plasmid was copied and transmitted through the division of *E. coli* cells for almost seven days so that the bacteria, when it was supplied with foreign nucleotides, become able to replace foreign bases with natural ones [107–115].

If we try to interpret a book written in four letters, we will not be able to tell many interesting stories. But if the same book contains more letters, then it would already be possible to construct new words and find completely new forms to use those words, and thus probably tell more such stories. According to Ross Thyer [116], Romesberg's work is a big step forward that tells us all we can do. Many skeptics thought Floyd's results were impossible, says Stephen Benner [117], because chemical reactions specific for DNA, like replication, are extremely susceptible to mutations. Still, Benner sees nothing strange because foreign cells in our organisms will soon become our reality. There are no more borders. If we go back in time and repeat evolution across four billion years, we might build a completely different genetic system. However, designing a completely synthetic organism is a big task in itself, although many claims that the organism has no links to unnatural DNA. Creating a new organism simply will certainly not happen for some time to come, because there is still much to be learned about DNA, which is integrated into every aspect of cell life. However, DNA has been present for about a billion years, which does not mean that scientists could not do something much more perfect than it.

When Stephen Benner decided to reproduce new genetic molecules, he didn't just mean to DNA, although the first thing he noticed was its unusual design because its skeleton contains repetitive, negatively charged phosphate groups, which are mutually repulsed. Therefore, it is much more difficult to intertwine the two DNA helices that form a double helix by pairing the bases of adenine (A) with thymine (T) and cytosine (C) with guanine (G), both of which are abundant in each cell and attracted by weak hydrogen bonds which is easy to break with water. It was simply fascinating how our precious genetic heritage is transmitted through these weak connections in the water, even though it seems almost impossible. And yet, life has found such a structure, in which it is possible in spite of everything [117].

This inspired Benner to take on the challenge of finding the right key to a similar or completely different solution, developing a lavish nursery of exotic base pairs outside the natural array of base pairs that would merge and copy in a similar way as natural base pairs. In addition to undoubted success, such work brought a number of problems because it soon became clear that only several artificial base pairs could be embedded into DNA because cells cannot fully absorb unnatural biochemistry. Even so, the DNA and RNA produced in this way, with unnatural base pairs, are already being applied to register the virus, finding many other applications in medicine. Following their inspired visions, scientists are expected to soon develop an organism with an extended genetic code, which will be able to memorize more information. Numerous limitations are related to insufficient knowledge of the structure of gene molecules, as well as a clear answer to the question of whether natural bases are required for life, or they represent only one of the many possible similar approaches [118,119].

Obviously, the Earth has evolved in a certain way, but that does not mean that there are no other ways to achieve similar goals, which is confirmed in Benner's research. In 1986, by researching the structure of DNA, he realized that what is look like an unequivocal lack in the DNA's skeleton was, in fact, its perfect feature because when a skeleton made of negatively charged phosphates was replaced with a neutral chemical group, each strand was found to be longer than expected by about a dozen units, and bent in itself, probably due to the repulsion of the charge, to keep the molecules in a stretched state, because the base pairs are subject to inter-twining. Benner then created base pairs that were similar to natural ones, which were themselves linked by a hydrogen bond, after which his team tested two new pairs: isoC and isoxy xanthosine, to finally determine that polymerase enzymes are such pairs incorporated by copying DNA or transcribing it into RNA. It clearly follows that they are able to read DNA containing unnatural base pairs, inserting complementary features into the growing DNA or RNA. As biological machines that translate RNA into a protein, ribosomes can also read RNA fragments containing isoC and apply them to attach to an artificial amino acid in the resulting protein. Interestingly, the researchers did not observe any problem with this mechanism because hydrogen atoms tended to circulate in a ring, while iso-G often converts to different shapes and pairs with T instead of izo-C [118,119].

Eric Kool tried to develop unnatural base pairs with a fixed arrangement of hydrogen bonds, making a base analogous to the natural T base with fluorine instead of oxygen atoms. This new base named difluoro toluene, labeled F, mimics the T base extremely faithfully, although it shows unexpected hydrogen spikes. DNA copying faithfully inserts A versus F, and vice versa, showing that the polymerase fitted correctly as long as the base had the regular form, although some scientists were very skeptical about it. They asked whether it is possible that hydrogen bonds are not necessary for DNA replication. Knowing that humans are so accustomed to hydrogen bonds that it is difficult for them to imagine their alternative, this question becomes very important. In this case, instead of forming hydrogen bonds, which are usually joined with hydrophilic or aqueous molecules, F and other forms of bases were hydrophobic, causing water to repel them, helping them to stabilize within the double helix. Following this discovery, Floyd Romesberg enlarged his set of hydrophobic bases, beginning from compounds like benzene and naphthalene, to produce a number of barely conceivable derivatives, all of which took them very far from anything resembling natural base pairs. During testing, researchers found that key base pairs during the replication process must be hydrophobic for enzymes that are inbuilt into base pair inside DNA, but they must also be able to accept hydrogen bonds so that enzymes can continue to copy DNA strands. Therefore, Romesberg base pairs MMO2 and SICS possess both hydrophobic and hydrophilic properties [107–114,120].

Ichiro Hirao came up with the unusual idea to design something like a coil created by arranging natural base pairs, adding a negatively charged, electron-rich chemical group that repels natural base pairs, to finally synthesize unnatural DNA in 2011 as hydrophobic base pairs Ds and Diol1-Pk, which can copy its DNA with 99.77%–99.92% fidelity at each replication. In the same year, Benner showed that artificial base pairs P and Z, which bind by hydrogen bonds, reached an accuracy at each replication of 99.8%, and finally, Romesberg reported a replication efficiency rate of 99.66%–99.99% for optimized forms of his bases, named Nam and 5SICS, which overlap with the lowest artificial DNA replication rate. All these researches have shown that their "best case approaches the worst natural case," from the point of view of fidelity of the obtained copies of DNA with unnatural bases [121].

Obviously, unnatural bases should be improved to a much greater extent in order to confirm such research in its entirety because, during experiments, it was presented that polymerases can copy a maximum of four bases paired in a row. To solve the problem of copying efficiency, Philip Holliger [122] used nucleic acids called XNA, with sugars usually present in DNA, while RNA was changed by alternative ring structures, creating during his research billions of mutants of natural polymerase, thus enabling it to evolve, by inducing selective pressure on it, with the aim of converting it to XNA. After that, the most effective mutants were compared to identify the most successful among them. The site on the polymerase in contact with DNA has been found to have an enzyme that acts as a regulator and has been adjusted so that the contact is as close as possible to ensure correct replication during the XNA copying process. In addition, Holliger's team constructed an enzyme that can convert XNA back to DNA [121].

Such research has already been confirmed in vitro, which is why researchers rightly hope that such a result will be confirmed in vivo. Philip Marlier and his associates replaced the T base with chloro-uracil by replacing the hydrogen atom with chlorine in the RNA of uracil bases, which was then introduced into *E. coli*, which could not produce thymine, and after approximately 5 months, they noticed that certain bacteria could no longer survive without chloro-uracil, because they had eliminated about 90% of the thymine from their genome during that period. Benner, Romesberg, and Hirao tried to persuade the cells to receive their base pairs. Although this automatically meant they would have to devise numerous ghosts, venturing boldly into the heart of the problem of extremely sophisticated re-examination of genetic material, Marline's team tried to replace all natural bases with artificial ones until Romesberg was focused on the development of organisms with only hydrophobic bases, which also seemed almost impossible because the cells contain do many ingredients that are optimized to work with natural bases [107–114].

The artificial base pairs still have practical application, as many large companies already use the Bener isoC and iso® pair to enhance the identification and control of viral infections, while Siemens [114] applies a set of linked DNA sequences binding to HIV-1 RNA in a patient's blood specimen. Inserting an artificial base into a certain sequence prevents its binding to a random DNA sequence, which binds to HIV-1 RNA, enabling a detection of HIV-RNA-1 even at very low concentrations. In addition, DNA and RNA can catalyze some cellular reactions, which is why they are used as drugs. Of course, programmers can also improve the performance of certain sequences by binding the appropriate chemical groups to base pairs, while artificial bases then may easily target particular sites in the sequence before each of the cytosine or guanine molecules is fully saturated. Romesberg's team added an inserter to groups of unnatural bases embedded in DNA, which allows them to precisely attach various molecules [117–119].

Of particular intrigue is the question of how early forms of life on the Earth evolved on the existing natural genetic script. Was the reason for that the limited number of available potential constituents? Adenine, for instance, is easily made from hydrogen cyanide, which was attended when life appeared. What is certain is that once organisms formed a set of base pairs, they, metaphorically speaking, locked them in the very heart of their existence. And albeit RNA is generally thought to have preceded DNA, and while it may not have been the best solution to support life. It was probably the only possible solution in prebiotic Earth, Benner asks himself, if nucleic acids would have formed on another planet on their own whether their base pairs would have the same as these on Earth? [117–119].

For that to really happen, it would obviously be necessary for organisms to be subject to the same restrictions because otherwise, the very root of life would be fundamentally different. In doing so, it is possible to apply some universal rules, which are related to the existence of DNA skeletons with repetitive charges, which were initially considered mandatory, although they depress bending, influencing that coils with different base sequences conduct alike in the replication process. Although some investigators have been successful with an alternative skeleton of DNA, most of their attempts have failed because the molecules attached to it were too rigid or too fragile to make a helix. All this points to a possible limitation in the chemical variations that can be introduced into the game, despite the imagination whose limits are not foreseen, because the question of why the chemistry of living organisms is exactly as it is, remains unanswered. Is it because it is the only possible answer, if any answer exists at all, or is the only way to come up with an answer to finally show how the genome as a whole is formed? [107–122].

It is already enough to take only eight small DNA strands at least 60 nucleotides long, combined with a master mixture of enzymes and reagents, incubated at 50°C for an hour, so that researchers can synthesize, in just five days, the mouse mitochondrial genome, a factory organelle energy for the plant and animal cells. In addition, this technique gives synthetic biologists the simplest possible tool for designing and building the genetic sequence of a synthetic vaccine or completely new drugs and converting a microbial cell into an alternative energy source [107–122].

Some researchers at various institutions have developed the synthesis of the entire bacterial genome, experimenting with about 1,100 bases of long DNA segments ordered from sequencing companies, introducing them into a yeast cell, and stacking them on top of each other. Thanks to that, we now have a method that allows us to make a 1-kilobase of parts even ourselves, says Daniel Gibson [123], who led this study. The main problem in realizing such a complex task, according to Gibson, was related to the purchase of sufficiently long pieces of DNA from appropriate companies, which deal with sequencing, because no one managed to string such long nucleotide chains on top of each other, with high-enough precision. The resulting errors embedded in the final product can only be avoided by sequencing all DNA segments, which significantly slows down the genome creation process. Sequence verification in smaller units is much easier, which is why the researchers switched to studies with shorter, 60-nucleotide pieces because they allow genome fragments to be unmistakably linked, mixing them together or adding 75 double-stranded pieces before mixing. After such processing, the resulting sequences are introduced into the *E. coli* bacterium through a well-designed automated process. The resulting clones are sequenced, extracting the correct genome sequences, which allow the realization of *E. coli* bacteria. This has no defects in its DNA, since this way, true copies can be obtained in all major segments, which enables the creation of a final product of 16.3 kilobases through the above three successive steps [107–122].

According to Ron Weiss, a much more efficient technique for building relatively large DNA complexes without error is the so-called viral replacement technique [124]. This approach significantly expands the possibilities of building such complexes, which, according to Gibson, can be used to obtain vaccines using synthetic biology (SB). For example, it is known that the flu virus easily mutates, which is why we need a new vaccine every other year. With this method, it is possible to monitor mutations and synthesize a new virus very quickly to make a new effective vaccine against the flu virus. The experiments began with the mitochondrial genome because errors in its sequences lead to many diseases for which there is currently no real treatment. However, although we have already managed to create a synthetic mitochondrial genome, unfortunately, it is still not clear whether it is functional. However, it is obvious that it can correct disorders in cells with mitochondrial defects, which opens the way for designing treatments for various diseases [123–126].

The French writer Georges Perek [124] wrote a whole novel without the use of the letter e. Still, it was clear that communicating with a limited number of letters was a frustrating job. Moreover, the genetic messages of cells are contained in only four letters, encrypted in four types of small molecules, called bases, which make up DNA. Recently, certain American scientists have shown the ability to write and send the message in a foreign language to normal cells, which means that the possibilities of the genetic code are significantly increased [123–126].

It has long been known that four DNA bases, adenine, thymine, cytosine, and guanine, labeled A, T, C, and G, appear in particular sequences along the DNA backbone wrapped in a double helix, forming a code that direct the cell how to assemble the given protein, which they need it to work, once again, for who knows how many times repeating it as a refrain, because that was a key fact. It is the essence of our existence. And then, when that clear fact finally found its place in the minds of his listeners, a torrent of words came that explained the depth of the genius of nature, which created everything from that seemingly very scarce material. Moreover, although protein molecules are complex in structure, made up of 20 components, they can be translated into the language of the four-letter genetic code because DNA bases are read in groups of three bases, with 64 permutations [123–126].

Despite the remarkable perfection of the natural genetic code, investigators have many causes to be curious about creating a new genetic code with various characters since DNA with unnatural base pairs may be tougher to chemical degradation than natural DNA. In addition, unnatural genes inbuilt into the genome can affect the way the natural genes translate into a protein, explaining how conventional bases combine in pairs that are condensed into double strands in the form of a double helix, with their mutual stickiness conditioned by the action of hydrogen bonds. It is very selective because A only sticks to T, and C only to G, and earlier efforts by researchers to introduce new hydrogen bonds between bases in DNA were hampered by the tendency of bases to stick together and thus be as little separated as possible [123–126].

Floyd Romesberg and his associates started from a different approach, incorporating into DNA a basic pair that interacts through hydrophobic interactions. Opposite to conventional DNA bases, a new type of base was relatively insoluble in water, allowing its units to be attracted to each other and grouped, making globules coalesce in water. This has proven to be a good solution, as artificial bases do not tend to pair with natural, hydrogen-bond bases. They have previously shown that a specific pair of synthetic bases, labeled 7AI and ICS, is efficient when mounted on a double helix backbone. In order to realize similar functions as DNA bases, such artificial constructions must communicate with protein enzymes that process DNA inside cells. Such bases should be accepted as true components of DNA from double helix-forming enzymes, such as DNA polymerase, because if the enzyme accepts these new bases, then a mutated version of DNA, composed of multiple segments, may be constructed and copied as desired. Both synthetic bases, 7AI and ICS, have been shown to be acceptable as building blocks of a specific type of bacterial DNA polymerase. And although these constructions show obvious advantages, they are far from perfect because each of them has a pronounced tendency to pair with another copy of itself, forming pairs 7AI:7AI and ICS:ICS, besides the desired pair 7AI:ICS. To avoid this problem, Romesberg used modified versions of his unnatural bases, which are produced in various variants, after which they try to find ideal pairs, which would be selectively bound. Such research paves the way for synthesizing an unnatural pair, which DNA polymerase would efficiently and selectively incorporate into the DNA double helix. This would result in a new type of pairing, labeled PP:MICS, which is the first insertion to the genetic code, allowing new genetic letters to be introduced without too many errors [123,124].

Nonlinear genetic data provides a new challenge of perceiving living cells and their internal functioning extremely exciting. Sentences woven into any language represent information in a line, the content of which is assembled so that words can be read in one direction, from left to right. DNA similarly encodes information, overlapping them in both directions. DNA is a double-stranded molecular helix, which has one strand parallel, which is placed in the opposite direction to the other complementary strand so that both DNA strands of the helix contain information useful to the cell. It is reminiscent of a book that is read upside down and backward after reading the book from right to left, up and forward in parallel [105],

Additionally, genomic sentences are dynamic wholes of the cellular machine that capture, divide, and combine phrases and clauses as needed. Each of the trillion cells of the human body would contain 6 m of unbent DNA strands if all 46 chromosomes were unpacked, unwrapped, and attached to each other. If we imagine something so unimaginable as real, it is clear that only a smart being can write a library on such a small surface of the material, which is completely invisible. Only super-intelligence can double the density of this information by writing, which is basically a modular type of sentence that can be read in both directions [105,106].

Now, human genome writers are trying to describe another new dimension of coded information, explaining how cellular machines within a single-celled nucleus dynamically access its DNA following their three-dimensional arrangement. Yu Heng Lau [127], who deals with three-dimensional genome analysis, believes that only in the last few years have we begun to understand how the folding genome affects the way it is expressed and controlled because some specific genes, even specific exons, are close to the bending site because the 3D structure of the genome shows an effect on gene fusion [105,106].

Segments encoded by exons [124] and proteins are inside of genes. Using a long string to present a gene, metaphorically speaking, dots inside that string, as dots in a sentence, represent exons, so that the genome functions as a dynamic library that opens and closes certain chapters and pages when necessary, and then folds particular strings, translating them into certain words and phrases for reading and copying in RNA. Although it is still unclear to anyone why this happened, these parts, which are bent outwards, expose the selected information of the given cellular machinery to the environment. The internal, bent parts hide data that are not needed at that moment. The statement that the 3D DNA arrangement affects the way cells are read and genes are interpreted, including how is the best way to approach certain exons, discourages a person who is trying to understand how a cell works inside, but also inspires it in trying to understand how a cell came into being. This powerful evidence suggests that a truly divine supergene was involved in its construction [106,124].

The ability to adapt life to perform an "unnatural" function flourishes to unlimited possibilities. By applying this technology, developing organisms that feed on waste oils and oil, after their outpouring, or cells that migrate to certain tissues and organs, and then secrete drugs only in those tissues is possible. It has come a long way before all this has become realistically possible. Although the four letters of the genetic code were discovered in the late 19th century, until 1950, scientists failed to understand the structure of DNA and the way life was created in its myriad forms. Since then, scientists have been preoccupied with the idea of finding ways to create the genetic code themselves, driven by their imaginative visions. It's an incredible turnaround, since only a decade ago, life seemed to be on Earth, which has evolved over billions of years, is an eternal secret, and it is almost unbelievable that something more important in the genetic code will soon change. The first such success dates back to 1989, when a Swiss research group modified the letters of the genetic code chemically, although they failed to add new letters [105,106].

Finally, in 2014, Romesberg added two new letters to *E. coli*, after which these organisms died quickly, and after that, this process, after numerous new attempts, was successfully realized and described in detail in his researches. In order to succeed in his idea, it was necessary to design the bacterial immune system so that it eliminates DNA that does not contain new letters. With enough adopted X and Y letters, base pairs, the organism can keep the new letters in its genetic code and survive by passing on genetic information to future generations. Although we are still far from fulfilling Romesberg's real dream, it is certainly a very important step related to building organisms that can perform tasks that the natural organism cannot perform [107–114].

To do that, it needs to discover a way suitable for creation inside of the cells their X and Y letters, just as it does with their letters A, T, G, and C. After that, such information would have to enter the cellular machinery, which has been evolving for billions of years, to understand new genetic letters, not just to recognize them, and to use those letters to make building blocks of cell life, say, for example, proteins. And that's not all, because, in order to ensure the creation of such an unnatural species, it is necessary to simultaneously avoid all possible side effects, such as genes that would pass uncontrollably from one species to many other species, which is a key requirement of any genetic engineering. To better cope with such challenges, Romesberg founded a biotechnology company called Santrock [106].

X-rays, cosmic rays, chemical reactions, or similar mechanisms can change the base pair in the DNA chain, forcing the generation of cells mutation. Such a change can lead to the formation of a new protein or enzyme. The theory of evolution says that billions of such mutations have created all the living beings present today. From the initial self-replicating molecule, which is generated spontaneously, unicellular organisms are developed, which evolved into multicellular organisms, at the top of which pyramids are vertebrates, all the way to the man himself. In the process, DNA structures evolved from the asexual format of a single helix in bacteria to a double helix of the chromosomal type, typical in all higher life forms. The number of chromosomes has also increased over time. While fruit flies possess 5 chromosomes, mice 20, humans 23, and dogs 39, the evolutionary mechanism of mutations does not explain how the genome develops and how individual mutations can create new chromosomes or extend the DNA chain [100,105].

Romesberg and his team believe that we live in a time that indicates the beginning of creating new forms of life. Now, with practically unlimited possibilities of increasing information, semi-synthetic organisms have been optimized, which provide an appropriate model to form an organism with completely artificial properties and traits that are not found anywhere in nature. A synthetic DNA base pair can be inserted into a living organism and retained in it throughout its life [107–126].

10.3.2 Enormous Potential of CRISPR Technology and Its Ethic Controversies

A total of $2.7 billion has been spent over the course of 13 years on the human genome project, HGP. This project has enabled scientists to decode the human genome. It is a story about three billion base pairs, which gives us information about how our cells are built. Therefore, the natural desire of every scientist who deals with this field is to understand this story as much as possible, which is like the endless flow of a river, composed of an incredible number of various details, talking about not only how the human genome is built and how it works but also what are the interconnections and interactions of all these numerous elements that make up this exciting saga of ourselves [106].

Therefore, the exciting idea is to come to the heart of the story by rebuilding the genome. This idea has tremendous pragmatic and theoretical implications because it provides a much deeper understanding of how this complicated creation within us works. Such insight is possible to get only when we can disassemble a machine and assemble it into a harmonious whole. If we are not able to realize this, we cannot understand the essence of its functioning. According to Monya Becker, such a synthetic genome would be a unique source of new information. On the basis of them, the production of well-defined types of biological drugs built from large molecules through appropriate cultivation of cell lines in laboratories would be possible, despite the risk of infection with the virus during breeding, which can then jeopardize the entire chain of their reproduction [127,128].

SB, as a new discipline dealing with new genome printing, whose impressive results can lead to much more successful stem cell therapy without the risk of infecting a patient with other diseases, as often happens today, due to insufficiently sterile conditions in treatment with them. In addition, if only the good spirit of this idea is followed, it would be possible to produce microorganisms that can allow humans to generate some of the missing amino acids. Adding new letters to the letter of life allows researchers of SB to significantly extend the scale of proteins they can synthesize [106].

For billions of years, the history of life was written with only four letters: A, T, C, and G, which form the chains of DNA subunits found in all living organisms. Theses letters expanded with forming a living cell with two unnatural DNA constituents in its genome. Such a step would be a step toward the synthesis of cells which is capable of producing the drug and other useful molecules while increasing the possibility of this method, one day can lead to the realization of new cells without the usual four DNA bases, for all organisms on Earth, increasing their genetic information potential [129].

The current research aims to write down the entire human genome in the next 10 years. In this way, genome printing, from initial criticism and skepticism, has led to a phase of serious global debate, focusing on the fastest possible DNA printing through the organization of a similar project. Such ventures would enable people to produce hitherto unheard-of things, such as new nutrients obtained through photosynthesis, by awakening dormant genetic pathways, which would enable the development of so-called phototropic cells, which could help people survive a rigorous, long-term space trip [130].

From the exotics information, it is possible to resurrect some extinct species, such as mammoths, by editing their genes, to the open considering the possibility of genetic improvement of human beings; all become at our fingertips. Particularly important is the opportunity to treat severe and autoimmune diseases. We are entering an unimaginable age in which man will become the creator of his own evolution. In life, everything is based on the same genetic code, which makes up cytosine (C), guanine (G), adenine (A), and thymine (T). It is the programmatic language of life, which has not fundamentally modified much from the Earth's primordial chaos. Like any other language, it can be applied to write messages. DNA in all its combinations describes a genome that is 2 mm long in *E. coli* and 30 meters long in the blue whale [106].

Although the language of DNA was written billions of years ago, we learned to read it in the last few years. For the first time in 1970, DNA sequencing was performed, after which, in the years that followed, due to a long and painstaking work, the human genome was finally read. Especially popular is the quote of the great theoretical physicist Richard Feynman, which synthetic biologists repeat as a mantra: "What I can't do, I don't understand" [106,131].

Although Feynman did not say it literally, such a thought translated into the language of synthetic biologists means that we could understand how life as a whole works. To understand these maxima, it is needed to understand the process of assembling and writing DNA, which is the basic unit of existence. Carried by this great idea, scientists are trying to achieve genome synthesis, which indirectly means writing and printing entire unnatural genes, and not just copying current DNA, as in the process of cloning organisms, beginning with the rudimentary organisms, trying to fully understand what certain letters into the genetic chapter of life mean [100].

The early success came in 2010 by Craig Venter [132], one of the leaders on the human genome project, when he produced the first unnatural cell, writing the entire genome of tiny bacteria named *Mycoplasma mycoides*, which then succeeded to insert another bacterium, called *Synthia*, into an empty cell. It was a fascinating discovery when the *Synthia* genome was methodically broken down to the minimum number of genes needed to sustain it, revealing a reduction life to its basic form, the function of each of its genes [132].

Sofia Rosenfeld explained that to do something very simple, the goal is to avoid complexity, which requires serious engineering. It was obvious that even the simplest bacterial genome is much more complicated than scientists initially considered. In Venter's gene pairing to create a synthetic cell of 473 genes, the function of 149 genes was completely unknown. Given that this number represents almost a third of the total number of genes, it becomes clear how much knowledge is still needed to understand the genetic code whose sequences we already know, even when it comes to the simplest organisms. And what about much more complex organisms! It reminds us of Feynman's quote about the difference between knowing the name of something and knowing that [106,133].

In order to clearly understand everything that is important about the genetic code, researchers will have to sequence, synthesize, and program huge quantities of genetic data. The organism's creation is now performed over DNA synthesis using DNA gene-editing procedures and then through painstaking and persistent learning through the experiments that follow. Once the synthesis is performed, it must be repeated a large-enough number of times to understand what gene arrangements are made at every step to perfect the synthesis itself. According to Venter's synthetic cell, the researchers added and subtracted the gene simultaneously, trying to determine what changes in such a procedure causes changes in the organism itself. If the synthesized bacterium were to die, it was a reliable sign that this gene was crucial for life [134].

Jokanović [106] says that, it is the heart of SB, concluding that in the future, we will gain the necessary knowledge about genes faster because we will be able not only to design them faster but also to build and test them faster. Therefore, Carlson's curve [135] is significant because it shows that the gene knowledge base will advance in a similar way as the computer programming speed base. Like the Internet that led to a series of significant technological advances in the 1990s, some companies, which are now carriers of the world economy, such as Amazon and Google, are helping the popularization of DNA printing technology to gain increasing understanding in the industry. Apple's CEO and founder, Steve Jobs, told his biographer just before his death from cancer: "I think the greatest innovations of the 21st century will be at the crossroads of biology and technology. A new era is beginning" [106].

Many companies and investors are assured that SB could revive some of the basic aspects of living and doing jobs. Technologists employed at companies like Microsoft trust that DNA could be transferred to a silicon plate, as a storage form. The genetic code is just a tool for preserving and transmitting information about how life functions. Since DNA is an extraordinary dense medium, scientists have developed a theoretically capable method of storing all the world's data on DNA, which, unlike all existing media used today for physical imaging, would never become obsolete. That's why Twist started to cooperate with Microsoft to enhance the process of storing data on DNA. But that's not all, since synthetic biologists are not satisfied only with copying existing life forms because they primarily want to create something completely new without rushing to bring dead organisms back to life in time [106].

In the last few years, two new genetic technologies have caused a real research transformation. One, relatively well known, is the ability to decode information in our genes, while the other is our newly emerging ability to change genes at will. In addition, these researches provide us with the opportunity to anticipate hazards to our health, eliminating many deadly diseases and transforming ourselves and the whole nature around us at the same time. Such powerful technology raises very complex and important questions about ourselves, inevitably changing the essence of our current view of the world, since for the first time in our long history, we appear as creators of evolution, with endless possibilities not only of our change but also with all the possible dangers that such changes bring [106].

Thanks to reading the genetic code, different types of tests have been developed to identify genetic diseases. Along with the growing popularity of these tests, many ethical problems have come to the fore because the possibility of genetic testing changes the lives of future parents and accelerates the dilemmas of many people who now face doubts for the first time in their history, whether to leave the child to grow up spontaneously or to intervene at birth to prevent certain diseases or possible deviant behaviors. Some of these technologies are quite simple, such as a blood test to diagnose Down syndrome, or the Dor Yeshorim genetic database, which was originally designed to confirm the genetic heritage of Jews. This base allows persons to escape partners with whom they could have a child who is deadly affected by Tai-Sachs disease [106], which is especially pronounced in Ashkenazi. Both of these seemingly ordinary examples underline significant ethical issues because the Dor Yeshorim database and other similar initiatives, like genetic tests for sickle cell anemia, which greatly affect African Americans, have allowed us to intentionally modify the frequency of personal human genes inside of given community [106].

The capability to apply genetic testing in deciding whether or not to have a child is a clear form of soft eugenics, helping those affected by serious disorders to prevent problems caused by poor genetic inheritance, which at first glance seems humane, as it indirectly reduces human suffering. However, such a worldview carries a great danger that we get involved in natural processes beyond any measure while making unreasonable decisions out of excessive fear that things will get out of control [106].

Steven Heine [136] added that DNA is not only destiny but also the place where problems begin. Some diseases are completely genetically determined, as in Huntington's disease or Dushen's muscular dystrophy, which has the terrible consequence that we will eventually have the disease if we have such a defective gene. But for the tremendous majority of illnesses, our future is not contained in genes, so the outcomes of genetic tests can mislead us because Heine reveals that this type of risk is only about 2.1% higher than the risk of developing the same disease in another person. For this reason, a very obvious question arises: what does an increased risk of 2.1% in a given case mean? [106].

A special challenge is our capability to modify DNA sequences in humans and other beings, using a gene-writing technique, first used in human cells in 2013, very soon changing fundamental research in the biological sciences. In the book *Crack of Creation*, one of the first researchers of this technique, Nobel Prize winner Jennifer Doudna, together with Samuel Sternberg, explains the history of her discovery and its impact on overall science. Her book raises deep ethical questions in addition to scientific ones [137].

Obviously, the possibilities of this technique are enormous because such technology, for the first time in our history, allows us to cut an unwanted DNA sequence from an organism and then delete or change it as desired, using a technique similar to computer processing programs. In addition, it can be used to introduce various control elements related to gene construction by simply transmitting instructions using light stimulation, which allows scientists to deal with the consequences of gene expression at a certain point in the life of the organism or expression in a certain region. Of course, as with any other method, some important limitations should be kept in mind when using this method because sometimes it is hard to deliver an exclusively designed DNA sequence to some distinct cells to change its genes, while the other restrictions are ethical [137].

It is extremely important to determine where and when this technology should be used because, as you can imagine, it can be used for mass destruction, since due to its simplicity, it can be easily spread over the Internet, which would help anyone with average knowledge of biology can deal with gene-editing problems. This decisively influenced Doudna to take an extremely critical distance on the technology. Of course, it was never disputed that it would be immoral to legally prohibit a family from removing from the embryo a defective part of a gene that could cause a terrible disease, such as Huntington's disease, which condemns an adult to a terrible death. Perhaps these apocalyptic visions are exaggerated because each genetic ingredient is encoded by numerous genes, which has a very small effect on the whole, which is currently unknown, although we are making great strides toward creating a new CRISPR race [137].

Doudna and Sternberg's fears about the danger of uncontrolled application of their technique, through the wide availability of gene disks with imprinted genetic changes and the possibility of spreading such changes across different populations almost instantaneously over the Internet are more than realistic. When such a disk is applied, the probability of possible gene modification rises exponentially with each generation, rapidly overwhelming the entire human population [137].

Although this is the technology that can suggest the most effective way to sterilize all mosquitoes or prevent them from carrying malaria, a gift from heaven. As such, it could have a huge impact on the epidemiology of some of the deadliest diseases, saving many people by changing the mosquito genome. The problem with such perfect technology is high because it is a biological bomb which could have unforeseeable side effects, because if we make a mosquito unfriendly to malaria parasites, we could cause, as and in the case of overuse of antibiotics, the parasite to mutate in such a way as to avoid the effects of a genetic change, which would mean that this change can make it immune to our recent anti-malarial drugs. Such changes would also cause undesired effects in the ecology of a given environment, which would cause many other species to be irreversibly damaged, or even extinct because everyone knows how fragile and sensitive our ecosystems change[106].

Besides that, it is obvious that a solution in one part of the world can become a catastrophe in another part of the world because certain diseases and pests move freely across imaginary borders and get into a new ecosystem. And despite everything, regardless of whether these events inspire or amaze us, the possibility of manipulating genes in different organisms with the help of this technology gives a stunning insight into evolution. We can now replay such exotic questions as how fish fins were replaced by legs in terrestrial vertebrates about 400 million years ago. Thanks to this method, some diseases, such as cancer, will be easily curable. It will even enable the accelerated improvement of a drug for Dushen's muscular dystrophy, although there are still numerous technical ambiguities regarding the efficient delivery of CRISPR sequences to all muscle cells affected by the disease. All this tells us to enter a new, completely amazing world [105].

Because of this, Feng Zhang [138] believes that the current limitations on gene intervention will be quickly overcome and that fundamentally changed people will soon begin to be born. This technology will enormously improve human health. A much more vigorous approach combines CRISPR technology with stem cells, which will be soon used for laboratory production of male and female gametes. If we add to this the possibility of growing and multiplying stem cells in vitro, which is their essential advantage over embryonic cells, it is obvious that this makes such cells ideal candidates for creating altered offspring using CRISPR technology. The way this would be possible is very simple. Stem cell genes would be altered first. They would then be transferred to a male or female gamete, after which they would produce offspring. People have always wanted to control their bodies. Moreover, some parents would like to know when and how to have children and how healthy their children will be [106,138].

New technologies that enable genome modification will lead to an increased interest of many individuals not only having children but also healthier children; if a genetic disease is present in the family, such as a deadly Huntington's disease, Sinclair believes that by changing the appropriate gene, a deadly gene defect present inside the egg can be avoided. That is why it is most important to correct these mutations before a fetus is formed. It follows that if embryonic cells become part of medical practice, it will open the door to unpredictable changes in human behavior, with unforeseeable consequences for human life and human identity, which in itself opens up numerous ethical dilemmas and social challenges. In [106] admits that it is impossible to answer the question of whether this is good in itself. Of course, there are no unfounded fears that children will be the subject of experiments, because experiments on the embryo will encourage a flood of parental desires for the superior characteristics of their offspring. This will finally open the door wide to unimaginable changes and medical abuses [105,106].

Sterility treatment is one such story that restores hope. Unfortunately, more and more young people have azoospermia, a rare genetic defect, because between one million and six million DNA letters are missing in the Y chromosome. And although the potential of their gametes is not such that they can have offspring, even in such patients, there is hope because it is possible to turn the cells of his skin into stem cells, to then produce the correct DNA, and produce male gametes[106]. Thanks to that, it is possible to cure infertility because it is essentially possible to replace the damaged letters in DNA with letters without side effects, which makes this method very attractive to use [106].

This way, many serious diseases would be eliminated or alleviated. That opens the possibility for many people to feel the life benefits, which they did not even know before. Thus, it is clear that the technology of gene modification creates the possibility of the elimination of genes that have caused a difficult-to-cure or incurable disease and a significant improvement in the characteristics of each person. It is hypothesized that after such interventions on the embryo, humans could possess bones that are difficult to break, or that the risk of heart attack would be drastically reduced, while a change in the gene that regulates amyloid protein content could cure Alzheimer's disease and dementia, allowing people to keep their shrewdness even in old age. Of course, that would most directly affect the further evolution of man. Through favorable versions of genes, vaccines could be made against certain most dangerous diseases we face today, like SARS Cov-2. However, despite everything, the issue of genetically modified babies remains extremely controversial because while some researchers think it is unacceptable, others believe that it is the only opportunity to improve the human race because the human genome is not defect less [105,106].

No wonder then that Nick Bostrom [139], through his book *Super-intelligence*, alerted the public to the risks of AI when humans use reproductive technologies to improve human intelligence because the way in which genes influence intelligence is not yet well known. However, such a reality is no longer speculation from the standpoint of high-tech eugenics. What if everyone could be a little smarter, we wonder. If some people were much smarter, you should be aware of the fact that even a small number of such super-intelligent persons would certainly completely transform the world through their pronounced inventiveness, discoveries, and innovations that everybody else would apply? In Bostrom's opinion, genetic advancement is a significant long-term task, as is the issue of climate change or the issue of the world economy, because the ability of humans to solve such problems is a factor that we face every day [106,139].

For some researchers, the eruptive progress of embryo genetics and biotechnology is inevitable; the scientists recognize that CRISPR is extraordinarily powerful. But that is exactly why many problems are opening up because we have to be absolutely sure that this technology will be carefully applied. If we change the embryo that defines us as humans, is it finally a change in human evolution? In [106] describes genetically modified plants, including roses that glow in the dark. In his fantasy of the future, clients will play with the genetic codes of plants and animals, designing new creatures on their computers. Anyone in the world who has little money will create various creatures and change the game, thus creating a whole new world, think [106].

In Ref [140] author uses a laser to create various types that have altered DNA for large pharmacies. His idea is the democratization of creation with minimal rules, which intimidates bioethicists. It is no longer uncommon for color-changing flowers to appear on the market, tests that detect the disease from a single drop of blood, and a pill that tells doctors if you have taken the drags. The technical and ethical limits of science are being pushed to the margins; Heinz believes that after 10 or 20 years, people will design their children digitally. At the last conference in Vienna, he spoke about how to create completely new organisms that never lived before, talking about the incredible possibilities of this fundamentally new technology, which is already opening the door to a new industrial revolution. That technology means printing life [106].

We need to take critically a businessman like Austin Heinz, who speaks that we need to change future generations of human beings and advance the human race. DNA editing has become much cheaper in the past decade, especially thanks to the innovations of Austin Heinz, a leading researcher at Cambrian, who modified or built completely new machines, such as the genomic 3D laser, which prints DNA, making DNA change much cheaper and faster. According to Heinz, machines create a DNA chain by lowering DNA segments one by one, many of which contain errors. Using his approach, millions of DNA coils are created simultaneously, with the largest number of them with errors, while only a few coils are correct [141].

This is the biggest human achievement in history because that is what speaks most strongly about us who we are, who, in turn, are the product of DNA, which is composed of four letters A, C, T, and G. When Heinz's company accepts an order for a specific gene, it adds DNA components in small beads which are then deposited as a thin film on a glass plate. The machine adjusts color to each DNA component. The following step is crucial. Lasers programmed to analyze color combinations ignore the wrong coils and "print" the correct ones by distinguishing them correctly from the others. The latest product is placed on a small plastic plate as a powder that the customer puts inside the cells of the organism [141].

However, although it is a security question, we do not want the industry to be regulated. So we need to figure out how to democratize creation without killing. Heinz and other researchers with similar ideas have years of technical obstacles in front of them in which they need to clarify all the contradictions that this technology brings before they create life, especially since these achievements are extremely important because they enable individuals to eliminate future suffering, caused by serious mutations in DNA that cause terrible things, such as Down syndrome and cystic fibrosis [106], which essentially turns life on the Earth into hell, which is why everything should be done to avoid such things.

In addition, it should be recognized that this technology, like many others so far, can be used for exotic purposes, such as changing the DNA of the embryo to obtain the desired color of the eyes or hair. According to Heinz, the only problematic dimension of the application of this technology is related to the creation of new, never before existing living creations, which requires some more time, although it seems to us an idea that is on the verge of science fiction [141].

Heinz believes that diversity and multiplicity are the essences of our nature, trying during the last years of his life to help a creation of thousands of various unusual products, which aim not only to reconstruct grandly damaged limbs but also to fight viruses developing various alternatives to the widespread use of antibiotics. Maybe one day, he was convinced, researchers will print DNA even on Mars, and while all this sounds too fantastic, it will all happen sooner or later [106].

10.4 Conclusions

The use of AI in CRISPR applications is advancing rapidly, with prediction tools constantly evolving. Although the goal of most AI programs is to enhance the efficiency of CRISPR-Cas9 trials, each of them shows some specific variation. Some programs/algorithms are elementary and adapted to the kind of organisms and cell types, while others collect a wider range of data, such as epigenetic information, and show differences in CRISPR efficacy in different environments.

AI provides an increasingly clear insight into biological mechanisms and is becoming more comprehensive, reducing the potential for human bias. The tools available have room for enhancement in the accuracy of prediction or the introduction of various experimental parameters, although they sometimes lead to misclassification of targets.

In this chapter, particular care is dedicated to the ethical issue of wide-spreading of SB combined with AI.

References

1. Jinek M, East A, Cheng A, Lin S, Ma E, Doudna J. RNA-programmed genome editing in human cells. *Elife*. eLife Sciences Publications Limited; 2013;2:e00471.
2. Jiang W, Bikard D, Cox D, Zhang F, Marraffini LA. RNA-guided editing of bacterial genomes using CRISPR-Cas systems. *Nat Biotechnol*. Nature Publishing Group, a division of Macmillan Publishers Limited; 2013;31:233–239.
3. Cong L, Ran FA, Cox D, Lin S, Barretto R, Habib N, et al. Multiplex genome engineering using CRISPR/Cas systems. *Science* 2013;339:819–823.
4. Nishimasu H, Ran FA, Hsu PD, Konermann S, Shehata SI, Dohmae N, et al. Crystal structure of Cas9 in complex with guide RNA and target DNA. *Cell*. 2014;156:935–949.
5. Jinek M, Jiang F, Taylor DW, Sternberg SH, Kaya E, Ma E, et al. Structures of Cas9 endonucleases reveal RNA-mediated conformational activation. *Science (80)*. 2014;343:1247997–1247997.
6. Cho SW, Kim S, Kim JM, Kim J-S. Targeted genome engineering in human cells with the Cas9 RNA guided endonuclease. *Nat Biotechnol*. Nature Research; 2013;31:230–232.
7. Abadi S, Yan WX, Amar D, Mayrose I. A machine learning approach for predicting CRISPR-Cas9 cleavage efficiencies and patterns underlying its mechanism of action. *PLoS Comput Biol*. 2017;13(10):e1005807.
8. Hwang WY, Fu Y, Reyon D, Maeder ML, Tsai SQ, Sander JD, et al. Efficient genome editing in zebrafish using a CRISPR-Cas system. *Nat Biotechnol*. Nature Publishing Group; 2013;31:227–229.
9. Trivedi TB, Boger R, Kamath G, Evangelopoulos G, Cate J, Doudna J, Jack Hidary J. Crispr2vec: Machine Learning Model Predicts Off-Target Cuts of CRISPR systems, bioRxiv preprint. 2020. doi: 10.1101/2020.10.28.359885.
10. Alavi A, Ruffalo M, Parvangada A, Huang Z, Bar-Joseph Z. A web server for comparative analysis of single-cell RNA-seq data. *Nat Commun*. 2018;9(1):1–11.
11. Alipanahi B, Delong A, Weirauch MT, Frey BJ. Predicting the sequence specificities of DNA- and RNA-binding proteins by deep learning. *Nat Biotechnol*. 2015;33(8):831–838.
12. Gandhi S, Haeussler M, Razy-Krajka F, Christiaen L, Stolfi A. Evaluation and rational design of guide RNAs for efficient CRISPR/Cas9-mediated mutagenesis in Ciona. *Dev Biol*. 2017 May 1;425(1):8–20.
13. Sinclair JC, Davies KM, Venien-Bryan C, Noble ME. Generation of protein lattices by fusing proteins with matching rotational symmetry. *Nat. Nanotechnol*. 2011;6:558–562.
14. Bianchini F. The problem of prediction in artificial intelligence and synthetic biology. *Complex Systems*. 2018;27:249–265.
15. Zenil H, Schmidt A, Tegnér J. "Causality, information, and biological computation: an algorithmic software approach to life, disease, and the immune system," *From Matter to Life: Information and Causality* (S. I. Walker, P. C. W. Davies and G. F. R. Ellis, eds.), New York: Cambridge University Press, 2017, pp. 244–279.
16. Shipman SL, Nivala J, Macklis JD, Church GM. CRISPR–Cas encoding of a digital movie into the genomes of a population of living bacteria. *Nature*. 2017;547:345–349.
17. Church GM, Regis E., *Regenesis: How Synthetic Biology Will Reinvent Nature and Ourselves*, New York: Basic Books, 2012.

18. Hsu PD, Lander ES, Zhang F. Development and applications of CRISPR-Cas9 for genome engineering. *Cell*. 2014;157(6):1262–1278.
19. Goldman N, et al. Towards practical, high-capacity, low-maintenance information storage in synthesized DNA. *Nature*. 2013;494:77–80. doi: 10.1038/nature11875.
20. Moravec H. *Mind Children: The Future of Robot and Human Intelligence*, Cambridge, MA: Harvard University Press, 1988.
21. European Group on Ethics in Science and New Technologies. "Statement on Artificial Intelligence, Robotics and 'Autonomous' Systems." (Aug 14, 2018), ec.europa.eu/research/ege/pdf/ege_ai_statement_2018.pdf.
22. Xie S, Shen B, Zhang C, Huang X, Zhang Y. sgRNAcas9: a software package for designing CRISPR sgRNA and evaluating potential off-target cleavage sites. *PLoS One*. 2014 Jun 23;9(6):e100448.
23. Cho SW, Kim S, Kim Y, Kweon J, Kim HS, Bae S, Kim JS. Analysis of off-target effects of CRISPR/-Cas-derived RNA-guided endonucleases and nickases. *Genome Res*. 2014 Jan;24(1):132–141.
24. Frock RL, Hu J, Meyers RM, et al. Genome-wide detection of DNA double-stranded breaks induced by engineered nucleases. *Nat Biotechnol*. 2015;33(2):179–186.
25. Morsy SG, Tonne JM, Zhu Y, Lu B, Budzik K, Krempski JW, Ali SA, El-Feky MA, Ikeda Y. Divergent susceptibilities to AAV-SaCas9-gRNA vector-mediated genome-editing in a single-cell-derived cell population. *BMC Res Notes*. 2017 Dec 8;10(1):720.
26. Wang X, Wang Y, Wu X, et al. Unbiased detection of off-target cleavage by CRISPR-Cas9 and TALENs using integrase-defective lentiviral vectors. *Nat Biotechnol*. 2015;33(2):175–178.
27. Wang J., Zhang X., Cheng L., Yonglun L. An overview and metanalysis of machine and deep learning-based CRISPR gRNA design tools. *RNA Biol*. 2020;17(1):13–22.
28. Cradick TJ, Fine EJ, Antico CJ, Bao G. CRISPR/Cas9 systems targeting β-globin and CCR5 genes have substantial off-target activity. *Nucleic Acids Res*. Oxford University Press; 2013;41:9584–9592.
29. Tsai SQ, Zheng Z, Nguyen NT, Liebers M, Topkar VV, Thapar V, et al. GUIDE-seq enables genome-wide profiling of off-target cleavage by CRISPR-Cas nucleases. *Nat Biotechnol*. 2014; 33:187–197.
30. Doench JG, Fusi N, Sullender M, Hegde M, Vaimberg EW, Donovan KF, et al. Optimized sgRNA design to maximize activity and minimize off-target effects of CRISPR-Cas9. *Nat Biotechnol*. Nature Publishing Group, a division of Macmillan Publishers Limited. All Rights Reserved; 2016;34:184–191.
31. Montague TG, Cruz JM, Gagnon JA, Church GM, Valen E. CHOPCHOP: a CRISPR/Cas9 and TALEN web tool for genome editing. *Nucleic Acids Res*. 2014;42:W401–W407.
32. Pliatsika V, Rigoutsos I. "Off-Spotter": very fast and exhaustive enumeration of genomic lookalikes for designing CRISPR/Cas guide RNAs. *Biol Direct*. 2015 Jan 29;10:4.
33. Naito Y, Hino K, Bono H, Ui-Tei K, S.F A, S B, et al. CRISPRdirect: software for designing CRISPR/Cas guide RNA with reduced off-target sites. *Bioinformatics*. Oxford University Press; 2015;31:1120–1123.
34. Zhou T, Yang L, Lu Y, Dror I, Dantas Machado AC, Ghane T, et al. DNAshape: a method for the high-throughput prediction of DNA structural features on a genomic scale. *Nucleic Acids Res*. 2013;41:W56–W62.
35. Lei Y, Lu L, Liu H-Y, et al. CRISPR-P: a web tool for synthetic single-guide RNA design of CRISPR-system in plants. *Mol Plant*. 2014;7(9):1494–1496.
36. Breiman, L. Bagging predictors. *Machine Learning*. 1996;26(2):123–140.
37. Gratz SJ, Ukken FP, Rubinstein CD, et al. Highly specific and efficient CRISPR/Cas9-catalyzed homology-directed repair in Drosophila. *Genetics*. 2014;196:961–971.
38. Sanjana NE, Shalem O, Zhang F. Improved vectors and genome-wide libraries for CRISPR screening. *Nat Methods*. 2014;11(8):783–784.
39. Peng D, Tarleton R. EuPaGDT: a web tool tailored to design CRISPR guide RNAs for eukaryotic pathogens. *Microb Genom*. 2015;1(4):e000033.

40. Molla KA, Yang Y. Predicting CRISPR/Cas9-Induced Mutations for Precise Genome Editing. Trends Biotechnol. 2020 Feb;38(2):136-141
41. Salman R, Kecman V. Regression as classification. In: *2012 Proceedings of IEEE Southeastcon*, Orlando, 2012.
42. Huang K, Xiao C, Glass LM, Critchlow CW, Gibson G, Sun J. Machine learning applications for therapeutic tasks with genomics data. *Patterns (N Y)*. 2021 Aug 9;2(10):100328.
43. Gao Y, Chuai G, Yu W, et al. Data imbalance in CRISPR offtarget prediction. *Brief Bioinform*. 2019;35(16):2783–2789.
44. Hruscha A, Krawitz P, Rechenberg A, et al. Efficient CRISPR/-Cas9 genome editing with low off-target effects in zebrafish. *Development*. 2013;140:4982–4987.
45. Mao Z, Bozzella M, Seluanov A, et al. Comparison of nonhomologous end joining and homologous recombination in human cells. *DNA Repair*. 2008;7(10):1765–1771.
46. He H, Garcia EA. Learning from imbalanced data. *IEEE Trans Knowl Data Eng*. 2009;21(9): 1263–1284.
47. LeCun Y, Bengio Y. *The Handbook of Brain Theory and Neural Networks* (M. A. Arbib, ed.), MIT press, 1998, pp. 255–258.
48. Ng AY. Feature selection, L1 vs. L2 regularization, and rotational invariance. In: *Proceedings of the 21st International Conference on Machine Learning*, Banff, 2004.
49. Hastie T, Tibshirani R, Friedman J. Introduction. In: *The Elements of Statistical Learning*. Springer Series in Statistics. New York, NY: Springer, 2009. doi: 10.1007/978-0-387-84858-7_1
50. Leinonen R, Sugawara H, Shumway M, et al. The sequence read archive. *Nucleic Acids Res*. 2011;39(suppl 1):D19–D21.
51. Rauscher B, Heigwer F, Breinig M, et al. Genome CRISPR—a database for high-throughput CRISPR/Cas9 screens. *Nucleic Acids Res*. 2017;45(D1):D679–D686.
52. Kim HK, Min S, Song M, et al. Deep learning improves prediction of CRISPR-Cpf1 guide RNA activity. *Nat Biotechnol*. 2018;36 (3):239–241.
53. Kleinstiver BP, Pattanayak V, Prew MS, et al. High-fidelity CRISPR-Cas9 nucleases with no detectable genome-wide off-target effects. *Nature*. 2016;529(7587):490–495.
54. Ren X, Yang Z, Xu J, et al. Enhanced specificity and efficiency of the CRISPR/Cas9 system with optimized sgRNA parameters in Drosophila. *Cell Rep*. 2014;9(3):1151–1162.
55. Sternberg SH, Doudna JA. Expanding the biologist's toolkit with CRISPR-Cas9. *Mol Cell*. 2015;58(4):568–574.
56. Liu Y, Schmidt B, Maskell DL. CUSHAW: a CUDA compatible short read aligner to large genomes based on the Burrows–wheeler transform. *Bioinformatics*. 2012;28(14):1830–1837.
57. Bae S, Park J, Kim JS. Cas-OFFinder: a fast and versatile algorithm that searches for potential off-target sites of Cas9 RNA-guided endonucleases. *Bioinformatics*. 2014;30(10):1473–1475.
58. McKenna A, Shendure J. FlashFry: a fast and flexible tool for large-scale CRISPR target design. *BMC Biol*. 2018;16(1):74.
59. Listgarten J, Weinstein M, Kleinstiver BP, et al. Prediction of off-target activities for the end-to-end design of CRISPR guide RNAs. *Nat Biomed Eng*. 2018;2(1):38–47.
60. Liu Q, He D, Xie L. Prediction of off-target specificity and cell-specific fitness of CRISPR-Cas System using attention boosted deep learning and network-based gene feature. *PLoS Comput Biol*. 2019 Oct 28;15(10):e1007480.
61. Luo J, Chen W, Xue L, et al. Prediction of activity and specificity of CRISPR-Cpf1 using convolutional deep learning neural networks. *BMC Bioinform*. 2019;20(1):332.
62. Haeussler M, Schönig K, Eckert H, et al. Evaluation of off-target and on-target scoring algorithms and integration into the guide RNA selection tool CRISPOR. *Genome Biol*. 2016; 17(1):148.
63. Chuai G, Ma H, Yan J, et al. DeepCRISPR: optimized CRISPR guide RNA design by deep learning. *Genome Biol*. 2018;19(1):80.
64. Ran FA, Cong L, Yan WX, et al. In vivo genome editing using staphylococcus aureus Cas9. *Nature*. 2015;520(7546):186–191.

65. Jinek, M. et al. A programmable dual-RNA-guided DNA endonuclease in adaptive bacterial immunity. *Science*. 2012;337: 816–821.
66. Kim D, Kim S, Kim S, et al. Genome-wide target specificities of CRISPR-Cas9 nucleases revealed by multiplex Digenome-seq. *Genome Res*. 2016;26(3):406–415.
67. Kuan PF, Powers S, He S, et al. A systematic evaluation of nucleotide properties for CRISPR sgRNA design. *BMC Bioinform*. 2017;18(1):297.
68. Chari R, Mali P, Moosburner M, et al. Unraveling CRISPR-Cas9 genome engineering parameters via a library-on-library approach. *Nat Methods*. 2015;12(1):823–826.
69. Zhou H, Zhou M, Li D, Manthey J, Lioutikova E, Wang H, Zeng X. Whole genome analysis of CRISPR Cas9 sgRNA off-target homologies via an efficient computational algorithm. *BMC Genomics*. 2017 Nov 17;18(Suppl 9):826. doi: 10.1186/s12864-017-4225-1.
70. Hall MA, Smith LA. Feature selection for machine learning: comparing a correlation-based filter approach to the wrapper. In: *Proceedings of the Twelfth International Florida, Artificial Intelligence Research Society Conference*, Orlando, Florida, 1999.
71. Eder Y, Todros K. Robust Two-sample location testing via probability measure transform. *IEEE Transactions on Signal Processing*. 2021;69:4724–4739. doi: 10.1109/TSP.2021.3092380.
72. Middleton D. *Statistical Communication Theory*, New York: McGraw-Hill, 1960.
73. Trunk GV. A problem of dimensionality: a simple example. *IEEE Trans Pattern Anal Mach Intell*. 1979;1(3):306–307.
74. Hughes GP. On the mean accuracy of statistical pattern recognizers. *IEEE Trans Inform Theory*. 1968;14(1):55–63.
75. Chari R, Yeo NC, Chavez A, et al. sgRNA scorer 2.0: a species independent model to predict CRISPR/Cas9 activity. *ACS Synth Biol*. 2017;6(5):902–904.
76. Hartenian, E, Doench JG. Genetic screens and functional genomics using CRISPR/Cas9 technology. *FEBS J*. 2015;282:1383–1393.
77. O'Brien AR, Burgio G, Bauer DC. Domain-specific introduction to machine learning terminology, pitfalls and opportunities in CRISPR-based gene editing. *Brief Bioinform*. 2021;22(1): 308–314.
78. Kaur K, Gupta AK, Rajput A, et al. ge-CRISPR—an integrated pipeline for the prediction and analysis of sgRNAs genome editing efficiency for CRISPR/Cas system. *Sci Rep*. 2016;6(1):30870.
79. Wong N, Liu W, Wang X. WU-CRISPR: characteristics of functional guide RNAs for the CRISPR/Cas9 system. *Genome Biol*. 2015;**16**(1):218.
80. Chari R, Mali P, Moosburner M, et al. Unraveling CRISPR-Cas9 genome engineering parameters via a library-on-library approach. *Nat Methods*. 2015;12(1):823–826.
81. Peng H, Zheng Y, Blumenstein M, et al. CRISPR/Cas9 cleavage efficiency regression through boosting algorithms and Markov sequence profiling. *Bioinformatics*. 2018;34(18):3069–3077.
82. Zhu H, Liang C. CRISPR-DT: designing gRNAs for the CRISPRCpf1 system with improved target efficiency and specificity. *Bioinformatics*. 2019;35(16):2783–2789.
83. Wang J, Xiang X, Cheng L, et al. CRISPR-GNL: an improved model for predicting CRISPR activity by machine learning and featurization. bioRxiv 2019;605790.
84. Xue L, Tang B, Chen W, et al. Prediction of CRISPR sgRNA activity using a deep convolutional neural network. *J Chem Inf Model*. 2019;59(1):615–624.
85. Chuai G, Wang QL, Liu Q. In silico meets in vivo: towards computational CRISPR-based sgRNA design. *Trends Biotechnol*. 2017;35(1):12–21.
86. Cui Y, Xu J, Cheng M, et al. Review of CRISPR/Cas9 sgRNA design tools. *Interdiscip Sci*. 2018;10(2):455–465.
87. Yan J, Chuai G, Zhou C, et al. Benchmarking CRISPR on-target sgRNA design. *Brief Bioinform*. 2018;19(4):721–724.
88. Moreno-Mateos MA, Vejnar CE, Beaudoin JD, et al. CRISPRscan: designing highly efficient sgRNAs for CRISPR targeting *in vivo*. *Nat Methods*. 2015;12(10):982–988.
89. Pranckevicius T, Marcinkevicius V. Comparison of naïve Bayes, random forest, decision tree, support vector machines, and logistic regression classifiers for text reviews classification. *Balt J Mod Comput*. 2017;5(2):221–232.

90. Lin J, Wong KC. Off-target predictions in CRISPR-Cas9 gene editing using deep learning. *Bioinformatics*. 2018;34(17):i656–i656.
91. Wilson LOW, Reti D, O'Brien AR, et al. High activity target-site identification using phenotypic independent CRISPR-Cas9 Core functionality. *CRISPR J*. 2018;1(2):182–190.
92. Mali P, Yang L, Esvelt KM, Aach J, Guell M, DiCarlo JE, et al. RNA-guided human genome engineering via Cas9. *Science*. 2013;339: 823–6. doi: 10.1126/science.1232033 - DOI - PMC - PubMed.
93. Doench JG, Fusi N, Sullender M, et al. Optimized sgRNA design to maximize activity and minimize off-target effects of CRISPR-Cas9. *Nat Biotechnol*. 2016;34:184–9.
94. Breiman L, Cutler A. Random forests. *Mach Learn*. 2001;45(1):5–32.
95. Hastie T, Tibshirani R, Friedman J. *The Elements of Statistical Learning*, 2nd edn, New York: Springer, 2009.
96. Slaymaker IM, Gao L, Zetsche B, Scott DA, Yan WX, Zhang F. Rationally engineered Cas9 nucleases with improved specificity. *Science*. 2016;351(6268):84–88.
97. Stemmer M, Thumberger T, del Sol Keyer M, Wittbrodt J, Mateo JL. CCTop: an intuitive, flexible and reliable CRISPR/Cas9 target prediction tool. *PLoS One*. 2015;10(4):e0124633.
98. Tsai SQ, Nguyen NT, Malagon-Lopez J, Topkar VV, Aryee MJ, Joung JK. CIRCLE-seq: a highly sensitive in vitro screen for genome-wide CRISPR-Cas9 nuclease off-targets. *Nat Methods*. 2017;14(6):607.
99. Yin H, Song CQ, Dorkin JR, Zhu LJ, Li Y, Wu Q, Park A, Yang J, Suresh S, Bizhanova A, et al. Therapeutic genome editing by combined viral and nonviral delivery of CRISPR system components in vivo. *Nat Biotechnol*. 2016;34(3):328–333.
100. Jokanović V. *Nanomedicine, the Greatest Challenge of 21st Century*, Belgrade, Serbia: Data Status, 2012.
101. Bostrom N. Are we living in computer simulation. *Philos Q*. 2003;53(211):243–255.
102. Armstrong S, Bostrom N, Shulman C. Racing to the precipice: a model of artificial intelligence development. *AI & Soc*. 2016;31:201–206.
103. Lello L, Raben TG, Hsu SD. Sibling validation of polygenic risk scores and complex trait prediction. *Sci Rep*. 2020;10:13190.
104. Impey C. "Mars and beyond: the feasibility of living in the solar system," *The Human Factor in a Mission to Mars: An Interdisciplinary Approach* (K. Szocik, ed.), Springer, 2019.
105. Jokanović V. *How Dead and Live Our Cells*, Beograd: "Vinča" Institute of Nuclear Science, 2013.
106. Jokanović V. *Man of Fifth Dimension*, Novi Sad: Prometej, 2021.
107. Liu J, Luo C, Smith PA, Chin JK, Page MG, Paetzel M, Romesberg FE. Synthesis and characterization of the arylomycin lipoglycopeptide antibiotics and the crystallographic analysis of their complex with signal peptidase. *J Am Chem Soc*. 2011;133(44):17869–17877.
108. Nicolaou KC, Ellery SP, Rivas F, Saye K, Rogers E, Workinger TJ, Schallenberger M, Tawatao R, Montero A, Hessell A, Romesberg F, Carson D, Burton D. Synthesis and biological evaluation of 2′, 4′- and 3′, 4′-bridged nucleoside analogues. *Bioorg Med Chem*. 2011;19(18):5648–5669. DOI: 10.1016/j.bmc.2011.07.022.
109. Lavergne T, Malyshev MD, Romesberg FE, Major groove substituents and polymerase recognition of a class of predominantly hydrophobic unnatural base pairs. *Chemistry*. 2012;18(4): 1231–1239.
110. Malyshev DA, Dhami K, Quach HT, Lavergne T, Ordoukhanian P, Torkamani A, Romesberg FE. Efficient and sequence-independent replication of DNA containing a third base pair establishes a functional six-letter genetic alphabet. *PNAS*. 2012;109(30):12005–12010.
111. Zhang Y, Lamb BM, Feldman AW, Zhou AX, Lavergne T, Li L, Romesberg FE. A semisynthetic organism engineered for the stable expansion of the genetic alphabet. *PNAS*. 2017;114(6):1317–1322.
112. Seo YJ, Hwang GT, Ordoukhanian P, Romesberg FE. Optimization of an unnatural base pair toward natural-like replication. *J Am Chem Soc*. 2009;131(9):3246–3252.
113. Betz K, Malyshev DA, Lavergne T, Welte W, Diederichs K, Dwyer TJ, Ordoukhanian P, Romesberg FE, Marx A. KlenTaq polymerase replicates unnatural base pairs by inducing a Watson-Crick geometry. *Nat Chem Biol*. 2012;8(7):612–614.

114. Betz K, Malyshev DA, Lavergne T, Welte W, Diederichs K, Romesberg FE, Marx A. Structural insights into DNA replication without hydrogen bonds. *J Am Chem Soc.* 2013;135(49):18637–18643.
115. Ast M, et al. Diatom plastids depend on nucleotide import from the cytosol. *Proc Natl Acad Sci USA.* 2009;106(9):3621–3626.
116. Shroff R, Cole AW, Diaz DJ, Morrow BR, Donnell I, Annapareddy A, Gollihar J, Ellington AD, Thyer R. Discovery of novel gain-of-function mutations guided by structure-based deep learning. *ACS Synth Biol.* 2020;9(11):2927–2935
117. Zumrut HE, Batool S, Argyropoulos KV, Williams N, Azad R, Mallikaratchy PR. Integrating ligand-receptor interactions and in vitro evolution for streamlined discovery of artificial nucleic acid ligands. *Mol Ther Nucleic Acids.* 2019 Sep 6;17:150–163.
118. Zumrut H, Yang Z, Williams N, Arizala JDR, Batool S, Benner S, Mallikaratchy P. Ligand Guided Selection (LIGS) with artificially expanded genetic information systems against TCR-CD3ε. *Biochemistry,* 2020;59(4): 552–562.
119. Singh I, Laos R, Hoshika S, Benner SA, Georgiadis MM. Snapshots of an evolved DNA polymerase pre- and post-incorporation of an unnatural nucleotide. *Nucleic Acids Res.* 2018;46(15):7977–7988.
120. Kent T, Rusanov TD, Hoang TM, Velema WA, Krueger AT, Copeland WC, Kool ET, Pomerantz RT. DNA polymerase θ specializes in incorporating synthetic expanded-size (xDNA) nucleotides. *Nucleic Acids Res.* 2016;44(19):9381–9392.
121. Betz K, Kimoto M, Diederichs K, Hirao I, Marx A. Structural basis for expansion of the genetic alphabet with an artificial nucleobase pair. *Angewandte Chemie.* 2017;56(39):12000–12003.
122. Wachowius F, Porebski BT, Johnson CM, Holliger P. Emergence of function from single RNA sequences by Darwinian evolution, bioRxiv preprint doi: 10.1101/2021.03.03.433769.
123. Gibson DG, Glass JI, Lartigue C, Noskov VN, Chuang RY, Algire MA, Benders GA, Montague MG, Ma L, Moodie MM, Merryman C. Creation of a bacterial cell controlled by a chemically synthesized genome. *Science.* 2010;329(5987):52.
124. Chavez A, Scheiman J, Vora S, Pruitt BW, Tuttle M, Iyer EP, Lin S, Kiani S, Guzman CD, Wiegand DJ, Ter-Ovanesyan D. Highly efficient Cas9-mediated transcriptional programming. *Nat Methods.* 2015;12(4):326–328.
125. Purnick PE, Weiss R. The second wave of synthetic biology: from modules to systems. *Nat Rev Mol Cell Biol.* 2009;10(6):410–422.
126. Tversky A, Kahneman D. Extensional versus intuitive reasoning: The conjunction fallacy in probability judgment, *Psychological Review.* 1983;90(4):293–315.
127. Lau YH, Stirling F, Kuo J, Karrenbelt MA, Chan YA, Riesselman A, Horton CA, Schäfer E, Lips D, Weinstock MT, Gibson DG. Large-scale recoding of a bacterial genome by iterative recombineering of synthetic DNA, *Nucleic Acids Res.* 2017;45(11):6971–6980.
128. Baker M. Repositories share key research tools. *Nature.* 2014;505(7483):272.
129. Jokanović V. The short story of the miraculous daily resurrection of life in us. *Zastita Materijala.* 2021;62(2):136–140.
130. Wang X, Ranellucci S, Katz J. Global-Scale Secure Multiparty Computation, Session A1: Multi-Party Computation 1 CCS'17, October 30-November 3, 2017, Dallas, TX, USA.
131. Jokanović V. *Bridge between Nanophysics and Alternative Medicine a New Energetic Approach to the Human Cell Treatment and Their Healing,* LAP LAMBERT Academic Publishing, 2016.
132. Hutchison CA, Chuang RY, Noskov VN, Assad-Garcia N, Deerinck TJ, Ellisman MH, Gill J, Kannan K, Karas BJ, Ma L, Pelletier JF. Design and synthesis of a minimal bacterial genome. *Science.* 2016;351(6280):aad6253.
133. Baker M. The next step for the synthetic genome. *Nature.* 2011;473:403–408.
134. Ellis T, Adie T, Baldwin GS. DNA assembly for synthetic biology: from parts to pathways and beyond. *Integr Biol.* 2011;3:109–118.
135. El Karoui M, Hoyos-Flight M, Fletcher L. Future trends in synthetic biology—a report. *Front Bioeng Biotechnol.* 2019;7(175):1–8.
136. Dar-Nimrod I, Heine SJ. Genetic essentialism: on the deceptive determinism of DNA. *Psychol Bull.* 2011;137(5):800–818.

137. Doudna JA, Sternberg SH. *A Crack in Creation: Gene Editing and the Unthinkable Power to Control Evolution*, Mariner Books, 2017.
138. Ran FA, Hsu PD, Wright J, Agarwala V, Scott DA, Zhang F. Genome engineering using the CRISPR-Cas9 system. *Nat. Protocols.* 2013;8(11):2281–2309.
139. Thorn PD. Nick Bostrom: superintelligence: paths, dangers, strategies. *Minds and Machines.* 25(3):285–289.
140. http://synbio.info/display/synbio/SynBio+Beta.
141. https://www.alienelement.com/ 2015/07/28/dna-printing-here-living-creature-create/.

11

Innovation and Emerging Computer Vision and Artificial Intelligence Technologies in Coronavirus Control

Mukesh Carpenter
Department of Surgery, Alshifa Multispecialty Hospital, Okhla, India

Vinod Kumar Jangid
Department of Respiratory Medicine, Medical College Kota, Rajasthan, India

Dharmendra Carpenter
Department of Anaesthesiology & Critical Care, Narayana Multispecialty Hospital, Jaipur, India

Lavanya Sharma
Amity Institute of Information Technology, Amity University, Noida, India

CONTENTS

DOI: 10.1201/9781003244165-14

11.1 Introduction

Coronavirus is the most infectious disease which belongs to a large family of viruses. This new virus and disease were unknown until the outbreak began in Wuhan, China, in December 2019. This virus may cause illness in humans or animals and therefore can be viewed as a pandemic. In humans, mutants of the Covid-19 virus cause a respiratory infection that varies from common cold to severe disease such as severe acute respiratory syndrome (SARS) and the Middle East respiratory syndrome (MERS). In today's scenario, the biggest to the society is the ongoing outbreak of this deadly virus which is also termed Covid-19. In this outbreak computer vision (CV) as a subset of artificial intelligence (AI) plays a very important role in solving several complex problems in the emergency department and seems to have great potential in controlling this outbreak [1–6].

India reported its first case of deadly virus infection in Kerala, India, in February 2020. A young female was admitted to the emergency department in General Hospital, Kerala, having cough and sore throat only with no other symptoms such as fever and shortness of breath. She disclosed that she returned to India from Wuhan on January 23, 2020, owing to virus outbreak situation there. She was asymptomatic and disclosed that she traveled from Wuhan to Kunming by train where she noticed a few persons with a respiratory infection on a railway station and on the train. In India, approximately 29 million were reported through June 4, 2021, as shown in Figure 11.1. About 354,000 casualties are reported as per the Worldometers database [2] as shown in Figure 11.2. As per the data recovery rate of the patient is 27,504,126 (99%) and the total number of deaths is 353,557 (1%).

CV is a promising tool for supporting the emergency department. In literature, various studies have shown that AI techniques can manage clinical care by lowering the administrative demands in hospitals. AI in the emergency department enables healthcare workers to manage a huge amount of data and transform it into lifesaving information. I is used to help healthcare works and stakeholders to manage the vast data and transform them into potentially life-saving information. These techniques have various applications in laboratories and hospitals such as surgery, disease diagnosis, prognosis, therapies, medical imaging, and drug discovery. However, it has various challenging issues [7–10]. Despite the advantages, AI applications continue to face serious challenges in healthcare. In this review, we will focus on the role of AI in patient low management as well as in predicting a patient's admission to a hospital [11–15].

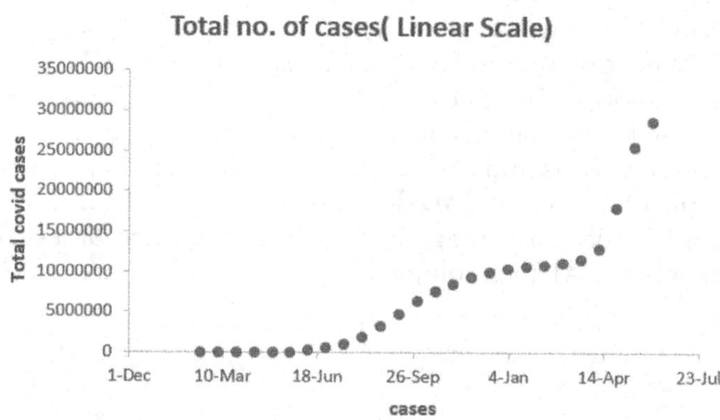

FIGURE 11.1
Total number of cases reported in India.

11.2 Background

Covid-19 is an infectious disease caused by a new strain of Coronavirus. This pandemic outbreak led to an invisible threat that is a very challenging task to detect and contain. This pandemic has posed critical challenges for governments, medical departments, and public health across the world [16]. The use of emerging AI tools and technologies can provide great help to the fight against the invisible threat to the humanoid. The first case was reported in the Wuhan city of China at the end of 2019 and soon it spread to other countries also. Presently, various variants or mutants of this Covid-19 virus such as the UK variant, South Africa variant, Indian variant, Brazilian variant, and California variant are diagnosed in different countries (Figure 11.3). In case of an outbreak in India, while it is unknown if the variant is driving the large Indian wave, the World Health Organization (WHO) officially has said that there is some available information to suggest increases

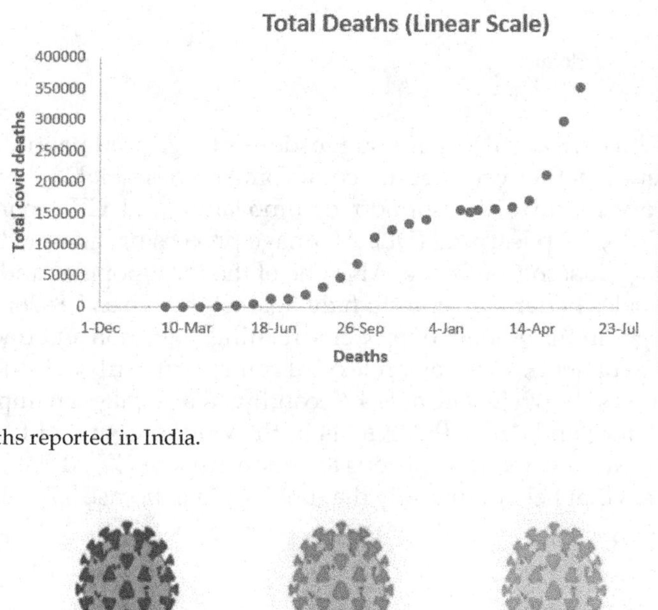

FIGURE 11.2
Total number of deaths reported in India.

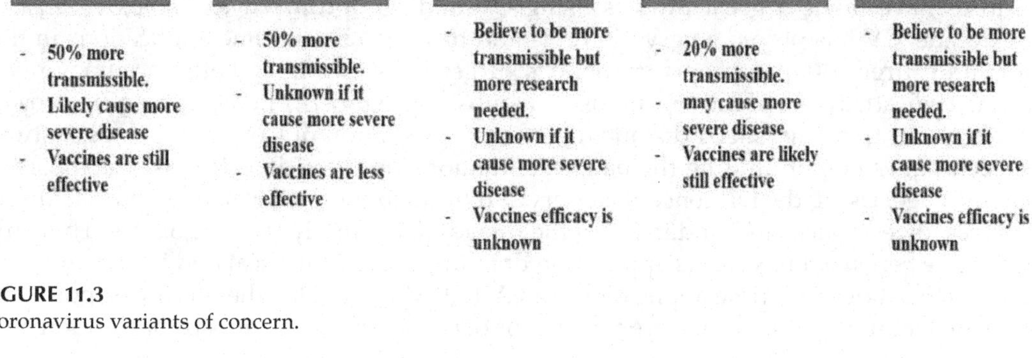

FIGURE 11.3
Coronavirus variants of concern.

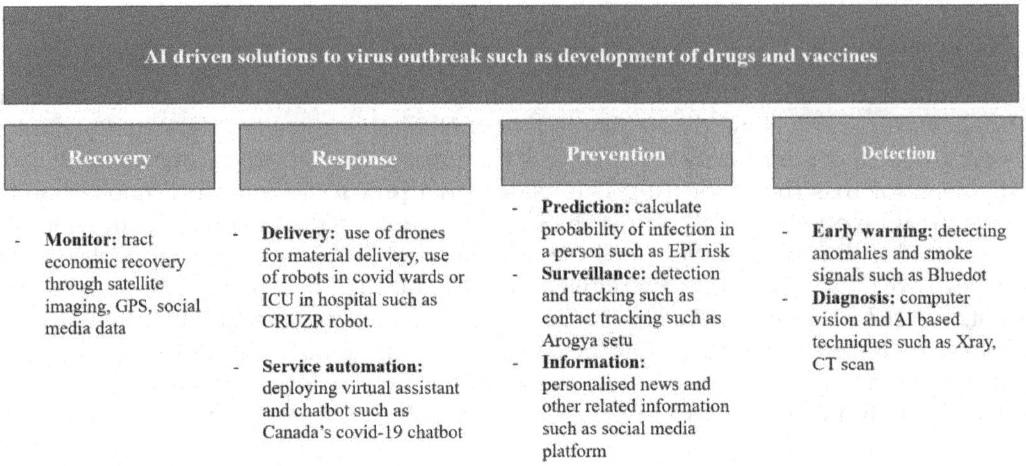

FIGURE 11.4
AI-driven solutions to the pandemic.

transmissibility [17]. There are also some suggestions of "reduced neutralization," but it is too early to comment whether vaccines are compromised or not [6].

Healthcare delivery requires the support of innovative and CV technologies such as machine learning (ML), deep learning (DL), AI, image processing, and medical imaging (CT scan, X-ray) to fight against this outbreak. AI is one of the most popular and innovative technologies which provides better outcomes to fight against this virus [18–26]. This technology has several advantages in the domain of proper screening, detection and tracking, prediction of existing and future patients. The major role of AI comes with early detection and diagnosis of this virus infection as shown in Figure 11.4. Secondly, AI also plays an important role in the development of vaccines and drugs that results in the work pressure of frontline worker or medical team. In the past few months, various techniques using CV, AI, ML, and DL are done by several researchers that help in the early diagnosis of this disease [27–36].

11.3 Computer Vision (CV) Technology

In order to fight against this deadly virus researchers and governments of different countries have made a great process using CV and its optimized technologies. In the past decade, CV has played a very vital role and has several real-time applications in the domain of surgery (laparoscopic or robotics surgery that results in better accuracy rate), clinical trial attrition reduction, disease diagnosis, prognosis, medical imaging, object detection and tracking, social distancing, drones, prediction of Covid-19. CV algorithms and techniques help in finding the pattern and anomalies in visions to get a diagnosis at an early stage. Using the DL concept (supervised or unsupervised learning approach), CV identifies, investigates, and understands the images accordingly [15,18,31,37–40]. This will prove to be beneficial in various application domains such as medical and healthcare.

Due to the above real-time applications of CV (AI, ML, and DL), there is a vast potential for vision technologies to fight Coronavirus for drug discovery, diagnosis, prevention and

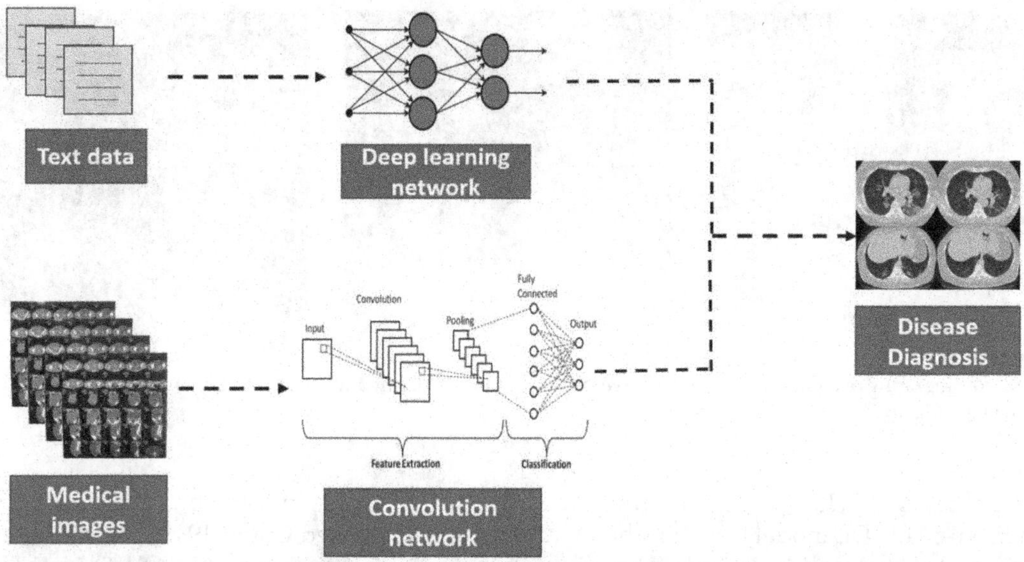

FIGURE 11.5
Application of deep learning for disease diagnosis.

control, and Coronavirus management. In the next sections, we will provide an overview of some of the important applications domain of vision-based technologies. The classification, localization, detection, and tracking of an object refer to the identification of different objects in a frame. With the growth of CV technologies, a huge amount of data is generated, and recent medical advancement has been producing a lot of digital data [17,34,35,41–44]. Using large datasets, CV models can acquire numerous pattern recognition abilities from the physical level to medical perception using DL, where the disease is diagnosed using DL convolution neural network (CNN) and data collected is in the form of both textual format and images (see Figure 11.5) [18,31].

11.4 Computer Vision for Covid-19 Diagnosis

The emerging CV and AI technologies such as X-ray radiography (CXR) and computed tomography (CT) play a very important role in the diagnosis of Covid-19. The main goal of diagnosis is to make a judgment about the exact feature of the disease and prognosis so that patient recovers. These technologies are fast, reliable, and widely available and also provide an affordable diagnosis of the disease which seems to be essential to fight against this disease [1,4,5,15,45–53].

11.4.1 X-Ray Radiography (CXR)

Digital chest CXR helps in the diagnosis of disease at a very affordable price and is widely available. It is mainly used for detections of chest pathology as compared to CR scans. Healthcare workers used CXR for the Covid detection and other disease detection such as cancer, tumors, pneumonia, and cardiac disease. In order to detect Covid-19 disease, some

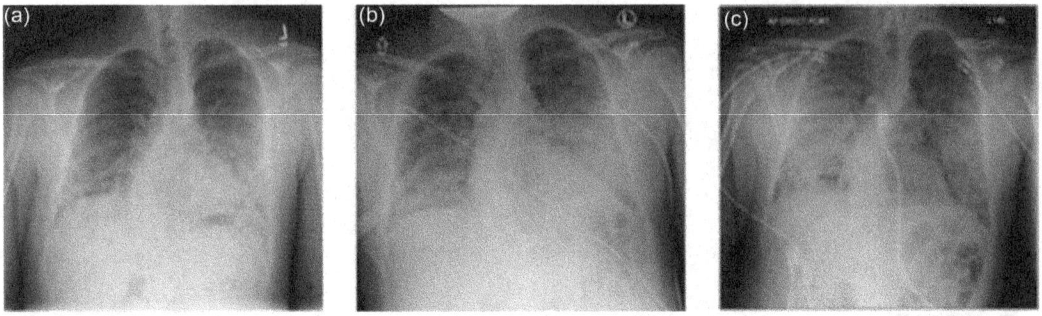

FIGURE 11.6
X-ray radiography of a patient (row wise): (a) early stage (day 3), (b) mediatory stage (day 4), and (c) highly infection stage (day 6).

preprocessing techniques are required to enhance the visuals in X-ray imaging. As shown in Figure 11.6, DL models exist in the literature for X-ray-based Covid-19 diagnosis. One of the most popular models is developed by Darwin AI, Canada namely COVID-Net. This model is trained with COVIDx dataset which is based on 13,645 patient chest X-ray cases. The accuracy of this model is 92.4% for this disease diagnosis. Based on the diagnosis clinicals can decide about the treatment plan of a patient [1,16,18,54,55].

11.4.2 Computed Tomography (CT)

This technique is also known as a noninvasive test which is used to generate a precise vision of a patient's chest with a radiology examination. This method provides better vision and more detailed visual information as compared to X-ray as shown in Figure 11.7. This technique provides complete visual information about fats, muscles, bones, and other body organs that will help in the diagnosis of tumors, cancer, surgery, anomalies in the heart, lungs, and liver. With respect to Covid-19 diagnosis, it clearly identified CT features of pneumonia such as ground-glass opacification and consolidation affecting the lower lobes with other added characteristics to be detectable depending on the stage of Covid-19. Using DL and image segmentation infected area can be identified between a normal person and an infected person. One of the hospitals in China namely "Renmin Hospital of Wuhan University" used this trained model for diagnosis of new Covid-19 strain. This model is based on approximately 46,096 CT images of both normal and infected persons that have been investigated by radiologists. This dataset comprises CT images of 106 patients, where 51 are Covid-19 infected. This model is based on the "UNet++" sematic segmentation model which is used to filter valid regions in the image. The accuracy of this method is 95.24% per patient with a positive predictive value of 84.62% and a negative predictive value of 100% [1,5,12].

Another method for diagnosis of Covid-19 is "COVNet" which is freely available on the internet source. Basically, extract visual information from volumetric chest CT using transfer learning based on a convolutional neural network that is 50 layers deep known as RESNET50. Segmentation of lungs was done using the U-Net model as a processing task. In this model, a dataset is used which contains 4,356 chest CT images from 3,322 patients retrieved from six hospitals. The rate of sensitivity and specificity is 90% with a 95% confidence interval. The dataset provided by the University of San Diego contains 349 CT visuals or images with clinical findings of this disease. An AI model is trained to demonstrate its potential achieved 85% accuracy [1,5].

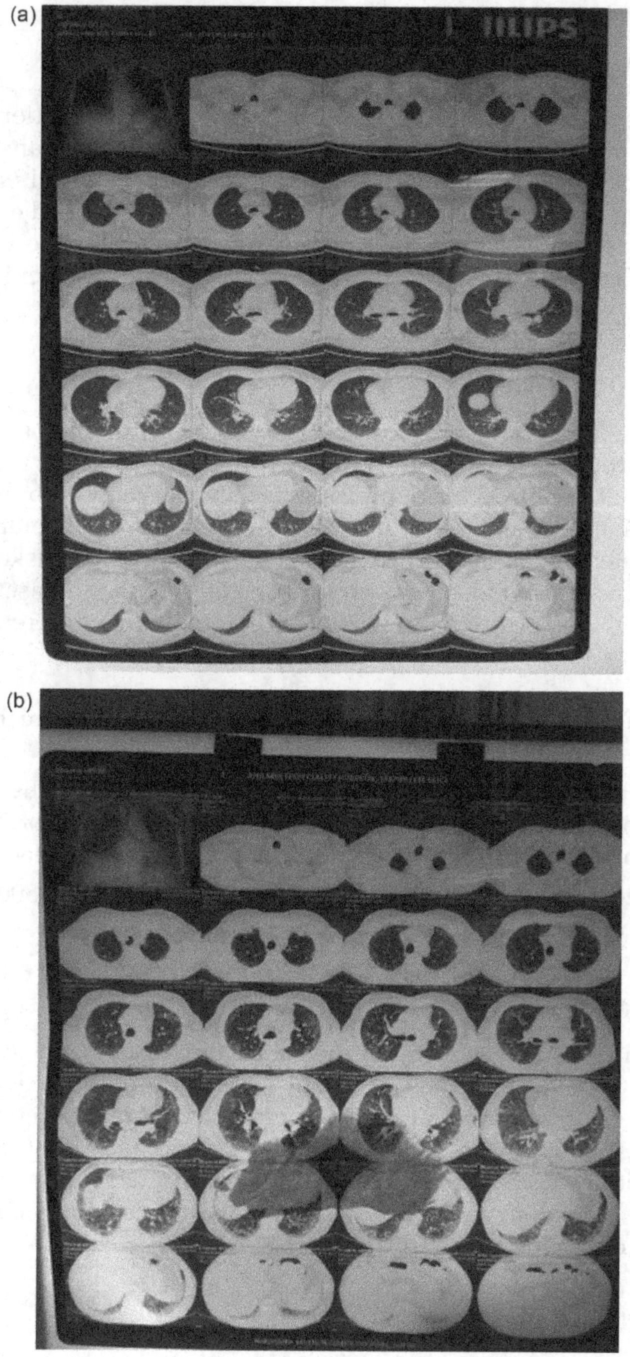

FIGURE 11.7
Computed tomography: (a) healthy patient and (b) Covid-19 patient.

11.5 Computer Vision for Covid-19 Prevention

The WHO provides detailed guidelines on Covid-19 infection prevention and control strategies. The main aim of this disease prevention is to restrict the transmission in health care locales with the implementation of standard precautions for all patients, healthcare workers, clinicals, early recognition, sources, and environmental and engineering control [54–59]. CV and AI techniques seem to be useful in implementing these strategies such as pandemic drones, face mask detection, thermography, and germ screening. In this section, some important CV techniques are discussed.

11.5.1 Face Mask Detection

A major strategy that can be useful in the prevention of this disease is the use of face masks and other protective equipment to restrict the spread of this disease. Most of the countries throughout the globe have implemented this strategy to prevent the spread of the virus (Figure 11.8). CV can be very useful to facilitate its implementation.

A masked face detection method which is based on a multilevel masked face recognition model achieved an accuracy rate of 95% on a masked face image dataset which is publicly available for research perspectives [11–13,20,26]. There are three different types of masked face datasets which are listed below:

- This dataset is used to train a masked face detection model which is the starting point for the successive masked face detection task [12].
- The real-world masked face recognition dataset is one of the world's largest real-world masked face datasets which consists of 5,000 images of 525 persons with face masks and 90,000 images of the same 525 subjects without face masks [11].
- The simulated masked face recognition dataset consists of 500,000 images of approximately 10,000 persons [13].

11.5.2 Thermal Imaging

This technique is also known as infrared thermography where an object is detected from the heat produced from its body. This technique is very useful in the detection of Covid-19 people such as fever screening at an early stage. This thermal imaging strategy can be applied to airports, shopping complexes, public places, and night vision applications (Figure 11.9). Infection screening can be done using infrared thermography and a CCD camera to capture optical signals with high-sensitive for contactless vital symptoms by using music feature matching algorithms [1–6,11–13,16].

11.5.3 Drones in Covid-19

In view of the pandemic situation, drones seem to be very useful in the prevention and control of these situations. Drones use remote sensing and digital imagery to detect infected persons as shown in Figure 11.10. There are several real-time applications such as aerial monitoring, vision-guided robot control for three-dimensional object detection, and remote life sign monitoring in the Covid-19 pandemic. The reports from the media and other available sources have discovered four major use cases of pandemic drones in Covid-19 [14,-21–26]. These are mentioned below:

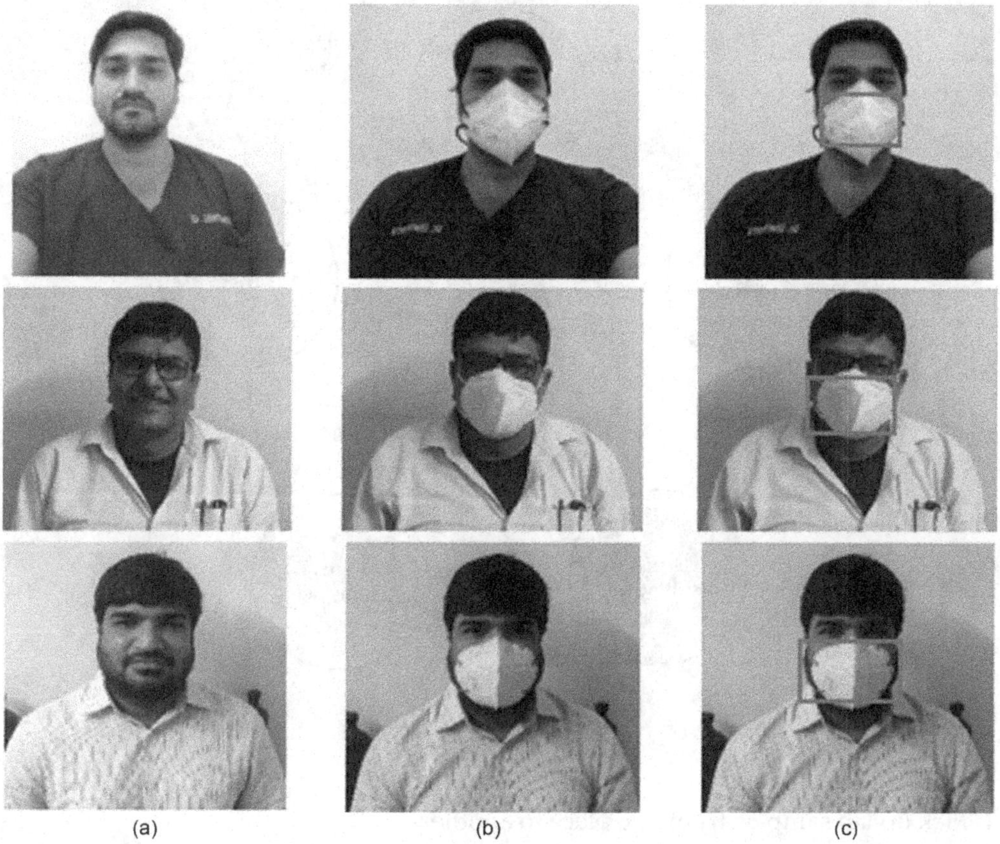

FIGURE 11.8
Face mask detection: (a) without a face mask, (b) with a face mask, and (c) with a detected face mask.

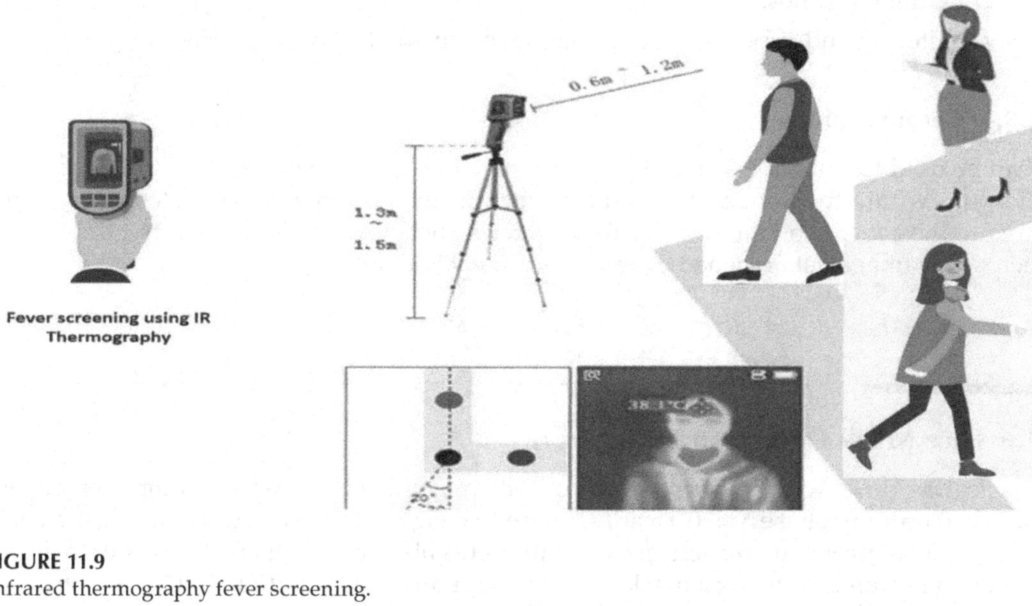

FIGURE 11.9
Infrared thermography fever screening.

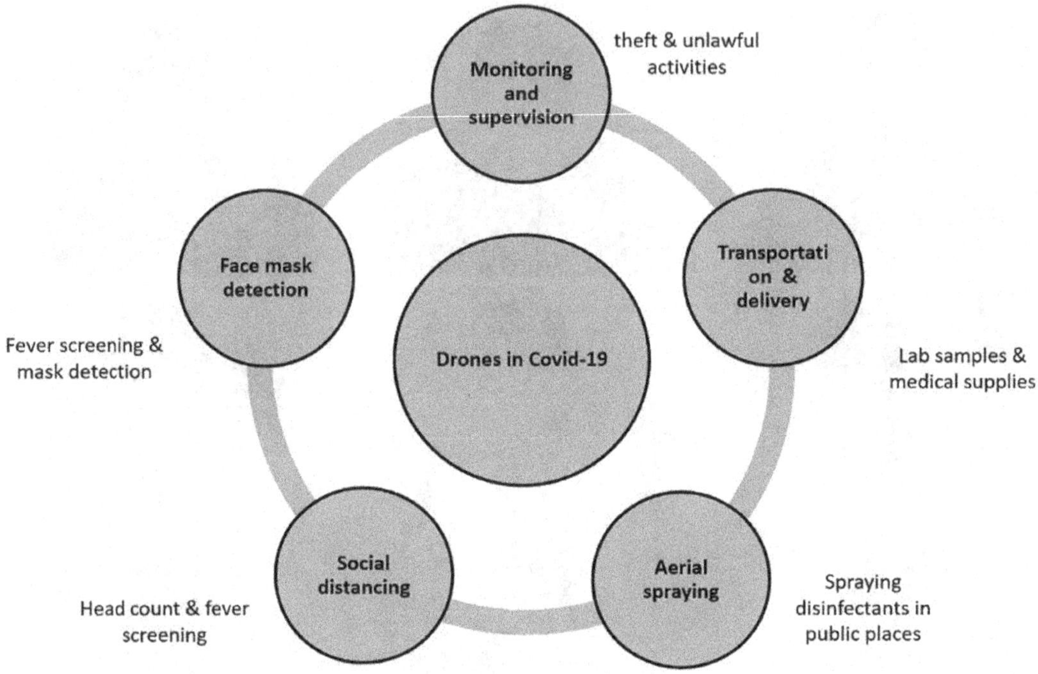

FIGURE 11.10
Drones during Covid-19.

1. Pick up lab samples from one place to another.
2. Transportation and delivery of medical supplies or equipments to reduce time duration that results in a minimum amount of exposure to infection.
2. Aerial spraying of public regions, hospital regions, and societies to disinfect the containment zones.
3. Monitoring and supervision of public space during lockdown period.

11.5.4 Germ Scanning

It can be used to combat this pandemic situation. CNN becomes standard for object classifications within the images due to their generalizability and accurate outcomes as compared to the conventional methods. To scan germ such as detection of bacteria light-sheet microscopy image data reached over 90% accuracy [1,12,16].

11.6 Face Mask Detection Framework

This section deals with the framework of face mask detection which comprises of two phases: (i) training phase and (ii) testing phase (see Figure 11.11). In this framework, a basic CNN model is created using TensorFlow with Keras library and OpenCV in order to detect whether a person is wearing a mask or not in this pandemic situation. In the first phase of

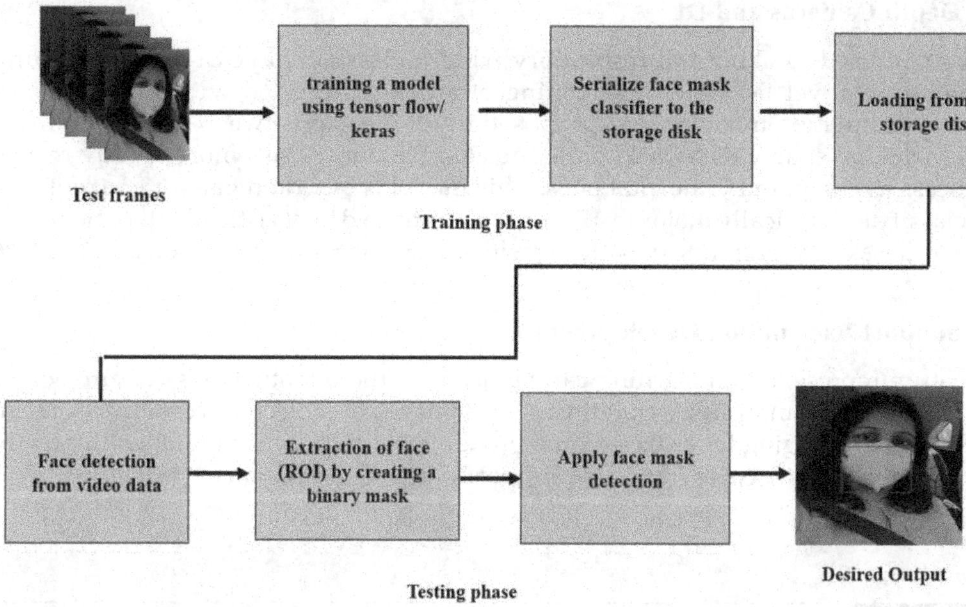

FIGURE 11.11
Face mask detection frameworks.

this framework, a dataset is created, and then modeling is done using TensorFlow/Keras [13]. Then, serializing the face mask detector to the storage disk and loading from the storage is done. A region of interest is a portion of an image that needs to filter out or operate on in a certain manner. In the next phase, extraction of the face by creating a binary mask is done. After this step, face mask detection is applied, and the desired output is obtained as a result (mask or no mask).

11.7 AI Vision for Covid-19 Treatment

In the present scenario, there is no specific treatment for this situation but most of the symptoms can be treated if diagnosed at an early stage. High fever, headache, breathing problem, loss of smell, and sore throat are some of the early signs of infection. The overall treatment depends on the clinical condition.

11.7.1 Disease Progression Score

This score also helps in the treatment of this disease. Clinicals can classify patients based on the severity of this disease. CV can be helpful in identifying the person who is very serious or critically ill to immediate medical attention. This score helps in the categorization of infected patients. This score is calculated by measuring infected zones from CT imaging to check the progression of patients from time to time and identify the most critically ill patient [1,12].

11.7.2 Depth Cameras and DL

These can be used as abnormal respiratory pattern classifiers to carry out an accurate and unassuming, yet larger-level screening of a person infected with this virus. The respiratory simulation model is first proposed to fill the gap between a huge volume of training data and sparse real-world data. Infected patients have a more speedy respiration process as compared to normal ones. This model is based on gated recurrent units NN to classify six clinically major respiratory patterns to detect critically ill patients. The model can classify the respiratory patterns with a 94.5% accuracy rate [1–6,11–14,16,18–26].

11.7.3 Support Vaccination Development

Deep feature representation learning can be used for the quantitative structure activity relationship (QSAR) analysis by combining 360° images of molecular conformations into DL. QSAR analysis using DL is based on a unique molecular image input technique that can be used for drug discovery and also supports vaccine development [1,3,60].

11.8 AI Vision for Ventilation Management in Intensive Care Unit (ICU)

In United Kingdom, a trial is going on to explore the AI treatment to monitor lung physiology of Covid-19 patients to optimize ventilation treatments. Covid-19 patients may develop SARS and MERS conditions due to which lungs do not provide enough amount of oxygen required. This condition is life-threatening and sometimes mechanical ventilators are required in ICU. The selection of incorrect ventilator settings can damage the patient's lungs. The researchers at Imperial College London are doing research on how AI can help personalize ventilator settings to a specific patient. The system can seem to be helpful for the next Covid-19 wave. In the second wave also several mutants of Covid-19 infections can be seen in different countries. This system can help to set the ventilators appropriately regardless of the cause. This new AI device will be known as the Beacon Care system which utilizes mathematical calculations to suggest the ventilator setting as per the need and requirement of a patient. Medical practitioners input data on a patient's condition, and the AI device advises the healthcare team on how to treat the patient. This device also checks the aeration of lungs and blood flow and captures oxygen for a better response to treatment. It can also predict whether patients are going to be worst or respond to different maneuvers [61].

A BEACON Care system is an ICU ventilation and nutrition assist system. This system will provide doctors with two possible solutions that are listed below:

1. Ventilator-dependent "open-loop" advice for optimizing ventilation in ventilated adult Covid-19 patients.
2. "Open-loop" advice for optimizing nutrition in ventilated adult Covid-19 patients.

This system takes inputs which are measured by BEACON and data inputs from the ventilator and manual data entry by the medical practitioner into a computer program that has been designed to assist the decisions to make changes in setting for ventilator patients undergoing mechanical ventilation (see Figure 11.12). The main motive of this system is to present assistance for ventilator and nutrition adjustment in graphical format [62].

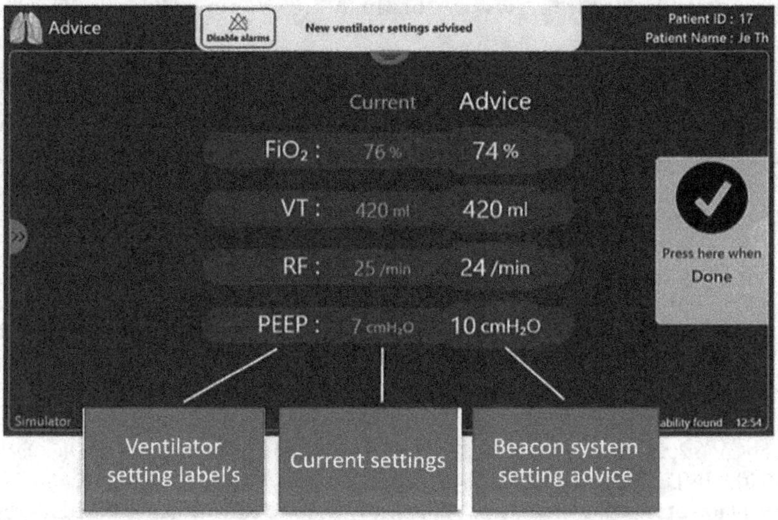

FIGURE 11.12
BEACON care system.

11.9 Future of Emerging AI Technologies

Due to the versatile nature of CV and its optimized technologies, they have the potential to support teams that are working in several disciplines and industries to combat Covid-19 challenges. AI vision advancement is becoming popular day by day, and new applications are currently being introduced and tested on a very large scale.

11.10 Conclusion

There are various real-time applications of CV present in literature. There is a vast potential for vision technologies to fight Coronavirus for drug discovery, diagnosis, prevention and control, and Coronavirus management. This chapter provides a brief discussion of diagnosis, detection, and prevention of the Covid-19 virus using CV and AI technologies. This chapter also presents a framework of face mask detection using CNN with TensorFlow and Keras to protect ourselves from infection. Furthermore, the future of emerging technology and ventilator management is also discussed.

References

1. Computer vision applications. Available at: https://viso.ai/applications/computer-vision-applications-for-coronavirus-control/ [accessed on 10 July 2021].
2. Covid-19. Available at: https://www.worldometers.info/coronavirus/country/india/.

3. Covid-19. Available at: https://www.ncbi.nlm.nih.gov/pmc/articles/PMC7530459/ [accessed on 10 July 2021].
4. Covid-19 Vaccines. Available at: https://www.france24.com/en/live-news/20210517-pfizer-moderna-vaccines-effective-against-indian-variants-study [accessed on 10 July 2021].
5. Covid-19. Available at: https://www.indiatoday.in/coronavirus-outbreak/story/who-names-labels-mutant-strains-corona-india-covid-strain-delta-1809235-2021-05-31 [accessed on 10 July 2021].
6. Covid-19 Variants. Available at: https://www.cdc.gov/coronavirus/2019-ncov/variants/variant-info.html#Concern [accessed on 10 July 2021].
7. E. Callaway. The race for coronavirus vaccines: a graphical guide. *Nature.* 580(7805), 576 (2020).
8. Y. Wang, D. Zhang, G. Du, et al. Remdesivir in adults with severe COVID-19: a randomised, double-blind, placebo-controlled, multicentre trial. *Lancet.* 395(10236), 1569–1578 (2020).
9. ClinicalTrials.gov. COVID-19, Recruiting Studies. https://clinicaltrials.gov/ct2/results?term=COVID-19&Search=Apply&recrs=a&age_v=&gndr=&type=&rslt=.
10. A. Das, M. Wasif Ansari, R. Basak. "Covid-19 Face Mask Detection Using TensorFlow, Keras and OpenCV", *2020 IEEE 17th India Council International Conference (INDICON)*, 2020, pp. 1–5, DOI: 10.1109/INDICON49873.2020.9342585.
11. Covid-19. Dataset Available at: https://paperswithcode.com/dataset/rmfd [accessed on 10 July 2021].
12. Covid-19. Available at: https://arxiv.org/abs/2003.09093 [accessed on 10 July 2021].
13. Face mask detection Covid-19. Available at: https://github.com/akashkam559/Covid19-Face-Mask-Detector [accessed on 10 July 2021].
14. Drone usage in Covid-19. Available at: https://www.unicef.org/supply/media/5286/file/%20Rapid-guidance-how-can-drones-help-in-COVID-19-response.pdf [accessed on 10 July 2021].
15. S. Singh, L. Sharma, B. Kumar. A machine learning based predictive model for coronavirus pandemic scenario. *J. Phys.: Conf. Ser.* 1714, 012023 (January 2021).
16. Covid-19. Available at: https://ui.adsabs.harvard.edu/abs/2020arXiv200309093W/abstract [accessed on 10 July 2021].
17. World Health Organization. Coronavirus disease 2019 (COVID-19): situation report. 156 (2020). www.who.int/docs/default-source/coronaviruse/situation-reports/20200624-covid-19-sitrep-156.pdf?sfvrsn=af42e480_2.
18. Sharma L., Garg P. (Eds.) (2020). *From Visual Surveillance to Internet of Things*, Chapman and Hall/CRC, New York. DOI: 10.1201/9780429297922.
19. R. Punia, L. Kumar, M. Mujahid and R. Rohilla, "Computer Vision and Radiology for COVID-19 Detection", *2020 International Conference for Emerging Technology (INCET)*, 2020, pp. 1–5, DOI: 10.1109/INCET49848.2020.9154088.
20. A. Anand, V. Jha, L. Sharma. An improved local binary patterns histograms techniques for face recognition for real time application. *Int. J. Recent Technol. Eng.* 8(2S7), 524–529 (July 2019).
21. L. Sharma, D.K. Yadav. Histogram based adaptive learning rate for background modelling and moving object detection in video surveillance. *Int. J. Telemed. Clin. Pract.* (June, 2016). ISSN: 2052-8442, DOI: 10.1504/IJTMCP.2017.082107.
22. L. Sharma, N. Lohan. Performance analysis of moving object detection using BGS techniques in visual surveillance. *Int. J. Spat. Data Sci.* February, 2019
23. L. Sharma, N. Lohan. (March, 2019). Internet of things with object detection. In: *Handbook of Research on Big Data and the IoT*, IGI Global, pp. 89–100. ISBN: 9781522574323, DOI: 10.4018/978-1-5225-7432-3.ch006.
24. L. Sharma. Introduction. In: Sharma L., Garg P. (eds) *From Visual Surveillance to Internet of Things*, Taylor & Francis, CRC Press, Boca Raton, FL, Vol.1, pp. 14.
25. L. Sharma, P. K. Garg. Block based adaptive learning rate for moving person detection in video surveillance. In: *From Visual Surveillance to Internet of Things*, Taylor & Francis, CRC Press, Boca Raton, FL, Vol. 1, pp. 201.
26. S. Makkar, L. Sharma. (2019). A face detection using support vector machine: Challenging issues, recent trend, solutions and proposed framework. In: Singh M., Gupta P., Tyagi V.,

Flusser J., Ören T., Kashyap R. (eds) *Advances in Computing and Data Sciences*. ICACDS 2019. Communications in Computer and Information Science, vol. 1046, Springer, Singapore. DOI: 10.1007/978-981-13-9942-8_1.
27. L. Sharma, P. K. Garg. IoT and its applications. In: *From Visual Surveillance to Internet of Things*, Taylor & Francis, CRC Press, Boca Raton, FL, Vol. 1, pp. 29.
28. C. Wang, P.W. Horby, F.G. Hayden and G.F. Gao. A novel coronavirus outbreak of global health concern. *The Lancet*, 395, 10223, 470–473 (2020).
29. E. Gorbalenya. Severe acute respiratory syndrome-related coronavirus–The species and its viruses a statement of the Coronavirus Study Group. *BioRxiv*, 2020.
30. L. Sharma, A. Singh, D.K. Yadav. Fisher's linear discriminant ratio based threshold for moving human detection in thermal video. *Infrared Physics and Technology*, 78, 118–128 (March, 2016) Elsevier.
31. L. Sharma (Ed.) (2021). *Towards Smart World*, Chapman and Hall/CRC, New York. DOI: 10.1201/9781003056751.
32. L. Sharma. (December 2020). Human detection and tracking using background subtraction in visual surveillance. In: *Towards Smart World*, Chapman and Hall/CRC, New York, pp. 317–328. DOI: 10.1201/9781003056751.
33. L. Sharma, D.K. Yadav, S.K. Bharti. "An improved method for visual surveillance using background subtraction technique,", *IEEE, 2nd International Conference on Signal Processing and Integrated Networks (SPIN-2015)*, Amity University, Noida, India, Feb. 19–20, 2015.
34. D.K. Yadav, L. Sharma, S.K. Bharti. "Moving object detection in real-time visual surveillance using background subtraction technique", *2014 14th International Conference on Hybrid Intelligent Systems*, Gulf University for Science and Technology, Kuwait, December 14–16, 2014.
35. L. Sharma, D.K. Yadav, M. Kumar. "A morphological approach for human skin detection in color images", *2nd national conference on "Emerging Trends in Intelligent Computing & Communication"*, GCET, Gr. Noida, India, 26–27 April 2013.
36. L. Sharma, S. Sengupta, B. Kumar. An improved technique for enhancement of satellite images. *J. Phys.: Conf. Ser.* 1714, 012051 (January 2021).
37. G. Jha, L. Sharma, S. Gupta. Future of augmented reality in healthcare department. In: Singh P.K., Wierzchoń S.T., Tanwar S., Ganzha M., Rodrigues J.J.P.C. (eds) *Proceedings of Second International Conference on Computing, Communications, and Cyber-Security*. Lecture Notes in Networks and Systems, vol. 203. Springer, Singapore, 2021. DOI: 10.1007/978-981-16-0733-2_47.
38. G. Jha, L. Sharma, S. Gupta. E-health in Internet of Things (IoT) in real-time scenario. In: Singh P.K., Wierzchoń S.T., Tanwar S., Ganzha M., Rodrigues J.J.P.C. (eds) *Proceedings of Second International Conference on Computing, Communications, and Cyber-Security*. Lecture Notes in Networks and Systems, vol. 203. Springer, Singapore, 2021. DOI: 10.1007/978-981-16-0733-2_48.
39. S. Kumar, P. Gupta, S. Lakra, L. Sharma, R. Chatterjee. "The zeitgeist juncture of "Big Data" and its future trends", *2019 International Conference on Machine Learning, Big Data, Cloud and Parallel Computing (COMITCon)*, 2019, pp. 465–469. DOI: 10.1109/COMITCon.2019.8862433.
40. S. Sharma, S. Verma, M. Kumar, L. Sharma. "Use of motion capture in 3D animation: Motion capture systems, challenges, and recent trends", *2019 International Conference on Machine Learning, Big Data, Cloud and Parallel Computing (COMITCon)*, 2019, pp. 289–294. DOI: 10.1109/COMITCon.2019.8862448.
41. H. Cai. Sex difference and smoking predisposition in patients with COVID-19. *Lancet Respir. Med.* 8(4), e20 (2020).
42. W. Guan, Z. Ni, Y. Hu, et al. Clinical characteristics of coronavirus disease 2019 in China. *N. Engl. J. Med.* 382(18), 1708–1720 (2020).
43. S. Ruan Likelihood of survival of coronavirus disease 2019. *Lancet Infect. Dis.* 20(6), 630–631 (2020).
44. R. Verity, L.C. Okell, I. Dorigatti, et al. Estimates of the severity of coronavirus disease 2019: a model-based analysis. *Lancet Infect. Dis.* 20(6), 669–677 (2020).
45. N. Yu, W. Li, Q. Kang, et al. Clinical features and obstetric and neonatal outcomes of pregnant patients with COVID-19 in Wuhan, China: a retrospective, single-centre, descriptive study.

Lancet Infect. Dis. 20(5), 559–564 (2020); A.A. Kelvin, S. Halperin. COVID-19 in children: the link in the transmission chain. *Lancet Infect. Dis.* 20(6), 633–634 (2020).
46. J-M. Kim, Y-S. Chung, H.J. Jo, et al. Identification of coronavirus isolated from a patient in Korea with COVID-19. *Osong. Public Health Res. Perspect.* 11(1), 3–7 (2020).
47. A.C. Walls, Y.J. Park, M.A. Tortorici, A. Wall, A.T. McGuire, D. Veesler. Structure, function and antigenicity of the SARS-CoV-2 spike glycoprotein. *Cell.* 18(2), 281–292 (2020).
48. Q. Wang, Y. Qiu, J.Y. Li, Z.J. Zhou, C.H. Liao, X.Y. Ge. A unique protease cleavage site predicted in the spike protein of the novel pneumonia coronavirus (2019-nCoV) potentially related to viral transmissibility. *Virol. Sin.* 20, 1–3 (2020).
49. R. Lu, X. Zhao, J. Li, et al. Genomic characterisation and epidemiology of 2019 novel coronavirus: implications for virus origins and receptor binding. *Lancet.* 395(10224), 565–574 (2020).
50. R.A. Khailany, M. Safdar, M. Ozaslan. Genomic characterization of a novel SARS-CoV-2. *Gene Rep.* 100682, 1–6 (2020).
51. Mousavizadeh L, Ghasemi S. Genotype and phenotype of COVID-19: Their roles in pathogenesis. *J Microbiol Immunol Infect.* 54(2), 159–163 (2021 Apr).
52. B. Tang, N.L. Bragazzi, Q. Li, S. Tang, Y. Xiao, J. Wu. An updated estimation of the risk of transmission of the novel coronavirus (2019-nCov). *Infect. Dis. Model.* 5, 248–255 (2020).
53. J. Cai, W. Sun, J. Huang, et al. Indirect virus transmission in cluster of COVID-19 cases, Wenzhou, China, 2020. *Emerg. Infect. Dis.* 26(6), 1343–1345 (2020).
54. N. van Doremalen, T. Bushmaker, D.H. Morris, et al. Aerosol and surface stability of SARS-CoV-2 as compared with SARS-CoV-1. *N. Engl. J. Med.* 382(16), 1564–1567 (2020); Y. Chen, L. Chen, Q. Deng, et al. The presence of SARS-CoV-2 RNA in feces of COVID-19 patients. *J. Med. Virol.* 92(7), 833–840 (2020).
55. C.T. Huang, H.H. Lin, S.Y. Ruan, et al. Efficacy and adverse events of high-frequency oscillatory ventilation in adult patients with acute respiratory distress syndrome: a meta-analysis. *Crit. Care.* 18(3), R102 (2014).
56. J. Hindson. COVID-19: faecal-oral transmission? *Nat. Rev. Gastro. Hepat.* 17(5), 259–259 (2020).
57. F. Bénézit, P.L. Turnier, C. Declerck, et al. Utility of hyposmia and hypogeusia for the diagnosis of COVID-19. *Lancet Infect. Dis.* 20(9), 1014–1015 (2020).
58. A.K. Winter, S.T. Hegde. The important role of serology for COVID-19 control. *Lancet Infect. Dis.* 20(7), 758–759 (2020).
59. H.C. Metsky, C.A. Freije, T.S. Kosoko-Thoroddsen, P.C. Sabeti, C. Myhrvold. CRISPR-based COVID-19 surveillance using a genomically-comprehensive machine learning approach. *bioRxiv* (2020).
60. World Health Organization. Draft landscape of COVID-19 candidate vaccines. www.who.int/publications/m/item/draft-landscape-of-covid-19-candidate-vaccines.
61. Covid-19. Available at: https://www.bloomberg.com/news/articles/2020-07-21/u-k-scientists-tap-ai-for-better-ventilation-for-covid-patients [accessed on 20 May 2021].
62. Covid-19. Available at: https://www.mermaidcare.com/beacon-caresystem [accessed on 20 May 2021].

12

State of the Art of Artificial Intelligence in Dentistry and Its Expected Future

Vukoman Jokanović

ALBOS doo, Innovative Company, Belgrade, Serbia, Vinča Institute of Nuclear Sciences, Belgrade, Serbia

M. Živković and S. Živković

School of Dentistry, University of Belgrade, Belgrade, Serbia

CONTENTS

12.1 Introduction

The main part of any NN is an unnatural neuron, virtually presented by the corresponding nonlinear model that mimics a human neuron's functions. Arranging and connecting virtual neurons by corresponding mathematical operations creates an artificial neural network (ANN), which aims to solve specific tasks, such as the classification of radiographic images that clearly show the health state of a particular tooth. The term "deep learning" (DL) refers to multilayer architectures of NNs, which are suitable for accepting and processing complex data structures, primarily images, with all their topographic specifics. Given a series of mathematical limitations, NNs can present each function and each input to provide the desired output. During data processing, classified data are subjected to appropriate software and interactively optimized to diminish a prediction errors (disagreement between actual and expected outcomes). Based on such analyzes, similar conclusions can be drawn even for data that were not the subject of such analysis [1–4].

Image diagnostics, which occupies a central place in many diagnostic procedures in various health fields, is particularly suitable for the application of artificial intelligence (AI) and for overcoming the dependence of the obtained result on the examiner. It enables an increase in the efficiency of diagnostics while reducing its costs because it eliminates the need for performing routine tasks by medical staff. In addition, it enables much more

DOI: 10.1201/9781003244165-15

successful systematization of heterogeneous data, enabling quality integration of heterogeneous data groups, such as disease reports, socio-demographic, clinic, image and bio-molecular data, etc. Also, it facilitates various forms of research, and through administrative time savings, increases the time for direct contact between doctors/dentists and their patients [5–6].

AI humanizes the patient care process itself through aloud expression, articulation, and text identification, enabling physicians to shorten the required time to keep records, making better health care more personally, and helping patients be more actively involved in data collection process. Continual non-invasive observation of patients' healthiness and bearing enables a much better understanding of the cause of the disease in a patient, with reduced diagnostic and treatment costs, thus reducing the burden of the health systems in treating the aging society which is characterized by an increasing number of severe and chronic patients. It helps easily solve any problems of labor shortage through rationalization of the work of health workers, thus supporting the World Health Organization (WHO) to achieve its sustainable development goals.

In dentistry, convolutional neural networks (CNNs) are used mainly in dental radiography, since radiographic images play an important role in dental diagnostics, starting from dental screening to treatment, planning and implementation. In addition, dentists regularly use different imaging attachments, from the certain anatomical area of a given person, followed by non-imaging information such as clinic reports and basic and dental information and the possible existence of systemic diseases in a given patient and medication uses. Such information is frequently gathered at whole time intervals to be effectively integrated and networked with the assistance of AI to improve the quality of disease identification, recovery prognosis, and requested intervention [1,7].

Although numerous dental diseases (caries, periodontitis, etc.) are significantly widespread, dentistry has still only partially adopted AI technologies. This is probably because medical and dental information are not accessible to the extent that some other data are available due to medical rules of the information protection and the corresponding administrative barriers. Data are frequently locked in separate, interoperable systems. In addition, the sets of different data are not correctly classified, and are often unrepresentative, compared to other data sets in the AI area. Finally, the data related to all patients are complicated, multidimensional, and delicate, with terminated possibilities for their triangulation and validation. Medical and dental information, like electronic medical history, show an insufficient level of integrity, in which some important data are often systematically missing. Data choosing often conditions the bias of choice, requiring that applications developed based on such data are also biased [1,8].

In addition, data processing and measurement, and validation of the results derived from them are often insufficiently reproducible and rough. If such data are used for learning and verifying, it induces the bias in assessing the data [1,2,8].

12.2 The Main Challenge in the Use of Virtual Reality in Dentistry

New technologies, such as AI technology, try to mimic the human brain. The AI is an area of knowledge and engineering that deals with intelligent computer interpretation of reality and the creation of objects, like the behavior of the human brain. Applications of these technologies include areas such as expert systems and especially image recognition and robotics. Therefore, they become an integral part of everyday life. Finally, this technology has significantly improved the knowledge in recent medicine and dentistry [9–10].

Today, almost all residents globally possess vast amounts of documented medical and genetic information. Many of them are irrelevant, while some of them that establish a link between genes and susceptibility to dental disease are very important. This connection between a disease and relevant genetic information increasingly shows the extent to which such information is irreplaceable when looking at the correct diagnosis and the optimal way of treating dental diseases. Due to this, inclusion of computers and software in medicine and dentistry has become a necessity. Clinical Decision Support System (CDSS), which relies on a dynamic (medical) knowledge base and inference mechanisms, based on a set of rules implemented through language-based medical logic modules, is suitable for efficient resolution of various doubts and problems that arise during clinical practice [10,11].

One of the AI systems used to predict genetic disorders is the genetic algorithm (GA), which mimics natural evolution during a search. The most complex part of this algorithm is structuring information, which allows to identify and extract important information from scanned documentation. The ability of computer algorithms to learn from stored data helps healthcare staff globally to identify better approaches to the treatment of diseases, from the diagnostic stage to the end of treatment. Furthermore, sharing these essential data with healthcare workers worldwide is crucial to the development of medicine and dentistry [12,13].

Dentistry assumes the collection of the significant volume of information in combination with well-developed clinic practices. As one of the AI techniques, it uses augmented reality, which offers the user a computer-generated image and view of the real world. Her discovery simplified the procedure of distributing an aesthetic prosthesis, according to the patient's desires, because with its help, the patient can imagine a virtual prosthesis and request to be modified in real time. Knowing that virtual simulation of a three-dimensional image is very close to practice solution, that evidence reduces patient anxiety and acts as an effective means of non-pharmacological pain control [14,15].

Augmented and artificial reality is mainly used in dental education to stimulate clinic practice and decrease the number of risks influenced by patient exercises. It creates conditions for exceptional improvement of the quality of feedback, which the virtual patient can offer to the students. At the same time, interactive communication between students and virtual assistants allows students to objectively evaluate their work, which creates conditions for high-quality training. Studies on the efficiency of these systems have shown that when using this method of learning, students achieve a significantly higher level of competencies in a shorter time than when using traditional simulation methods and aids [16].

As a virtual assistant, this software solves many usual problems in clinical practice with greater accuracy, with less employed staff and less personal errors than traditionally applied methods, starting with booking and coordinating regular checkups, informing patients and dentists about possible greater genetic susceptibility (patients with various types of dental disease, which are indicated by the results of periodontal screening (patients with diabetes) or screening for oral cancer). The virtual assistant also helps to maintain more efficient paperwork and thus makes a clinical diagnosis and treatment plan accurate by providing the dentist with details of the patients like allergies they are susceptible to. It also helps to establish a general history, because it provides real-time information about the patients' previous therapy, which includes the use of corresponding drugs for different disorders such as cardiovascular disorder, the provision of emergency remote assistance in cases where a dentist is not at hand, etc. [17].

Virtual reality software allows us to create a complete highly detailed and extremely accessible virtual database for each patient, allowing the dentist to quickly access all relevant information (much faster and more efficiently than the usual method of collection, with all relevant attachments, from dental records to radiography). Due to its unique learning ability, this software can be trained to perform many other important tasks, especially when it is associated to the imaging systems such as magnetic resonance imaging (MRI) and CBCT, helping to identify many invisible deviations of the teeth. This condition is also used to accurately locate landmarks on the radiograph in cephalometric diagnosis, because it is very easy to obtain the necessary information and then compare them with the established databases to support dentists in decision-making. It also gives excellent results in pathology because it allows simultaneous scanning of a large number of sections to find some characteristic details, which play a crucial role in final diagnostic and clinical treatment [18].

In orthodontics, these computer programs can carry out numerous analyses on the basis of radiographs improving medical diagnosis and treatment. With the invention of intraoral scanners and cameras, the usual behavior of dental patients has become a thing of the past, because digital impressions are faster and more accurate, eliminating many unnecessary laboratory analyses and significantly decreasing the errors of reading images. Thanks to convenient AI software, the computer guides the dentist during the process of creating a digital impression and helps him realize the ideal impression. Based on the data entered into the system, it is possible to foresee even the outcome of complete treatment [19,20].

With the assistance of corresponding software, AI helps the dentist design the most appropriate aesthetic prosthesis by applying patient anthropological calculations, based on the facial measurements, ethnicity, and the patient's wishes, when the prosthesis is made. The prosthesis itself is currently made using computer-aided diagnosis (CAD) CAM technologies, using subtractive milling and additive production technology such as 3D printing. This technology improves classical conventional casting, significantly reducing defects in the prosthesis. Such technology is also applied for making orthodontic plates. Furthermore, in the area of implantology and surgery, these algorithms help to plan operations to the superfine details before the actual operation. Besides, one of the most significant applications of AI is in the area of oral and maxillofacial robotic surgery [21,22].

The key challenge in the field of robotics is the simulation of the movement of the human body. AI has dramatically improved the field of surgery, which is now very widespread in the developed countries of the world. Besides, one of the most intriguing uses of AI is related to the bioprinting of tissue and even organs, by arranging thin cell layers that can be used to reconstruct lost oral hard and soft tissues caused by its pathological disorder [23].

12.3 Fundamentals of Artificial Intelligence and Its Performances in Dentistry

Intelligent systems, especially DL, are systems that successfully mimic the cognitive functions of people in problem-solving. They are based on software models, capable of thinking and operating logically, like the human brain, learning from the available data to predict possible solutions. AI is exceptionally efficient in radiology due to its capability to register anomalies in radiographic figures that are not noticeable with a dentist eye. In that way, fast recording and diagnosis reduce not only the workload of the radiologist, but also the risk of cognitive bias. AI systems thus complement other modalities and imaging in setting adequate diagnoses and dental treatment, while in the area of maxillofacial radiology, they enable a quality analysis of complex images, exact place of landmarks, determination of bone structure, risk assessment of oral cancer, and evaluation of metastatic lymph nodes, periapical and pathologies of the maxillary sinus [9,24].

Radiologists play the main role in clarifying 2D and 3D images. Until the advent of AI, the interpretation of images has been biased. The most often applied imaging procedures for detecting and diagnosing dental diseases are plain radiography (PR), computed tomography (CT), MRI, positron emission tomography (PET), and ultrasonography (US), with the apparently wider applications of these procedures for analysis greatly increasing the amount of data to be interpreted and thus the workload of the radiologist. AI, machine learning (ML), and DL exceptionally improved our approach to data. They are a powerful tool for radiologists in analyzing large amounts of diagnostic figures, which, with enhanced velocity and precision of diagnosis, successfully detect abnormalities on the teeth [24].

Since AI is the ability of machines to mimic the brain's cognitive functions, it is easy to conclude that it is based on learning, using computer algorithms to solve various real problems. ML uses improved iterative procedures to detect features that are legally repeated in data. It is basically a nonlinear process, explained over linear algebraic data arrangements. ML presents the capability of computers or smart equipment, such as robots, to explain brain-like intelligence based on its ability to learn, make optimal decisions based on its experience, and improve its experience. ML problems are related to the input data or consequences that are a product of the learning system itself. It allows us to enhance foresight performance and the output model following data changes in real time. In addition, it provides quantitative analysis of the data of diagnostic imaging, based on enhanced figure quality, shortening the period of the data collection, optimizing the course of the clinical procedure, and improving communication between the dentist and the patient [9,24,25].

DL is a group of AI systems consisting of the algorithms of multiple sheets mutually interacted and layered into numerous significant levels, forming an extensive ANN. Currently investigated DL software containing CNNs uses self-enhancing back-propagation algorithms, which are created directly from data through a process of continuous data delivery, over and over again, so that optimal prediction can eventually be realized. These software are often applied to process large and complex figures obtained by 2D radiography or 3D CT. This is the difference between DL and ML, which creates algorithms exclusively by focusing on specific data [26,27].

ANNs are software that mimic the human brain's behavior, consisting of nerve cells mutually interconnected in networks of nerves. The dendrites of these cells accept given input signals, transmitting them to their nucleus, known as a soma, where these signals are processed and transmitted to the axon, which transmits further signal to the synapse, after which the signal is transferred again to the dendrite of the next neuron. ANNs are similar to neural networks (NNs) in the human brain, but these networks are artificial, given in forms of various software. ANNs consist of at least three main sheets (input, hidden, and output) that are mutually interacted. The input sheets accept the input sign and submit it to other sheets without processing. The next sheet processes the input signal and then directs it to the next sheets to continue processing, so that connection spreads signals from one sheet to another, wherein the first exciting sheet is known as the perceptron. Hidden sheets of artificial neurons mimic neurons in the brain. The NN itself performs a calculation based on a set of input data, creating the output by using the specific algorithms. If necessary, the obtained output signal can be returned to the input, instead of the output signal being non-feedback, linear [26–28].

CNNs are basically DL networks, consisting of several constituents mutually harmonized, where the convolution sheets, i.e., the unification sheets, are entirely interconnected. The CNNs enhance their knowledge directly from the data, recognizing the features within the figures, and select the obtained result into appropriate groups based on the subject, which is observed. The first two sheets (convolutional and merged sheets) perform property extraction: fully interconnected sheets provide the end result, while the last excited sheet classifies the separated functions into specific groups. Learnable parameter sets (kernels) are used to each figure location, with such cores optimized by the convolution sheet learning process to minimize mismatch of output and baseline values, using an optimization algorithm known as feedback. Hyper-parameters, like the arbitrary number and kernel size, are requested before the CNN learning begins. The aggregation sheet processes from top to down sampling operations, decreasing the number of parameters to be learned, to map fully connected sheets after down sampling to the last output to which the excitation function is used, classifying the obtained information. Thus, by a CNN, for example, a tumor could be categorized as benign or malignant by reducing the number of parameters to be learned so that after sampling downwards, fully connected layers are mapped to the final output to which the activation function is applied, which classifies the obtained information into categories. Thus, by a CNN, a tumor could be categorized according to malignancy, so that after sampling downwards, fully interconnected sheets are mapped to the total output. On the basis of this output, the classification is done, and the belonging category is determined [10,29,30].

ANNs are learned and tested using different data sets (validation data set) before their practice implementation. During learning, each network excitation is evaluated by a numerical scale of values, which shows the connection strength of these values in relationship with output changes during learning. Multiple circles of learning and testing on various data sets help us enhance artificial networks' prediction performance. Increasing the number of data implies primarily converting the data in training process so that the network constantly improves its input during further iteration during the training to decrease the number of errors. CNNs very often request multiple circles of learning to reach requested diagnostic precision [1,31].

Recently three kinds of AI systems approaches are known: brain simulation, symbolic and sub-symbolic, and statistical access. Symbolic access assumes a cognitive simulation of logic, and knowledge-based intelligence. Various computer approaches design the base of sub-symbolic accesses. Cognitivism assumes the improvement of various rules-based algorithms known as expert systems, which mimic human intelligence on a machine, trying to solve challenging real-world tasks. Cognitive systems are based on probability. Their goal is to increase the efficiency of human intelligence related to very hard problems of finding the best outputs from unrelated data and directing them to create appropriate hypotheses to make the best possible recommendations or decisions [1,10,28–31].

The modern approach, known as connectionism (connectivity), is based on the so-called naive (insufficiently researched) programs spontaneously driven by data. Such programs are bound to each other in CNNs mimicking functions of the neurons and synapses in the brain. Such programs possess the ability to process big amounts of data. Therefore, CNNs possess the capability to learn faster than previously defined expert systems. New CNNs programs applied for DL contain multiple sheets of algorithms that extract typical features from the scattered, unorganized data. They include supervised and unsupervised, and unattended learning. Supervised learning is associated with the operation of calculating and then the error is regulated to obtain the best result, using a function derived from the manually marked data included in the training and which rely on a number of examples of such training. Moreover, these programs are trained to distinguish regular and irregular properties by analyzing hand-assigned figures, providing a corresponding classification, in which the output variable is, for example, whether or not there is disease [2,32].

Unattended learning is some kind of ML based on the entered data. Learning data are unmarked and sheared into various clusters, generating the so-called clustering algorithms. The main constrain is a lack of the explicit accuracy assessment of the results, which are different from those obtained by supervised learning. Consolidation of knowledge assumes results obtained through constant interaction with the surrounding to achieve certain aims, with an award for the correct result and a licking for the incorrect result. In contrast, unsupervised learning uses unmarked input data, from which the algorithms themselves infer which datum is essential. The most common algorithms used are data clustering algorithms and a priori algorithms that involve spotting and determining/learning the necessary rules, to process big data amounts. These algorithms use AI procedures to collect experience straight from the data. With the increased volume of data, the learning performance of these algorithms increases. It means that more images and scans with a clinically approved diagnosis induce much better machine identifying of pathology or abnormalities inside the figure [2,33].

There are several types of software learning algorithms. Some of them use linear regression, and the code establishes connections among dependent and independent variables. Software based on logistic regression gives the probability of an event occurring. A tree algorithm is a type of supervised learning algorithm that is used for problem-solving and classification. It is a flowchart-like structure, where each internal node represents a test of one of the attributes, while each branch represents the test outcome, with each node or node terminal having a class designation. In support vector algorithms, each data item is plotted as a point in n-dimensional space, where n represents the number of features. Then, these programs find a hyper-plane clearly selecting the dots associated with specific data. The size of the hyper-plane is related to the number of intake characteristics. If this number is 2, only one line is reduced hyper-plane. If the number is 3, then the hyper-plane is reduced to 2D plane, while the hyper-plane geometry becomes challenging to present when the number is greater than 3 [2,14,15].

Fuzzy logic is an approach to computing based on the degrees of truth, rather than the usual true or false (1 or 0) logic. Therefore, it mimics human behavior, giving a certain result as partially true or false. It assumes that each output belongs somewhere between true and false, with different shades of green and blue instead of a completely green and blue. Fuzzy learning is a kind of AI used in medicine to diagnose, for example, diabetic neuropathy, or determine the required dose of the drug, based on the calculation of the brain volume, using MRI images. Based on such an approach, it is also possible to characterize ultrasound images or CT images of certain organs of the human body. Because these information are between partially correct and partially incorrect values, this method has limited application [19,21,33].

Application of AI in radiology includes calculating the probability of the illness (yes or no) and scenarios of replacing the radiologist with AI due to a more precise, faster, and repeatable software result. Finally, additional scenarios assume the use of CNNs, with imaging findings to help radiologists interpret obtained results. In addition to such possible realization, clinical applicability implies an additional division aimed at detecting and identifying abnormalities in the figures invisible to the eyes. In addition, it includes automatic subdivision, which collects data related to the degree of potential disorder of tissues, severity of the illness that affected them, to save the radiologist's time due to the reduced need for manual segmentation. In addition, it enables the classification of the disease, in which the abnormality in the figure is grouped into a category (big or small risk). In this way, CNNs over DL help dentists in their diagnosis and treatment, gaining their applicability in the medical area, although only a few studies have this topic as a significant subject in dentistry [21,33,34].

CAD is the first medical application of AI in oral radiology. Earlier, it was shown that this method is very suitable for the detection of cancer in internal organs of the human body. The CAD method diagnoses the specific disease for which it is specially trained, and the characteristics of CAD algorithms can only be enhanced by adding new data sets. Nevertheless, although the performances of CADs have not advanced enough to completely replace radiologists, their output is often used as an important information in discovering complex pathology states. This means that the radiologist can agree or disagree with the computer output, but when radiologist is in doubt, CAD can enhance his diagnosis. The essential constraint of CAD is the high amount of false-positive detections, and in some cases, CAD can omit some disordered places. Today, AI-based CAD methods are replaced by self-learning, which diagnoses and treats problems using cognitive computational algorithms (CCA). Such systems reduce the workload of medical staff by eliminating the necessity of manual grouping, providing crucial information about functions of the human tissues, and giving the most common and very accurate information about the severity of the disease itself [21,33–36].

CNNs use automatic grouping for identification of segment-specific features in wide-scale data. U-net segmentation is the most popular method of CNNs grouping used in medicine to distinguish bone and soft tissues. Automatic radiographic findings enable the accurate localization of important details. This method is applied in CT and MRI to detect irregularity in figures unnoticed during visual observation. CNNs enable accurate identification of disease edges. Particular edge-based algorithms define disease regions well. CNNs are suitable for locating specific details in partly hidden areas that are overlapped and invisible to the eyes. CNN algorithms allow studying the whole area part by part, while knowledge-based algorithms discover new anatomical details very preciously [34–37].

Today, the use of computer technology is gradually increasing in order to help dentists in the treatment of oral diseases. Augmented reality (AR) technology is the first application of AI already broadly used. AR is based on an interactive knowledge in actual life, which allows transfer of a 3D virtual object to a 3D reality, while computer virtual content and the real environment overlap. AR method has numerous applications in various areas of medicine. This method is present in oral and maxillofacial dentistry, because it allows perfect operation planning with great precision. It is also used when placing dental implants, because it diminishes period and price of implantation. It is applied in osteotomy of facial skeletal proliferation and in medical staff's education. In addition, it facilitates the study of anatomy by live visualization of the operated body magnified by mirror figures [34–37].

12.4 Some Examples of the Application of Artificial Intelligence in Dentistry

ANNs particularly are important when the etiology of the illnesses is multifactorial, as in the case of recurrent aphthous ulcer, whose exact etiology is unknown. In this case, the diagnosis is based on its recurrence, excluding other factors or internal disorders, when radiological diagnosis is taken as a gold standard. In such cases, when the opinion of the dental surgeon is likened with a diagnosis given with the help of ANNs, the obtained results show high reliability. The situation is similar in diagnosing complicated oral diseases or radiology, where large quantities of patient data are registered by digital IOPA (intraoral peri-apical X-ray), 3D scanning, etc. In all these cases, AI provides a quick and reliable diagnosis and instructions for treating oral diseases. It is crucial for identifying patients at danger from developing oral cancer, especially in their early stages. Genetic software and ANNs are also successful in predicting the size of unaltered canines or premolars or predicting tooth surface loss [38,39].

Virtual dental assistants (VDA) based on AI can make different jobs in dental institutions with fewer errors and accuracy than the average trained operator. They make it easier to schedule a patient's meeting at the clinic, monitor the patient's previous dental history, take care of insurance, serve as aids to the dental surgeon in all his action and treatment, and warn dentists of common patient habits like smoking and drinking [39,40].

Recently, AI is applied in different stages of orthodontics, from dental diagnosis to dental treatment. 3D spots and virtual procedures are suitable for assessment of craniofacial and dental irregularities. By 3D scanning, straighteners can be printed and orthodontic procedure adjusted. After that, the aligner is printed; a specific algorithm enables to take decisions on how given tooth/teeth can be moved, and how much pressure is needed for this action. Pressure points are recognized for that particular tooth/teeth, which decreases the probability of error and speeds up the healing process [41,42].

Therefore, for the best possible patient prosthesis, there are several factors that the dentist must keep in mind, such as anthropological computations, facing measurements, aesthetics, and the patient's desires. The application of AI enables precise fitting of dentures, while systems based on CAD/CAM ensure finished dental prostheses with great precision [43, 44]. They are also used to design inleys, onlays, crowns, and bridges, thus replacing the conventional method of casting prostheses, while reducing production time and potential manufacturing errors [36,41,42,45].

The digital convolutional method uses 16 convolutional sheets and 2 completely connected sheets used for detection of the periodontitis of premolars and molars. It is effectively used in the categorization of patients who have severe and chronic periodontitis induced by their specific immune reaction. Additionally, AI is also present in the design of the dental chairs, which is strongly changed in comparison with a typical hydraulic. This chair possesses an electric and fully automated sensor for controlling chair movements, whereby it can also react to the dentist's voice, without requiring any physical effort. The future in which the dental chair will calculate the weight and other essential patient parameters during his sitting on it for treatment is expected [46–49].

12.5 Applications in Dental and Maxillofacial Radiology

Periapical granulomas, abscesses, and cysts (periapical lesions) are usually present in everyday oral clinical practice. All of them are visible on radiographs. However, there is a possibility that some of them will go unnoticed because the images do not have a good-enough contrast. It has been shown that even in such cases, AI can precisely identify areas of teeth prone to caries and other periapical pathologies, because they enable automated segmentation that defines more precise boundaries of lesions, which facilitates their differentiation. It is expected that these systems will be applied to implants in the near future to enable early detection of peri-implantitis and thus prevent the development of possible more serious complications in those places [50,51].

Maxillary sinusitis is an inflammation of the mucous membrane that is radio-graphically characterized by thickening of the mucosa >4 mm, lacking continuity of the bottom of the maxillary sinus, opacity, focal alveolar atrophy, and associated periodontal disease. CT and CBCT are the most suitable methods for detecting such indications. In conventional radiography, diagnostic inconveniences are influenced by the overlapping of the maxillary sinus, the sinus bone architectures of the face, which often gives false negative outcomes. DL algorithms enhance the diagnostic capability of conventional radiographic imaging, reducing the need for unnecessary frequent CT scanning of patients. In a diagnosis of maxillary sinusitis, DL systems show 87.5%–93% accuracy, which is better or at least comparable to the findings of experienced radiologists. Such help is especially valuable in diagnostic support to inexperienced radiologists [27,52].

Deep analysis tools help periodontists detect alveolar bone loss early and diagnose changes in bone density in the furcation region. The CNNs show exceptional diagnostic possibilities, reaching accuracy values significantly higher than those in usual clinical practice (81% compared with 76%). Such high diagnostic efficiency is achieved by applying specific automated systems, which are used as reliable auxiliary tools for the determination of periodontal bone loss on panoramic radiographs, thus reducing the diagnostic effort of radiologists. This method allows the identification of periodontal diseases more exactly, in much shorter period than classical procedures. Besides, ANNs are used to predict early risks of osteoporosis by analysis of bone structure and its mineral content (BMD), reducing the needs for frequent BMD analysis and uses of other densitometric methods [53,54].

AI is used for the early detection of head and neck cancers and their metastases in the lymph nodes, which significantly affects the kind of treatment and the prognosis of disease development in such persons. CT and MRI are imaging techniques often used to detect metastases in lymph nodes. The application of CNNs has improved these CT diagnosis. Such systems based on CNNs figure classification are characterized by high accuracy (78.2%), sensitivity (75.4%), and specificity (81.0%), which is on the level of findings of radiology experts. They use DL programs in calculation of the survival probability of patients with oral cancer. DL systems show much better diagnostic performance, from classical statistical models. They provide perfect recommendations, suggesting the most efficient treatments of patients with oral cancer, and address some diagnostic concerns among clinicians regarding whether an oral lesion is benign or potentially malignant, creating the preconditions for timely treatment intervention [55,56].

Osteoarthritis is a frequent degenerative disease of the temporomandibular joint (TMJ), following the destruction of articular cartilage and resorption of subchondral bone. Since a procedure for quantifying morphological disorders in the beginning of illness has not yet been developed, a non-invasive technique based on shape variation analysis has recently been used for this purpose, using ANNs to classify morphological disorders in 3D mandible condyle model, including the following categories: G1, normal; G2, almost normal; and G3–6, which describe different degrees of disorders. Rainbow mapping of 3D prototypes defines the real position of the morphological disorder on the condylar area. Thus, ANNs categorization improves clinicians' understanding of changes in temporomandibular joints [57,58].

Primary headaches, due to, for example, migraines and increased pressure, are not caused by diseases, while secondary headaches usually indicate an illness. Doctors give diagnoses of chronic headache only after clinical testing and on the basis of MRI images.

Based on such images, they usually suggest appropriate medical therapy, while in some patients, the diagnosis itself is much more complex and requires multiple consultations to resolve the permanent symptoms of the headache. For the precise diagnosis of headaches, using AI, an automated system for its classification has been developed, which uses an appropriate learning algorithm. Such a methodology shows exceptional accuracy (97.85%) and significantly reduced errors in the classification of headaches ($P<0.05$). It enables the management of headache triggers by automatically detecting them, allowing patients to balance their behavior to prevent consequences that cause headaches. Recently, the compression of vascular structures induced by different anatomic structures, such as elongation or medial/lateral deviation of the styloid and ponticulus posticus, is explored. It was found that they cause a deceleration of cerebral blood flow which induces cervicogenic headache. It was shown that CNNs are very efficient in clarification of vertebrobasilar deficiency which causes cervicogenic headache [59,60].

Sjögren's syndrome is a long-term autoimmune disease that affects salivary and lacrimal glands and often seriously affects other organ systems, such as the lungs, kidneys, and nervous system, with main associated symptoms of severe dryness of skin, mouth and eyes, joint pain and stiffness, persistent dry cough, etc. Today, various algorithms based on AI allow the analysis of large data sets from various groups of patients suffering from this disease. The investigations show that the diagnostic accuracy of DL was significantly more reliable than the accuracy of that of radiologists with insufficient experience [61].

In endodontics, the researchers approved that AI algorithms are extremely precious in locating the apical foramen and identifying vertical root fractures if CBCT images are used for its analysis. In addition, NNs have been found to improve proximal diagnosis of caries by 39.4% compared with bite radiography. Finally, AI algorithms help prosthetists enhance the crown's consistency and aesthetics by using appropriate templates. Prosthesis fabrication is performed using the corresponding software for design and production, such as ablative milling and application of additives, which diminish errors in the final products. AI helps orthodontists to determine the need for tooth extraction, monitor tooth mobility, etc. They help clinicians make more accurate digital impressions and thus save time by reducing unnecessary laboratory examinations [62–64].

AI has changed the area of oral and maxillofacial surgery based on the appearance of image-guided surgery. Before surgery, CT, MRI, and CBCT images are now parts of clinical practice in many large hospitals. Such recordings allow the desired procedure to be derived more exactly than before the introduction of these techniques. Surgical removal of the lower third molars is challenging due to the great proximity of the third mandibular molar (M3) and the inferior alveolar nerve (IAN). Such interventions may cause neurosensory injury of the chin and lower mouth. The automated segmentation of panoramic images prior to extraction of M3 determines the proximity of M3 to neurosensomotor tissues, thus preventing damages to the tissues. In the future, these types of interventions will have an increasing application in orthognathic surgery influenced by extraordinary accuracy of image recognition, which shows different dentofacial irregularities [65, 66].

Examination of skeletal remains is particularly significant in forensic tests for identification of victims. The teeth are significant for their identification, because they are very persistent to decay and stand unchanged after the disconnection between soft and skeletal tissues. ANNs have shown an accuracy of about 95% in identifying the sex on 900 anthropological skulls after reconstruction by CT. In addition, ANNs have been shown to improve the accuracy of age estimation methods. With multilayer NNs, the percentage success rate was up to 99.9%, from which it follows that methods based on NNs are reliable for determining the age and sex of skeletal remains [66].

Three-dimensional printing, a manufacturing process in which objects are made by layering by layer-by-layer printing, is one area of application of AI, which helps to optimize grids in a CAD file, estimating the most effective modality of printing, with detection of any printing errors. AI algorithms anticipate expected errors during printing and solve the problem by creating new sheets of support to remove all contradictions between sheets, looking for a solution and making final decision when people are unable to find it in real time, while solving the problem of filling layers of uniform density in a quality manner is not possible [1,2,4].

AI provides essential information about tissues' functional behavior and severity of given illness, facilitating early cancer screening and detecting patients at its risk. The dilemma is only will this technology be substitute for radiologists in near future, even though AI has diminished the workload of radiologists, because they still cannot completely replace human intelligence. In addition, such approaches request a serious knowledge base. When it is applied to images from a different context, they can be misinterpreted, resulting in false positives or false negatives. The optimal solution for radiologists is to be trained to be able to apply computer algorithms in practice. This includes their basic training and understanding of biometric and genomic databases so that they can improve the application of this technology [1,2,4].

Radiomics is a new translational area of investigations whose goal is to withdraw multidimensional data from radiographic figures by specific algorithms. The radiomic process can be sheared into different stages, with defined inputs and outputs, like image collection and reassembly, figure segmentation, modification and identification of its characteristics, and analysis and development of appropriate models. These programs can extract approximately 400 patterns from CT and MRI tumor images, like surface form, magnitude, and roughness, unnoticed by the naked eye. Therefore, this is a very efficient method for predicting the metastatic probability of tumors and their oncogene potential and assessing the patient's response to therapy, which is applied to him [37,38].

Biobank Images is an application that enables the storage of a large amount of data and images. The biobank for image processing contains a library of biological materials and interconnected data that are intended for a large population of patients in an organized manner. Such data can be used in algorithms to simulate various diseases' progression [368,39,67].

Hybrid intelligence can be defined as a combination of human and machine intelligence. It expands possibilities of the human capabilities. It assumes harmonious cooperation of people with intelligent machines in order to solve various complex problems. It is applied to merge multimodal images for better diagnosis and medical treatment. Boundary edges are important for any figure because the resulting input is detected and registered, which is processed within the appropriate NN [40,68].

12.6 Other Applications

In order to handle the huge increase in patient data, appropriate intelligent software for their processing and storage is needed. These data are very diverse and consist of data related to patients' history to their processing, which aims to diagnose the disease in patients. Because of its remarkable ability to store and process such data in the desired way, AI has found many applications in dentistry and medicine, although it still cannot substitute the functions of dental surgeons [40,68].

The ANNs predict the need of teeth extraction during the orthodontic regime of dental treatment. Analysis of data related to tooth restoration shows the differences in the behavior of various materials for tooth restoration, as an important factor determining the lifespan of their restoration [69].

Genetic algorithms are successfully applied for the solvation of hard dental problems. These stochastic investigation methods based on the principle of survival of the fittest include stages based on mutation, inheritance, selection, and crossbreeding in examination of optimal problem-solving. The essential improvement of genetic algorithms is their clear orientation to solving problems. They are a powerful tool for optimizing a possible solution, because they are based on simple rules, which facilitates their application. Such algorithms are used to optimize dental implants in order to diminish the possible problem of mechanical fracture of the implants and ensure its extended term of strength. Genetic algorithms are suitable for detecting caries in the beginning to avoid more serious tooth disease. It is also used to reconstruct missing parts of teeth, reconstruct their surface, and maintain the smoothness of teeth [45,70].

AI with its data surfing and identification capabilities possesses numerous advantages over clinic and traditional analytical decision methods. The obtained algorithms are more exact and thus help doctors to make more successful diagnosis and provide improved health care. This diagnosis includes the observance and testing of the patients, collection and interpretation of their data, knowledge and experience of medical staff, and formulation of diagnosis and treatment by medical staff. AI enables more orderly, more systematic, and structured data and also enables significantly more efficient care while reducing treatment costs [71,72].

AI also provides an additional opportunity to train dentists. Integrated with MRI and CBCT, it can notice the slightest deviations from usual appearance, which allows precise localization of smaller apical foramen by increasing radiographic accuracy and more successful diagnosis of proximal caries [36,73,74].

12.7 Conclusions

This chapter describes the AI systems used in dentomaxillofacial radiology. Their exceptional usability in all areas of dentistry has been shown in making a fast diagnosis and treatment for unusual problems that are very difficult to solve without their help. It has been shown that these systems have a remarkable effect on reducing the workload of radiologists, enabling them to improve the relevance and validity of their diagnoses. It has been shown that the clinical applications of AI systems are practically unlimited, although their application is still in their infancy. Their importance in various fields of dentistry and maxillofacial radiology is especially emphasized.

The way and the essential aspect of their functioning are shown. Various forms of AI used in all branches of dentistry are also presented, and their important specifics are shown. It has been shown that they represent the future in dentistry that is already ahead of us, because they are highly efficient and reliable in giving diagnostic recommendations, establishing a treatment plan for complex problems that are often unsolvable without its application.

In addition, these systems have been shown to save radiologists' time, enabling radiologists to enhance their competence and efficiency significantly. It is obviously necessary to educate radiologists about the use values and the way of manipulating the appropriate computer software to apply these methods in a safe way. It has been shown that the clinic applications of AI are essentially endless, although the development of AI for use in medicine and dentistry is still in their infancy.

References

1. Jokanović V. Chapter 4: Smart healthcare in smart cities, In: *Towards Smart World: Homes to Cities Using Internet of Things* (Sharma, L. ed.), First edition. Taylor and Francis Group, Boca Raton, FL, (2020) pp. 45–73.
2. Jokanović V, Jokanović B. Chapter 27: Brain-computer interface: state of art, challenges and future, In: *Artificial Intelligence Technologies, Applications, and Challenges* (Sharma, L., Garg, P. eds.), Taylor and Francis Group, Boca Raton, FL, (2021) pp. 25–38.
3. Schwendicke F, Samek W, Krois J. Artificial intelligence in dentistry: chances and challenges, *J Dent Res*, 99, (2020), (7), 769–774.
4. El Naqa I, Ruan D, Valdes G, Dekker A, McNutt T, Ge Y, Wu QJ, Oh JH, Thor M, Smith W, Rao A, Fuller C, Xiao Y, Manion F, Schipper M, Mayo C, Moran JM, Haken RT. Machine learning and modeling: data, validation, communication challenges. *Med Phys*. 45, (2018), (10), e834–e840.
5. Gianfrancesco MA, Tamang S., Yazdany J., Schmajuk G. Potential biases in machine learning algorithms using electronic health record data. *JAMA Intern Med*, 178, (2018), (11), 1544–1547.
6. Israni ST, Verghese A. Humanizing artificial intelligence. *JAMA*, 321, (2019), (1), 29–30.
7. Schwendicke T, Golla T, Dreher M, Krois J. Convolutional neural networks for dental image diagnostics: a scoping review. *J Dent*, 91, (2019), 103226.
8. Watt RG, Daly B, Allison P, Macpherson LMD, Venturelli R, Listl S, Weyant RJ, Mathur MR, Guarnizo-Herreño CC, Celeste RK, Peres MA, Kearns C, Benzian H. Ending the neglect of global oral health: time for radical action. *Lancet*, 394, (2019), (10194), 261–272.
9. Bahaa K, Noor G, Yousif Y. The Artificial Intelligence Approach for Diagnosis, Treatment and Modelling in Orthodontic. 2011, INTECH Open Access Publisher.
10. Khanna SS, Dhaimade PA. Artificial intelligence: transforming dentistry today. *Ind J Basic Appl Med Res*, 6, (2017), (3), 161–167.
11. Sutton RT, Pincock D, Baumgart DC, Daniel C, Sadowski DC, Richard N, Fedorak RN, Karen I, Kroeker KI. An overview of clinical decision support systems: benefits, risks, and strategies for success. *npj Digit Med*, 3, (2020), (17), 1–10, doi:10.1038/s41746-020-0221-y.
12. Dutã M, Amariei CI, Bogdan CM, Popovici DM, Ionescu N, Nuca CI. An overview of virtual and augmented reality in dental education. *Oral Health Dent Manage*, 10, (2011), 42–49.
13. Lobo FG, Goldberg DE. The parameter-less genetic algorithm in practice. *Inf Sci*, 167, (2004), 217–232.
14. Yeager D, Paging HAL. What will happen when artificial intelligence comes to radiology? *Radiology Today*, 17, (2016), (5), 12.
15. Xie X, Wang L, Wang A. Artificial neural network modeling for deciding if extractions are necessary prior to orthodontic treatment. *Angle Orthod*, 80, (2010), 262–266.
16. Muñoz-Saavedra L, Miró-Amarante L, Domínguez-Morales M. Augmented and virtual reality evolution and future tendency. *Appl Sci*, 10, (2020), 322–345.
17. Lei X, Tu G, Liu AX, Li C, Xie T. "The insecurity of home digital voice assistants - amazon alexa as a case study," CoRR, vol. abs/1712.03327, 2017.
18. Higaki T, Nakamura Y, Tatsugami F, Nakaura T, Awai K. Improvement of image quality at CT and MRI using deep learning. *Japanese J Radiol*, 37, (2019), 73–80.
19. Kattadiyil MT, Mursic Z, AlRumaih H, Goodacre CJ. Intraoral scanning of hard and soft tissues for partial removable dental prosthesis fabrication. *J Prosthet Dent*, 112, (2014), 444–448.

20. Bearn DR, Chadwick SM, Jack AC, Sackville A. Orthodontic undergraduate education: assessment in a modern curriculum. *Eur J Dent Educ*, 6, (2002), 162–168.
21. Vecsei B, Joós-Kovács G, Borbély J, Péter Hermann P. Comparison of the accuracy of direct and indirect three-dimensional digitizing processes for CAD/CAM systems – An in vitro study. *J Prosthet Dent*, 61, (2017), (2), 177–184.
22. Murray MD, Darvell BW. The evolution of the complete denture base. Theories of complete denture retention–a review. Part 1. *Aust Dent J*, 38, (1993), 216–219.
23. Gu GX, Chen CT, Richmond DJ, Richmond DJ, Buehler MJ. Bioinspired hierarchical composite design using machine learning: simulation, additive manufacturing, and experiment. *Mater Horizons*, 5, (2018), 939–945.
24. Wu J, Liu X, Zhang X, He Z, Lv P. Master clinical medical knowledge at certificated-doctor-level with deep learning model. *Nat Commun*, 9, (2018), (1), 4352.
25. Jain K, Chen H. Matching of dental x-ray images for human identification. *Pattern Recognit*, 37, (2004), (7), 1519–1532.
26. Jaiswa M, Gupta N, Sin A. Study on patient's awareness towards role of artificial intelligence in dentistry. *Int J Health Sci Res*, 9, (2019), (10), 35–39.
27. Nagi R, Aravinda K, Rakesh N, Gupta R, Pal A, Mann AK. Clinical applications and performance of intelligent systems in dental and maxillofacial radiology: a review. *Imag Sci Dent*, 50, (2020), (2), 81–92.
28. Jung SK, o Kim TW. New approach for the diagnosis of extractions with neural network machine learning. *Am J Orthod Dentofacial Orthop*, 149, (2016), 127–133.
29. Khanagar SB, Al-ehaideb A, Maganur PC, Vishwanathaiah S, Patil S, Baeshen HA, Sarode SC, Bhandi S. Developments, application, and performance of artificial intelligence in dentistry e A systematic review. *J Dent Sci*, doi:10.1016/j.jds.2020.06.019.
30. Ramesh AN, Kambhampati C, Monson JRT, Drew PJ. Artificial intelligence in medicine. *Ann R Coll Surg Engl*, 86, (2004) (5), 334–338.
31. Dutã M, Amariei CI, Bogdan CM, Popovici DM, Ionescu N, Nuca CI. An overview of virtual and augmented reality in dental education. *Oral Health Dent Manage*, 10, (2011), 42–49.
32. Kikuchi H, Ikeda M, Araki K. Evaluation of a virtual reality simulation system for porcelain fused to metal crown preparation at Tokyo medical and dental university. *J Dent Educ*, 77, (2013), 782–792.
33. Seol YJ, Kang HW, Lee SJ, Atala A., Yoo JJ. Bioprinting technology and its applications. *Eur J Cardio-Thoracic Surg*, 46, (2014), 342–348.
34. Tandon D, Rajawat J. Present and future of artificial intelligence in dentistry. *J Oral Biol Craniofac Res*, 10, (2020), 391–396.
35. Kareem SA, Pozos-Parra P, Wilson N. An application of belief merging for the diagnosis of oral cancer. *Appl Soft Comput J*, 51, (2017), 1105–1112.
36. Yaji A, Prasad S, Pai A. Artificial intelligence in dento-maxillofacial radiology. *Acta Sci Dent Sci*, 3, (2019), 116–121.
37. Kalappanavar A, Sneha S, Annigeri RG. Artificial intelligence: a dentist's perspective. *J Med Radiol Pathol Surg*, 5, (2018), 2–4.
38. Park WJ, Park JB, History and applications of artificial neural networks in dentistry. *Eur J Dent*, 12, (2018), (4), 594–601.
39. Tunjugsari V, Sabiq A, Sofro ASM, Kardiana A. Investigating CDSS success factors with usability testing. *Int J Adv Comput Sci Appl*, 8, (2017), (11), 548–554.
40. Tripathi P, Malathy C, Prabhakaran M. Genetic algorithms based approach for dental caries detection using back propagation neural network. *Int J Recent Technol Eng*, 8, (2019), ISSN: 2277-3878.
41. Mago VK, Mago A, Sharma P, Mago J. Fuzzy logic based expert system for the treatment of mobile tooth. *Soft Tools Algor Biol Sys*, 696, (2011), 607–614.
42. Albuha Al-Mussawi RM, Farid F. Computer-based technologies in dentistry: types and applications. *J Dent*. 13, (2016), 215–222.

43. Devito KL, de Souza Barbosa F, Filho WN. An artificial multilayer perceptron neural network for diagnosis of proximal dental caries. *Oral Surg Oral Med Oral Pathol Oral*, 196, (2008), 879–884.
44. Choi IGG, Cortes ARG, Arita ES, Georgetti MAP. Comparison of conventional imaging techniques and CBCT for periodontal evaluation: A systematic review. *Imag Sci Dent*, 48, (2018), 79–86.
45. Naylor CD. On the prospects for a (deep) learning health care system. *JAMA*. 320, (2018), (11), 1099–1100.
46. Joda T, Bornstein MM, Jung RE, Ferrari M, Tuomas T., Zitzmann NU. Recent trends and future direction of dental research in the digital era. *Int J Environ Res Public Health*, 17, (2020), 1987, doi:10.3390/ijerph17061987.
47. Joda T, Waltimo T, Pauli-Magnus C, Probst-Hensch N, Zitzmann NU. Population-based linkage of big data in dental research. *Int J Environ Res Public Health*, 15, (2018), 2357–2362.
48. Joda T, Gallucci GO, Wismeijer D, Zitzmann NU. Augmented and virtual reality in dental medicine: A systematic review. *Comput Biol Med*, 108, (2019), 93–100. doi: 10.1016/j.compbiomed.2019.03.012. Epub 2019 Mar 15. PMID: 31003184.
49. Iskander M, Ogunsola T, Ramachandran R, McGowan R, Al-Aswad LA. Virtual Reality and Augmented Reality in Ophthalmology: A Contemporary Prospective. *Asia Pac J Ophthalmol (Phila)*, 10(3), (2021), 244–252. doi: 10.1097/APO.0000000000000409. PMID: 34383716.
50. Glick M. Taking a byte out of big data. *J Am Dent Assoc*, 146, (2015), 793–794.
51. Miyazaki T, Hotta Y. CAD/CAM systems available for the fabrication of crown and bridge restorations. *Aust. Dent J*, 56, (2011), (Suppl. 1), 97–106; Wong SH, Al-Hasani H, Alam Z, Alam A. Artifcial intelligence in radiology: how will be affected? *Eur Radiol*, 29, (2019), 141–143.
52. Hosny A, Parmar C, Quackenbush J, Schwartz LH, Aerts HJ. Artifcial intelligence in radiology. *Nat Rev Cancer*, 18, (2018), 500–510.
53. Kim HG, Lee KM, Kim EJ, San Lee J. Improvement diagnostic accuracy of sinusitis recognition in paranasal sinus X-ray using multiple deep learning models. *Quant Imag Med Surg*, 9, (2019), (6), 942–951.
54. Lee JH, Kim DH, Jeong SN, Choi SH. Detection and diagnosis of dental caries using a deep learning-based convolutional neural network algorithm. *J Dent*, 77, (2018), 106–111.
55. Krois J, Ekert T, Meinhold L, Golla T, Kharbot B, Wittemeier A, Dörfer C., Schwendicke F. Deep learning for the radiographic detection of periodontal bone loss. *Sci Rep*, 8, (2019), 8995–9001.
56. Chan S, Siegel EL. Will machine learning end the viability of radiology as a thriving medical specialty? *Br J Radiol*, 92, (2019), 20180416.
57. Zhang Z, Sejdić E. Radiological images and machine learning: trends, perspectives, and prospects. *Comput Biol Med*, 108, (2019), 354–370.
58. Radcliff K, Raphael IJ, Rasouli MR, Kepler CK, Albert TJ, Parvizi J. Pelvic incidence in patients with hip osteoarthritis. *Spine J [Internet]*, 12, (2012), (9), 68–69.
59. Chen H, Zhang K, Lyu P, Li H, Zhang L, Wu J, Lee CH. A deep learning approach to automatic teeth detection and numbering based on object detection in dental periapical films. *Sci Rep*, 9, (2019), 3840–3851.
60. Wight S, Osborne N, Breen AC. Incidence of ponticulus posterior of the atlas in migraine and cervicogenic headache. *J Manipulative Physiol Ther*, 22, (1999), (1), 15–20.
61. Naik M, de Ataide ID, Fernandes M, Lambor R. Future of endodontics. *Int J Curr Res*, 8, (2016), 25610–25616.
62. Sun Z, Zhang Z, Fu K, Zhao Y, Liu D, Ma X. Diagnostic accuracy of parotid CT for identifying Sjögren's syndrome. *Eur J Radiol*, 81, (2012), 2702–2709.
63. Murata M, Ariji Y, Ohashi Y, Kawai T, Fukuda M, Funakoshi T, et al. Deep learning classifcation using convolutional neural network for evaluation of maxillary sinusitis on panoramic radiography. *Oral Radiol*, 35, (2019), 301–307.
64. Kim Y, Lee K, Sunwoo L, Choi D, Nam C-M, Cho J, Kim J., Bae YJ, Yoo R-E, Choi B., Jung C, Kim JH. Deep learning in diagnosis of maxillary sinusitis using conventional radiography. *Invest Radiol*, 54, (2019), 7–15.

65. Rueda S, Gil JA, Pichery R, Alcañiz M. Automatic Segmentation of Jaw Tissues in CT Using Active Appearance Models and Semi-automatic Landmarking, In *Medical Image Computing and Computer-Assisted Intervention-MICCAI 2006* (Larsen, R., Nielsen, M. & Sporring, J. eds.). Springer, Berlin Heidelberg, (2006) pp. 167–174.
66. Bewes J, Low A, Morphett A, Pate FD, Henneberg M. Artifcial intelligence for sex determination of skeletal remains: application of a deep learning artifcial neural network to human skulls. *J Forensic Leg Med*, 62, (2019), 40–43.
67. Shiraishi J, Li Q, Appelbaum D, Doi K. Computer-aided diagnosis and artificial intelligence in clinical imaging. *Semin Nucl Med*. 41, (2011), 449–462.
68. Sharma, L., Garg, P.K. (Eds.). *Artificial Intelligence: Technologies, Applications, and Challenges* (1st ed.). Chapman and Hall/CRC, New York, (2021). https://doi.org/10.1201/9781003140351.
69. Wang J, Suenaga H, Yang L, Kobayashi E, Sakuma I. Video see-through augmented reality for oral and maxillofacial surgery. *Int J Med Robot*, 13, (2017), (2), doi:10.1002/rcs.1754.
70. Wang S, Summers RM. Machine learning and radiology. *Med Image Anal*, 16, (2012), (5), 933–951.
71. Moghimi S, Talebi M, Parisay I. Design and implementation of a hybrid genetic algorithm and artificial neural network system for predicting the sizes of unerupted canines and premolars. *Eur J Orthod*, 34, (2011), 480–486.
72. Saghiri MA, Asgar K, Boukani K K, Lotfi M, Aghili H, Delvarani A, Karamifar K, Saghiri A M, Mehrvarzfar P, Garcia-Godoy F. A new approach for locating the minor apical foramen using an artificial neural network. *Int Endontic J*, 45, (2012), 257–265.
73. Wang CW, Huang CT, Lee JH, Li CH, Chang SW, Siao MJ, Lai TM, Ibragimov B., Vrtovec T, Ronneberger O., Fischer P, Cootes TF, Claudia Lindner C. A benchmark for comparison of dental radiography analysis algorithms. *Med Image Anal*, 31, (2016), 63–76.
74. Gligorijević V, Malod-Dognin N, Pržulj N. Integrative methods for analyzing big data in precision medicine. *Proteomics*, 16, (2016), 741–758.

13

Analysis of Machine Learning Techniques for Airfare Prediction

Jaskirat Singh, Deepa Gupta, and Lavanya Sharma
Amity Institute of Information Technology, Amity University, Noida, India

CONTENTS

13.1 Introduction

Social media is now an important part of everyone's life. As a result, buying and selling stuffs online is normal these days, people use applications and websites to purchase stuffs satisfying their needs by sitting at home using online payment methods and other methods with proper secure ways. Analysis of specific product ideas can inform companies of the level of satisfaction a customer got with their services. The system of buying

DOI: 10.1201/9781003244165-16

tickets is used to purchase flight tickets days before the flight schedules prebooking for the flights that are planned to make a trip between two destinations in which a buyer can select the date, time, and a flight with a stoppage or no stoppage. Some aviation lines don't agree with the procedure because a plane organization might change the prices depending on the availability of tickets and demand for tickets. It helps in the booking of tickets from anywhere through application programming interfaces and online procedures and payment to avoid the last-minute rush for tickets [1–5].

Most of the airlines use difficult ways and rules for managing revenue for the execution of typical systems for estimation. Mostly, customers try to buy the tickets in advance of the departure date so that they will not have to pay high prices of tickets as the date comes closer and demands rise for the left tickets. But sometimes this results in spending more money than expected for the same booking. The ideal aim of the airways is to gain profit on the tickets, whereas the customer searches for the minimum rate as per their budget. But this system confirms a ticket to persons so that they could travel according to the date they have selected. At the point of purchasing tickets, the middle-upper class people look for cheaper and under-budget tickets. The rate of airline tickets at less cost is consistently rising due to competition among different airlines [6–8]. Many websites have been introduced on the internet for booking tickets at a cheaper cost. Using machine learning algorithms they usually show different tickets by different airlines for the same route based on the previous airfare prices.

The data collected is in the form of raw data with values of more than one data type but when the training of data will take place the machine learning (ML) algorithm requires the data in numerical format (i.e., integers). The data is required to be changed into the numerical format for reducing time complexity. So, this is not as simple as we think it is. It includes a lot of working on raw data to understand data by methods such as data visualization and data modeling. Different algorithms will be applied and we will get the predictions with their accuracy percentages and also the parameters required [9–16].

13.2 Background

It is not an easy task for a customer to buy a plane ticket at a very low cost. These few tricks are being investigated to determine the time and date to get airline tickets with the lowest billing rate. Most of these frameworks use state-of-the-art frameworks which are known as machine learning. To make the right decision before buying airplane tickets Gini and Groves violated the partial least squares regression modeling. Details are collected at key locations for booking travel experiences for 2019 in the months of March, April, May, and June. Additional data were compiled and used to monitor the demonstration relationships of the final model. Janssen has developed an aspiring model using the linear quantile blended regression process for the San Francisco–New York study where daily flights are provided by www.infare.com. Two key points, for example, the number of departure days and whether the flight is over a week or a working day, are considered a model promotion. The model accurately balances flights ahead of time from the date of flight. In any case, the model does not interfere with the wide range of time frames, closing the date of flight [17–21].

Wohlfarth introduced a model used to improve time while purchasing a ticket online ticketing based on major preparations known as masked point processors, information mining frameworks (strategy and collection), and countless test frameworks. The plot is proposed to transform a variety of recreational activities into integrated themes of a game program that can support the test of individual integration. This relevant topic is included in the discussion based on a close assessment of the behavior. The progress model measures the appropriate change schemes. A tree-based investigation was used to select the best planned collection and a short time later it was considered a disagreement on the movement model. A study conducted by Dominguez-Menchero suggests that the optimal purchase time depends on a nonparametric isotonic strategy for rebellion in a particular course, senders, and duration. The model offers the most satisfying number of days before buying a plane ticket. The model looks at two types of variables: travel and acquisition date [22–29].

13.3 Looking into the Data by the Team

It is a piece of work of more than just one IT professional. A proper cycle works for the thing we just do by entering our details in minutes. Many data science field IT professionals collaborate to gain insights from the data. First, a dataset of previous bookings of tickets of different airlines is collected. Which consist of features such as timing, destination, stoppage, airline, and dates. Then a business analyst will look into the data and some insights. Then a data scientist/analyst will collaborate with the business analyst to decide from where to extract the data (such as from third-party application programming interfaces and databases). Then the data scientist must check if the data is in the right format or not because before applying machine learning algorithms the data should be in the right format to make the process less complex and more precise. To make it ready for machine learning al gorithms data preprocessing, data wrangling, data cleaning, feature encoding, and other methods are applied to the raw dataset. For these procedures mostly two programming languages are used Python and R. But in this report, I used Python and its libraries to perform every step of converting raw data into the right format for analysis and after all the mentioned steps we will apply different algorithms to the model [22–27,30].

13.4 Data Collection and Preprocessing

13.4.1 Dataset

The dataset consists of 200 rows and 11 columns with different features as mentioned, such as airline, date of journey, source, destination, route, departure and arrival time, duration, total stops, and price for meals (Figure 13.1). The dataset is prepared accordingly to fulfill the needs. It is totally raw data as entered in the database, and it shows planned flight tickets with their price and every detail. We will process this raw data through various steps and procedures. So, we have to analyze this raw data and change it to a suitable form because machine learning algorithms only understand the numerical values.

13.4.2 Understanding Data and Preprocessing

As I have named the dataset "Data_Flight.xlsx" to "train_data" to easily access and to implement codes on it so it should be assigned after importing Python libraries such as pandas, numpy, seaborn, and matplotlib.

	A	B	C	D	E	F	G	H	I	J	K
1	Airline	Date_of_Journey	Source	Destination	Route	Dep_Time	Arrival_Time	Duration	Total_Stops	Additional_Info	Price
2	Multiple carriers	24/06/2019	Delhi	Cochin	DEL → HYD → COK	07:10	16:10	9h	1 stop	No info	6723
3	Multiple carriers	21/03/2019	Delhi	Cochin	DEL → BOM → COK	08:00	15:30	7h 30m	1 stop	No info	8307
4	GoAir	21/04/2019	Delhi	Cochin	DEL → HYD → COK	05:55	13:35	7h 40m	1 stop	No info	6146
5	Jet Airways	27/05/2019	Delhi	Cochin	DEL → AMD → BOM	13:25	12:35 28 May	23h 10m	2 stops	No info	15129
6	Jet Airways	1/05/2019	Kolkata	Banglore	CCU → BOM → BLR	06:30	12:00	5h 30m	1 stop	No info	14388
7	Jet Airways	6/04/2019	Banglore	Delhi	BLR → DEL	06:00	08:45	2h 45m	non-stop	In-flight meal not included	4544
8	Jet Airways	12/04/2019	Banglore	Delhi	BLR → DEL	11:10	14:05	2h 55m	non-stop	In-flight meal not included	5198
9	IndiGo	18/05/2019	Delhi	Cochin	DEL → BOM → COK	09:15	01:30 19 May	16h 15m	1 stop	No info	7501
10	GoAir	01/03/2019	Banglore	New Delhi	BLR → GOI → DEL	15:45	20:15	4h 30m	1 stop	No info	12314
11	Air India	15/05/2019	Delhi	Cochin	DEL → BOM → COK	13:00	19:15	6h 15m	1 stop	No info	8372
12	Jet Airways	21/05/2019	Kolkata	Banglore	CCU → BOM → BLR	16:30	23:35	7h 5m	1 stop	No info	14781
13	Jet Airways	24/06/2019	Banglore	Delhi	BLR → DEL	19:50	22:50	3h	non-stop	No info	8016
14	IndiGo	12/03/2019	Chennai	Kolkata	MAA → CCU	07:55	10:15	2h 20m	non-stop	No info	6297
15	Jet Airways	1/05/2019	Kolkata	Banglore	CCU → BOM → BLR	08:25	09:20 02 May	24h 55m	1 stop	No info	14388
16	GoAir	1/06/2019	Delhi	Cochin	DEL → AMD → COK	07:25	13:35	6h 10m	1 stop	No info	6391
17	Jet Airways	9/06/2019	Kolkata	Banglore	CCU → BOM → BLR	21:10	10:05 10 Jun	12h 55m	1 stop	No info	14939
18	Jet Airways	21/05/2019	Kolkata	Banglore	CCU → BOM → BLR	08:25	19:50	11h 25m	1 stop	No info	14781
19	IndiGo	6/03/2019	Delhi	Cochin	DEL → BOM → COK	07:30	21:00	13h 30m	1 stop	No info	14871
20	SpiceJet	6/04/2019	Banglore	Delhi	BLR → DEL	09:30	12:20	2h 50m	non-stop	No check-in baggage include	4319
21	Jet Airways	15/05/2019	Kolkata	Banglore	CCU → DEL → BLR	17:00	09:45 16 May	16h 45m	1 stop	No info	13067
22	Jet Airways	01/03/2019	Banglore	New Delhi	BLR → BOM → DEL	16:55	08:15 02 Mar	15h 20m	1 stop	No info	26890
23	Jet Airways	15/03/2019	Banglore	New Delhi	BLR → BOM → DEL	07:00	13:15	6h 15m	1 stop	In-flight meal not included	9134
24	IndiGo	3/04/2019	Banglore	Delhi	BLR → DEL	21:15	00:15 04 Apr	3h	non-stop	No info	3941
25	Air India	24/06/2019	Delhi	Cochin	DEL → BOM → COK	07:00	19:15	12h 15m	1 stop	No info	8879
26	Jet Airways	24/05/2019	Kolkata	Banglore	CCU → BOM → BLR	18:55	12:00 25 May	17h 5m	1 stop	No info	14571
27	IndiGo	15/06/2019	Kolkata	Banglore	CCU → BLR	09:35	12:10	2h 35m	non-stop	No info	5644
28	Jet Airways	9/06/2019	Kolkata	Banglore	CCU → BOM → BLR	06:30	04:40 10 Jun	22h 10m	1 stop	In-flight meal not included	9663
89	Jet Airways	1/06/2019	Delhi	Cochin	DEL → NAG → BOM	06:45	04:25 02 Jun	21h 40m	2 stops	In-flight meal not included	10919
90	IndiGo	3/03/2019	Kolkata	Banglore	CCU → BLR	20:25	23:05	2h 40m	non-stop	No info	6038
91	IndiGo	21/05/2019	Delhi	Cochin	DEL → COK	05:35	08:50	3h 15m	non-stop	No info	5601
92	IndiGo	9/03/2019	Delhi	Cochin	DEL → BOM → COK	07:30	21:00	13h 30m	1 stop	No info	12351
93	Air India	15/04/2019	Delhi	Cochin	DEL → COK	05:10	08:00	2h 50m	non-stop	No info	6094
94	Multiple carriers	15/05/2019	Delhi	Cochin	DEL → BOM → COK	10:35	19:15	8h 40m	1 stop	No info	7575
95	Jet Airways	27/05/2019	Delhi	Cochin	DEL → BOM → COK	21:50	04:25 28 May	6h 35m	1 stop	No info	16079
96	Jet Airways	15/06/2019	Delhi	Cochin	DEL → BOM → COK	19:15	19:00 16 Jun	23h 45m	1 stop	In-flight meal not included	10262
97	Air Asia	9/06/2019	Delhi	Cochin	DEL → BLR → COK	16:45	22:25	5h 40m	1 stop	No info	13470
98	Jet Airways	27/06/2019	Banglore	Delhi	BLR → DEL	11:10	14:05	2h 55m	non-stop	No info	8016
99	Jet Airways	9/03/2019	Delhi	Cochin	DEL → BOM → COK	18:15	18:50 10 Mar	24h 35m	1 stop	No info	16289
00	Jet Airways	6/06/2019	Delhi	Cochin	DEL → BOM → COK	19:45	04:25 07 Jun	8h 40m	1 stop	In-flight meal not included	10262
01	IndiGo	27/06/2019	Banglore	Delhi	BLR → DEL	18:25	21:20	2h 55m	non-stop	No info	4823

FIGURE 13.1
Dataset based on previous flight bookings.

In this step, we have to deal with mostly two things:

A. **Missing Values**

First, we have to check the missing values in the raw dataset. Using Python libraries the command used for searching missing values, Program line to find missing values:

train_data.isna().sum()

Now we have to deal with the missing values. As per the above result, only Route and Total_Stops have one missing value; so, we have to apply a code line to correct it [13]. Program line to correct missing values:

train_data.dropna(inplace=True)
train_data.isna().sum()

B. **Data Cleaning to Make Our Data Ready for Analytics as well as Modeling**

First, we have to check the data type:

train_data.dtypes

Second, we have to change the data types of times. In order to change we have to declare a function in which the algorithm of changing data type is mentioned. So, the function is:

def change_into_datetime(col):

```
1  train_data.dtypes

Airline                    object
Date_of_Journey    datetime64[ns]
Source                     object
Destination                object
Route                      object
Dep_Time           datetime64[ns]
Arrival_Time       datetime64[ns]
Duration                   object
Total_Stops                object
Additional_Info            object
Price                       int64
dtype: object
```

FIGURE 13.2
Data types of each feature.

train_data[col]=pd.to_datetime(train_data[col])
Now, use loop for changing the data type of the whole column.
For i in ['Date_of_Journey',' Dep_Time', 'Arrival_Time']:
change_into_datetime(i)
After using this function run the code line train_data.dtypes again and you will get the results in which the data types of times will be changed (Figure 13.2).

13.4.3 Extracting Derived Features from the Data

As it is difficult for a machine learning algorithm to understand the dates and time entered in the raw data so we have to split it into days, months as well as years from the journey date, arrival time, duration, etc.

After applying the Python code for splitting up the column into the result is shown in Figure 13.3a.

This will help in reducing time and space complexity and more precisely the algorithm will work on this dataset. Similarly, we have modified the duration column for better understanding [5].

As we know the duration of flights is calculated in hours and minutes. Therefore, suppose the duration time for the flight from Delhi to Cochin is 19 hours then it will show 19 hours 0 minutes, because there may be many duration times which contain some minutes too. For example, the duration time of the flight from Bangalore to Delhi is 2 hours 50 minutes. So, now both the duration time is in the same format whether the duration time has any of the parameters as zero. It will be shown in the same format. An output after implementation is shown in Figure 13.3b.

And to perform this method on data we have used the split function to split the duration considering the duration time as a list which has two parameters—hours and minutes—where the values can vary. The algorithm behind splitting the duration is

IF: Length of duration=2 (after splitting), then the same value will be printed.
ELSE: IF there is an 'h' (for hours) in the length of duration.
THEN, duration=duration+'0m'

And now, the Duration column in the dataset is further divided into two columns 'Duration_hours' and 'Duration_minutes' for better understanding while the machine learning algorithm works on the dataset.

(a)

journey_month	journey_day	Dep_Time_hour	Dep_Time_minute	Arrival_Time_hour	Arrival_Time_minute
3	24	22	20	1	10
1	5	5	50	13	15
9	6	9	25	4	25
12	5	18	5	23	30
1	3	16	50	21	35

(b)

	Airline	Source	Destination	Route	Duration	Total_Stops	Additional_Info	Price	journey_month	journey_day	Dep_Time_hour	Dep_Time_minute
0	IndiGo	Banglore	New Delhi	BLR → DEL	2h 50m	non-stop	No info	3897	3	24	22	20
1	Air India	Kolkata	Banglore	CCU → IXR → BBI → BLR	7h 25m	2 stops	No info	7662	1	5	5	50
2	Jet Airways	Delhi	Cochin	DEL → LKO → BOM → COK	19h 0m	2 stops	No info	13882	9	6	9	25
3	IndiGo	Kolkata	Banglore	CCU → NAG → BLR	5h 25m	1 stop	No info	6218	12	5	18	5
4	IndiGo	Banglore	New Delhi	BLR → NAG → DEL	4h 45m	1 stop	No info	13302	1	3	16	50

FIGURE 13.3
(a) Separation of arrival and departure times in hours and minutes. (b) Duration column correction.

Every column containing more than two parameters is now divided into more columns according to the number of parameters the columns had (see Figure 13.4). This is all about data preprocessing and extracting features from the data to reduce the complexity to understand data by the machine learning algorithms. Now we will proceed to further steps and we have divided the duration column so now we have to remove the combined one.

For removal, code line:

```
drop_column(train_data, 'Duration') #to remove 'Duration' column
train_data.head() #to print the updated data
```

Dep_Time_hour	Dep_Time_minute	Arrival_Time_hour	Arrival_Time_minute	Duration_hours	Duration_minutes
22	20	1	10	2	50
5	50	13	15	7	25
9	25	4	25	19	0
18	5	23	30	5	25
16	50	21	35	4	45

FIGURE 13.4
Duration is calculated in hours and minutes.

and to change the data types of 'Duration_hours' and 'Duration_minutes' to integer we can use astype() function.

For conversion, code line:

```
train_data['Duration_hours']=train_data['Duration_hours'].astype(int)
train_data['Duration_minutes']=train_data['Duration_minutes'].astype(int)
```

13.4.4 Handling Categorical Data and Feature Encoding

Now, we will work on the categorical data. Firstly, we will deal with the columns containing objects as their values to check that we have to pass a code line that is:

```
cat_col=[col for col in train_data.columns if train_data[col].dtype == 'O']
cat_col
```

This will print the name of columns that contain data as an object in the form of a concatenated string.

Similarly, columns not having values an object will be categorized separately by passing code line that is:

```
cont_col=[col for col in train_data.columns if train_data[col].dtype!= 'O']
cont_col
```

We have just changed the condition here; the data type should not be an object. Hence, a concatenated string is printed.

Feature encoding is done with the categorized data. In simple words, it is processed to transform categorical variables into continuous variables or numbers and then use them in the model to achieve results and this results in better implementation of a methodology for prediction and getting less complex results [3,7].

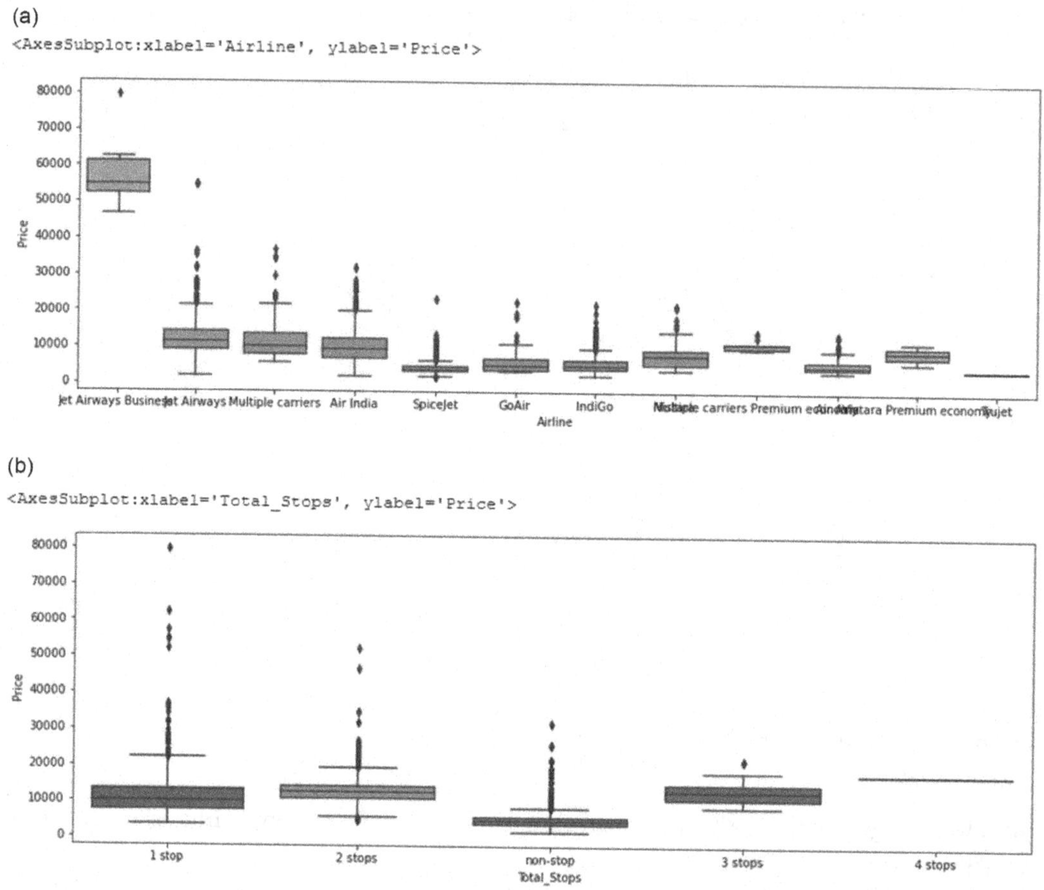

FIGURE 13.5
(a) Airline versus price graph. (b) Total stops versus price graph.

For a visual representation of data usually through graphs, we can run a code and get the graph with respect to two parameters. For example, if I want a graph on airlines w.r.t. their prices, the code line will be as follows:

plt.figure(figsize=(15,5)) #declared the size of figure we want

sns.boxplot(x='Airline', y='Price', data=train_data.sort_values('Price', ascending=False)), result is shown in Figure 13.5a.

Another example is shown in Figure 13.5b, only one changed parameters x='Total_Stops' as total stops is a column in the dataset.

Now we will use one hot encoding to convert categorical variables into integers. One hot encoding is a process of representing the categorical variables mapped as integer values because ML algorithms work on an integer value.

In pandas, we have to include get_dummies which is used for data manipulation which is a part of data preprocessing.

As we can see the categorical variables of column "Airline" are converted into integer format (Figure 13.6).

```
1 Airline = pd.get_dummies(categorical['Airline'],drop_first=True)
```

```
1 Airline.head()
```

	Air India	GoAir	IndiGo	Jet Airways	Jet Airways Business	Multiple carriers	Multiple carriers Premium economy	SpiceJet	Trujet	Vistara	Vistara Premium economy
0	0	0	1	0	0	0	0	0	0	0	0
1	1	0	0	0	0	0	0	0	0	0	0
2	0	0	0	1	0	0	0	0	0	0	0
3	0	0	1	0	0	0	0	0	0	0	0
4	0	0	1	0	0	0	0	0	0	0	0

FIGURE 13.6
Data with a numeric value of each feature (first five rows).

	Total_Stops	Route_0	Route_1	Route_2	Route_3	Route_4	Air India	AirAsia India	GoAir	IndiGo	Jet Airways	Multiple carriers	Multiple carriers Premium economy	SpiceJet	Vistara	Chennai
0	1	3	13	3	4	0	0	0	0	0	0	1	0	0	0	0
1	1	3	4	3	4	0	0	0	0	0	0	1	0	0	0	0
2	1	3	13	3	4	0	0	0	1	0	0	0	0	0	0	0
3	2	3	0	2	1	0	0	0	0	0	1	0	0	0	0	0
4	1	2	4	1	4	0	0	0	0	0	1	0	0	0	0	0

FIGURE 13.7
Ready data for implementation of algorithms.

13.4.5 Analysis of Dataset

Preparing the data for further steps is done firstly by analyzing the dataset, uncovering the useful hidden features inside other features. Also, some features have been derived from other features and some columns have been divided into more than one column as it contained more than one parameter and those parameters are important factors, analysis becomes necessary for recognizing good and bad features from the data [5].

After all data preprocessing and analysis of the data, the above image shows how prepared data look like (see Figure 13.7). All data are in numerical form and ready for implementing machine learning algorithms.

13.5 Analysis of Machine Learning Techniques

To develop a model that will predict the prices of airlines we have different machine learning algorithms to implement on the dataset after data preprocessing. They are linear regression, random forest, KNN algorithm, and decision tree. To evaluate the performance of this model, some definite parameters are examined as follows: (i) R-squared value, (ii) mean absolute error (MAE), and (iii) mean squared error (MSE) [4] formulas for the three parameters are the following:

$$R^2 = 1 - \frac{\sum_{n}^{t=1} \left(yi - \hat{y}i\right)^2}{\sum_{n}^{t=1} \left(yi - \bar{y}i\right)^2} \tag{13.1}$$

$$\text{MAE} = \frac{1}{n}\sum_{n}^{t=1} |yi - \hat{y}i| \tag{13.2}$$

$$\text{MSE} = \frac{1}{n}\sum_{n}^{t=1} (yi - \hat{y}i)^2 \tag{13.3}$$

13.5.1 Random Forest

It is another machine learning algorithm that contains multiple decision trees. This algorithm usually combines the version which is hard to predict to construct better predictive models for usage. It accumulates the lower version to make a massive version. Its capabilities are illustrated and passed to the trees without any substitution to attain the distinct selection trees which are cannot be correlated. The fundamental theory that distinguishes random forest algorithms from the selection tree is accumulated uncorrelated trees [4,14]. Random Forest has almost all the identical parameters similar to a decision tree. Low integration between models is the key. Just as low-correlated investments (such as stocks and bonds) come together to form a portfolio larger than the sum of its shares, inconsistent models can produce more accurate forecasts than any other prediction. The reason for this positive effect is that the trees protect each other from their own mistakes (as long as they do not always fail in one place). While some trees may not be good, many other trees will be good, so as a group the trees are able to move in the right direction. The requirements for a random forest to do well are as follows:

1. There needs to be a certain signal in our features for the models built using those features to perform better than simply guessing.
2. The predictions (and therefore errors) made by each tree need to have a low correlation with each other.

The dimensionality and nature of theta depend upon its usage withinside the improvement of a tree. After the creation of countless trees, they choose the most popular category. This technique is known as random forests [31] and the implemented result is shown in Section 13.6.

13.5.2 Linear Regression

To find out the relation between two parameters which are continuous linear regressions analysis is preferred to use. It is a method of modeling a value set as a target, and it is based on the number of independent variables we have in data. As well as how correlated dependent variables and independent variables are in the selected data. In this method of analysis, the number of independent variables is only one and the connection among the independent and the dependent variables can differ linearly. Before diving deep into machine learning the major concept we should know about are gradient descent and cost function.

$$y(\text{pred}) = b_0 + b_1 * x \tag{13.4}$$

Values in the above equation b_0 and b_1 are used so to reduce the error as much as possible. The squared value which is predicted and the actual value difference results in the error.

MSE is used for dealing with the negative data, as squaring of negative or positive turns the value to positive. In this equation, b_1 is called bias which shows the positive or negative relation among the variables x and y. MSE, MAE, and R-squared values are used to measure the accuracy of the linear regression problem [4,19]. The implemented results are discussed in Section 13.6.

13.5.3 Decision Tree

It is a tool that supports decisions it works using a tree-like model of the decisions. It is an algorithm that only has conditional controlling statements. In this, the tree count segregates the data or information which has been collected into small modules and also makes the data determined. The results at the end show that the tree has decision nodes as well as leaf nodes. Each node shows the test done on an extracted feature of data and each leaf node constitutes a label of the class (or class label) and the branches of the model show the conjunction between features leading to those class labels [14]. The decision might contain at least two branches w.r.t. rate. At first, the collected data is considered as root. There are two major parameters while calculating decision trees: Information Gain and Gini index [3]. Proportional change in entropy is known as Information Gain, whereas Gini Index is a component which is responsible for measuring how regularly an arbitrarily picked component may be differentiated wrongly, which results in errors. It implies that a characteristic with less Gini Index should be liked. For regression tree, the capacity of the cost could be a squared equation:

$$E = \sum \left(y - y^{\wedge}\right)^2 \tag{13.5}$$

In the above equation, y denoted the original value from the data and the other parameter y-cap is the value that is predicted. The presence of a class that contains the highest number of values of an assumed value acquired by the function used for splitting is known as Information Gain. If in case the value is kept splitting again and again with no cause at the leaf node [15,16]. The algorithm will be slow in completing, huge as well as over fitted. To overcome this, the slightest count on the example that is trained of the leaf node is allocated. The implementation results are shown in Section 13.6.

13.5.4 K-Nearest Neighbor (KNN) Algorithm

In the K-nearest neighbor relapse algorithm, the result is the mean of its K-nearest neighbors. Similar to Support Vector Machine, KNN is likewise a technique without parameters. Thinking about a couple of qualities, outcomes are registered to accomplish the perfect value. KNN algorithm is a managed order calculation that can likewise be utilized as a regressor. It relegates another information highlight the class. It is nonparametric because it doesn't take any presumption. It figures the distance between each preparation model and another information point [3,20]. To process this distance, following distance estimation techniques are utilized:

1. **Euclidean Distance**

$$\mathrm{ED} = \sqrt{\sum_{i=1}^{k} \left(x_i - y_i\right)^2} \tag{13.6}$$

2. **Manhattan Distance**

$$MD = \sum_{i=1}^{k} |x_i - y_i| \tag{13.7}$$

3. **Hamming Distance**

$$D_H = \sum_{i=1}^{k} |x_i - y_i| \tag{13.8}$$

K is the count of entries according to datasets that are selected by the training models that are nearly around fresh data points. The implementation of this algorithm [27] along with the results is shown in Section 13.6.

13.6 Algorithms Implementation and Evaluation

After data preprocessing and getting the data prepared to apply machine learning algorithms. We can easily apply the algorithms to the model by declaring a function and then importing a few algorithm functions from Scikit-learn (Figures 13.8a–c, 13.9a–d, 13.10a–d, 13.11a–d).

13.6.1 Random Forest Algorithm

CODE

(a)

```
In [356]:  from sklearn import metrics

In [357]:  def predict(ml_model):
               model = ml_model.fit(X_train,y_train)
               print('Training Score: {}'.format(model.score(X_train,y_train)))

               y_predictions=model.predict(X_test)
               print('Predictions are: {}'.format(y_predictions))
               print('\n')

               r2_score=metrics.r2_score(y_test,y_predictions)
               print('r2 score is : {}'.format(r2_score))

               print('MAE: ',metrics.mean_absolute_error(y_test,y_predictions))
               print('MSE: ',metrics.mean_squared_error(y_test,y_predictions))
               print('RMSE: ',np.sqrt(metrics.mean_absolute_error(y_test,y_predictions)))
               sns.distplot(y_test-y_predictions)


In [358]:  from sklearn.ensemble import RandomForestRegressor

In [359]:  predict(RandomForestRegressor())
```

FIGURE 13.8
(a) Random forest algorithm in Python with function for prediction. (b) Predictions and training scores of random forest. (c) Graph showing the density of price of tickets for random forest predictions.

(Continued)

OUTPUT

(b)

```
predict(RandomForestRegressor())
```

```
Training Score: 0.9345515592337842
Predictions are: [ 4785.3    5274.915  6902.7    5524.155  4265.46  12720.93   9845.685
 11641.665 10673.875 10201.49  10638.975 11634.695  5227.94  11167.275
  4269.08  11667.53  10280.295 11830.895  9894.545 12224.7   12074.585
 12495.33  12006.785 11437.04   6160.405 10847.27   6736.795 11056.735
  6978.53   8269.83   7331.13  11920.86  12344.29  12771.955 12056.965
 10375.07   6405.305 12495.33   4890.32  13260.385]


r2 score is : 0.5370799542772002
MAE:  1914.4497500000002
MSE:  4976586.14911
RMSE:  43.754425490457535
```

(c)

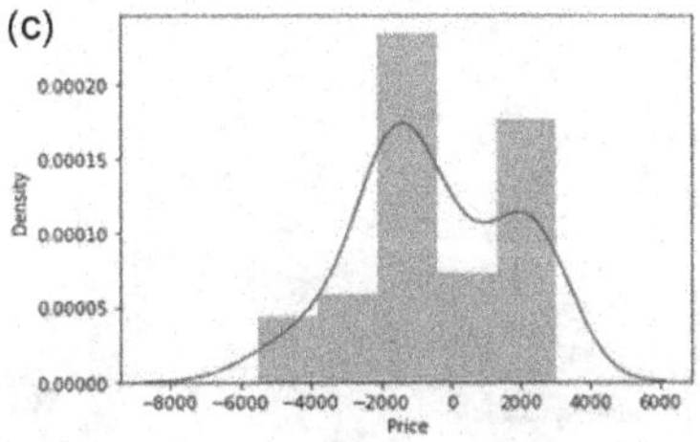

FIGURE 13.8 (*Continued*)
(a) Random forest algorithm in Python with function for prediction. (b) Predictions and training scores of random forest. (c) Graph showing the density of price of tickets for random forest predictions.

13.6.2 Linear Regression Algorithm

CODE

(a)

```
from sklearn import metrics
```

```
def predict(ml_model):
    model = ml_model.fit(X_train,y_train)
    print('Training Score: {}'.format(model.score(X_train,y_train)))

    y_predictions=model.predict(X_test)
    print('Predictions are: {}'.format(y_predictions))
    print('\n')

    r2_score=metrics.r2_score(y_test,y_predictions)
    print('r2 score is : {}'.format(r2_score))

    print('MAE: ',metrics.mean_absolute_error(y_test,y_predictions))
    print('MSE: ',metrics.mean_squared_error(y_test,y_predictions))
    print('RMSE: ',np.sqrt(metrics.mean_absolute_error(y_test,y_predictions)))
    sns.distplot(y_test-y_predictions)

```

FIGURE 13.9
(a) Function used for linear regression prediction. (b) Importing linear regression algorithm in Python. (c) Predictions and training scores of linear regression. (d) Graph showing the density of price of tickets for linear regression predictions.

(Continued)

(b)

```
1  from sklearn.linear_model import LinearRegression

1  predict(LinearRegression())

Training Score: 0.6802847199975024
Predictions are: [ 4884.2682849    5752.88517393  8054.72364265  5586.08056937
  3197.54538812 11478.18711196  8809.67680706 12522.81159282
 10546.0253592  10204.06059304  7121.32985375 10642.75191557
  6409.3872656   9439.12092236  4369.13237689 11214.65208597
  7507.90115324 12505.65336008 10615.96486075 12258.36782237
 12784.43667529 12021.64022373 10995.0232724  12059.84971341
  5607.41563467 12997.64253262  5941.46220511  9560.20056472
  4519.32775663  8682.06015656  5994.01968571 11954.20050314
 11920.11660391 12646.96519359 11806.56149146 11217.29756241
  6224.88119052 12021.64022373  3882.92739531 12934.17847648]


r2 score is : 0.5548648791291171
MAE:  1845.5846787204487
MSE:  4785389.728402108
RMSE:  42.96026860624185
```

OUTPUT

(c)


```
1  from sklearn.linear_model import LinearRegression

1  predict(LinearRegression())

Training Score: 0.6802847199975024
Predictions are: [ 4884.2682849    5752.88517393  8054.72364265  5586.08056937
  3197.54538812 11478.18711196  8809.67680706 12522.81159282
 10546.0253592  10204.06059304  7121.32985375 10642.75191557
  6409.3872656   9439.12092236  4369.13237689 11214.65208597
  7507.90115324 12505.65336008 10615.96486075 12258.36782237
 12784.43667529 12021.64022373 10995.0232724  12059.84971341
  5607.41563467 12997.64253262  5941.46220511  9560.20056472
  4519.32775663  8682.06015656  5994.01968571 11954.20050314
 11920.11660391 12646.96519359 11806.56149146 11217.29756241
  6224.88119052 12021.64022373  3882.92739531 12934.17847648]


r2 score is : 0.5548648791291171
MAE:  1845.5846787204487
MSE:  4785389.728402108
RMSE:  42.96026860624185
```


(d)

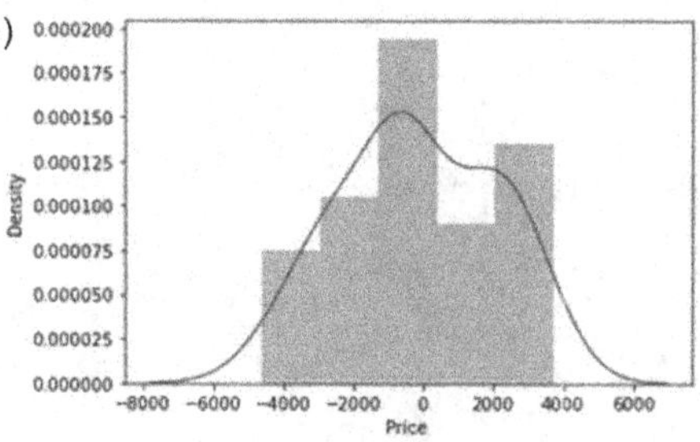

FIGURE 13.9 (*Continued*)
(a) Function used for linear regression prediction. (b) Importing linear regression algorithm in Python. (c) Predictions and training scores of linear regression. (d) Graph showing the density of price of tickets for linear regression predictions.

13.6.3 Decision Tree Algorithm

CODE

(a)

```
from sklearn import metrics
```

```
def predict(ml_model):
    model = ml_model.fit(X_train,y_train)
    print('Training Score: {}'.format(model.score(X_train,y_train)))

    y_predictions=model.predict(X_test)
    print('Predictions are: {}'.format(y_predictions))
    print('\n')

    r2_score=metrics.r2_score(y_test,y_predictions)
    print('r2 score is : {}'.format(r2_score))

    print('MAE: ',metrics.mean_absolute_error(y_test,y_predictions))
    print('MSE: ',metrics.mean_squared_error(y_test,y_predictions))
    print('RMSE: ',np.sqrt(metrics.mean_absolute_error(y_test,y_predictions)))
    sns.distplot(y_test-y_predictions)

```

(b)

```
from sklearn.tree import DecisionTreeRegressor
```

```
predict(DecisionTreeRegressor())
```

```
Training Score: 1.0
Predictions are: [ 3943.  4030.  6069.  4030.  6356. 14781.  7258.  9663.  9905. 10651.
 13470. 10577.  3943. 13977.  3850. 13029. 13470. 13470.  8266. 10844.
 11627. 14571. 13067. 14151.  4030. 11627. 12314.  9443.  8273.  9794.
  9794. 12242. 10844. 12898.  9899. 10577.  4030. 14571.  4823. 14939.]


r2 score is : -0.048379793669953264
MAE:  2725.575
MSE:  11270523.625
RMSE:  52.2070397551901
```

```
C:\Users\Inderjeet\anaconda3\lib\site-packages\seaborn\distributions.py:2551: FutureWarning: `dist
recated function and will be removed in a future version. Please adapt your code to use either `di
re-level function with similar flexibility) or `histplot` (an axes-level function for histograms).
  warnings.warn(msg, FutureWarning)
```

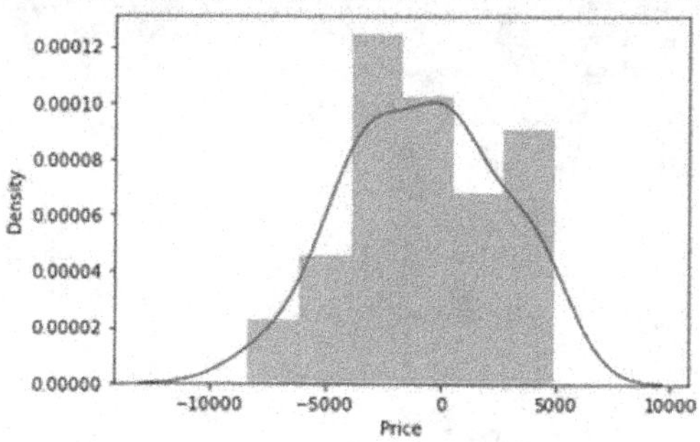

FIGURE 13.10
(a) Function used for decision tree prediction. (b) Importing decision tree algorithm in Python. (c) Predictions and training scores of a decision tree. (d) Graph showing the density of price of tickets for decision tree predictions.

(Continued)

OUTPUT

(c)

```
Training Score: 1.0
Predictions are: [ 3943.  4030.  6069.  4030.  6356. 14781.  7258.  9663.  9905. 10651.
 13470. 10577.  3943. 13977.  3850. 13029. 13470. 13470.  8266. 10844.
 11627. 14571. 13067. 14151.  4030. 11627. 12314.  9443.  8273.  9794.
  9794. 12242. 10844. 12898.  9899. 10577.  4030. 14571.  4823. 14939.]


r2 score is : -0.048379793669953264
MAE:  2725.575
MSE:  11270523.625
RMSE:  52.2070397551901
```

```
C:\Users\Inderjeet\anaconda3\lib\site-packages\seaborn\distributions.py:2551: FutureWarning: `d
recated function and will be removed in a future version. Please adapt your code to use either
re-level function with similar flexibility) or `histplot` (an axes-level function for histogram
  warnings.warn(msg, FutureWarning)
```

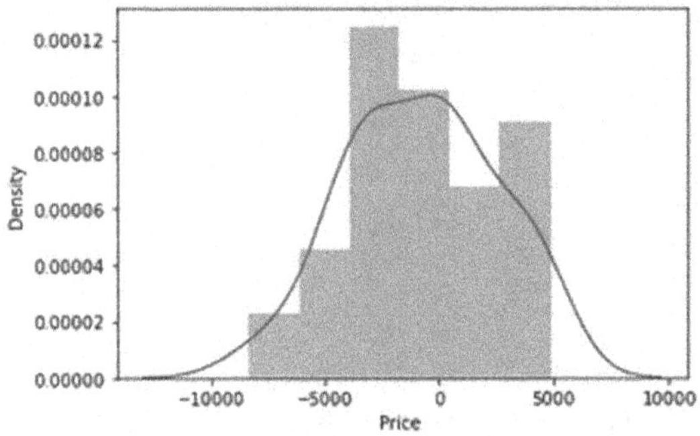

(d)

```
Training Score: 1.0
Predictions are: [ 3943.  4030.  6069.  4030.  6356. 14781.  7258.  9663.  9905. 10651.
 13470. 10577.  3943. 13977.  3850. 13029. 13470. 13470.  8266. 10844.
 11627. 14571. 13067. 14151.  4030. 11627. 12314.  9443.  8273.  9794.
  9794. 12242. 10844. 12898.  9899. 10577.  4030. 14571.  4823. 14939.]


r2 score is : -0.048379793669953264
MAE:  2725.575
MSE:  11270523.625
RMSE:  52.2070397551901
```

```
C:\Users\Inderjeet\anaconda3\lib\site-packages\seaborn\distributions.py:2551: FutureWarning: `d
recated function and will be removed in a future version. Please adapt your code to use either
re-level function with similar flexibility) or `histplot` (an axes-level function for histogram
  warnings.warn(msg, FutureWarning)
```

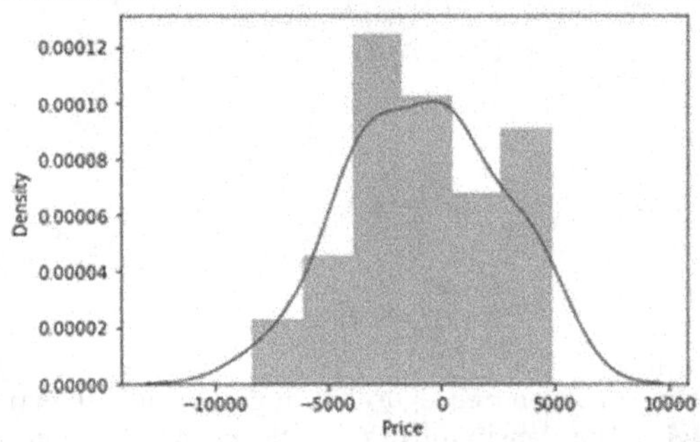

FIGURE 13.10 *(Continued)*
(a) Function used for decision tree prediction. (b) Importing decision tree algorithm in Python. (c) Predictions and training scores of a decision tree. (d) Graph showing the density of price of tickets for decision tree predictions.

13.6.4 K-Nearest Neighbor Algorithm

CODE

(a)

```
from sklearn import metrics
```

```
def predict(ml_model):
    model = ml_model.fit(X_train,y_train)
    print('Training Score: {}'.format(model.score(X_train,y_train)))

    y_predictions=model.predict(X_test)
    print('Predictions are: {}'.format(y_predictions))
    print('\n')

    r2_score=metrics.r2_score(y_test,y_predictions)
    print('r2 score is : {}'.format(r2_score))

    print('MAE: ',metrics.mean_absolute_error(y_test,y_predictions))
    print('MSE: ',metrics.mean_squared_error(y_test,y_predictions))
    print('RMSE: ',np.sqrt(metrics.mean_absolute_error(y_test,y_predictions)))
    sns.distplot(y_test-y_predictions)

```

(b)

```
from sklearn.neighbors import KNeighborsRegressor
```

```
predict(KNeighborsRegressor())
```

```
Training Score: 0.47382829052595044
Predictions are: [ 4725.8   8985.5   9810.    8925.6   6617.5   8347.1 11018.2 10830.3 10967.6
  9492.7 10448.6 10410.9  6183.9  9915.4  8639.3  5826.4  7554.2  8111.
 12017.4  9847.5 11367.1 11442.2 11645.    9568.1  8925.6 11648.4  7209.
 11090.4  5364.8  5767.5  8099.5  9545.1  8024.8  9874.8 11389.2  8723.3
  9909.  11442.2  5707.4  7585.6]


r2 score is : 0.27586239368938326
MAE:  2181.2974999999997
MSE:  7784783.767249999
RMSE:  46.704362751246265
```

```
C:\Users\Inderjeet\anaconda3\lib\site-packages\seaborn\distributions.py:2551: FutureWarning: `d
recated function and will be removed in a future version. Please adapt your code to use either
re-level function with similar flexibility) or `histplot` (an axes-level function for histogra
  warnings.warn(msg, FutureWarning)
```

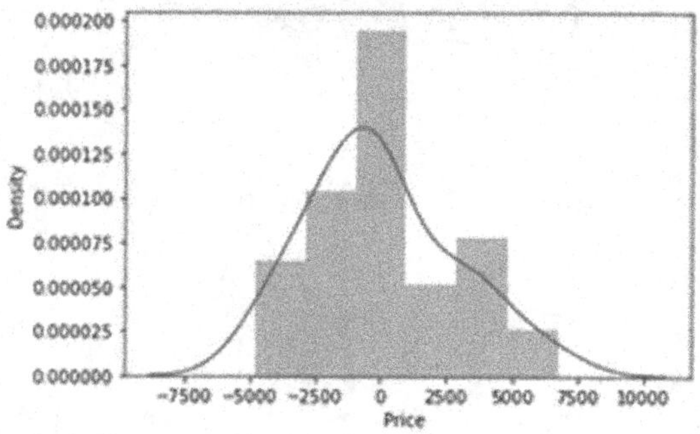

FIGURE 13.11
(a) Function used for K-nearest neighbor prediction. (b) Importing KNN algorithm in Python. (c) Predictions and training scores of KNN algorithm. (d) Graph showing the density of price of tickets for KNN predictions.

(Continued)

OUTPUT

(c)

```
1  from sklearn.neighbors import KNeighborsRegressor
```

```
1  predict(KNeighborsRegressor())
```

```
Training Score: 0.47382829052595044
Predictions are: [ 4725.8  8985.5  9810.   8925.6  6617.5  8347.1 11018.2 10830.3 10967.6
  9492.7 10448.6 10410.9  6183.9  9915.4  8639.3  5826.4  7554.2  8111.
 12017.4  9847.5 11367.1 11442.2 11645.   9568.1  8925.6 11648.4  7209.
 11090.4  5364.8  5767.5  8099.5  9545.1  8024.8  9874.8 11389.2  8723.3
  9909.  11442.2  5707.4  7585.6]

r2 score is : 0.2758623936893832 6
MAE:  2181.2974999999997
MSE:  7784783.767249999
RMSE:  46.70436275124626 5
```

```
C:\Users\Inderjeet\anaconda3\lib\site-packages\seaborn\distributions.py:2551: FutureWarning: `d
recated function and will be removed in a future version. Please adapt your code to use either
re-level function with similar flexibility) or `histplot` (an axes-level function for histogram
  warnings.warn(msg, FutureWarning)
```

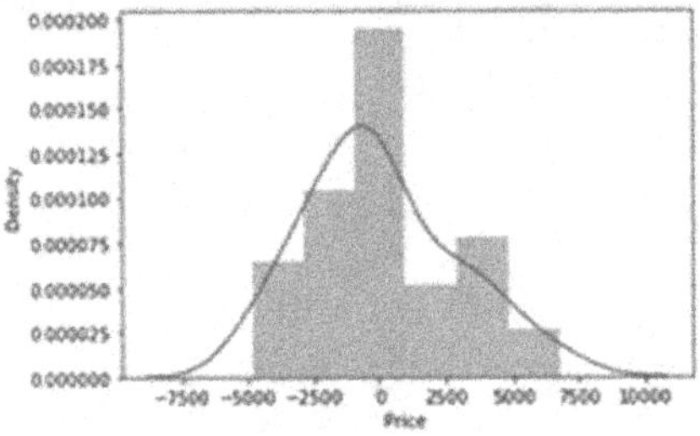

(d)

```
1  from sklearn.neighbors import KNeighborsRegressor
```

```
1  predict(KNeighborsRegressor())
```

```
Training Score: 0.47382829052595044
Predictions are: [ 4725.8  8985.5  9810.   8925.6  6617.5  8347.1 11018.2 10830.3 10967.6
  9492.7 10448.6 10410.9  6183.9  9915.4  8639.3  5826.4  7554.2  8111.
 12017.4  9847.5 11367.1 11442.2 11645.   9568.1  8925.6 11648.4  7209.
 11090.4  5364.8  5767.5  8099.5  9545.1  8024.8  9874.8 11389.2  8723.3
  9909.  11442.2  5707.4  7585.6]

r2 score is : 0.2758623936893832 6
MAE:  2181.2974999999997
MSE:  7784783.767249999
RMSE:  46.70436275124626 5
```

```
C:\Users\Inderjeet\anaconda3\lib\site-packages\seaborn\distributions.py:2551: FutureWarning: `d
recated function and will be removed in a future version. Please adapt your code to use either
re-level function with similar flexibility) or `histplot` (an axes-level function for histogram
  warnings.warn(msg, FutureWarning)
```

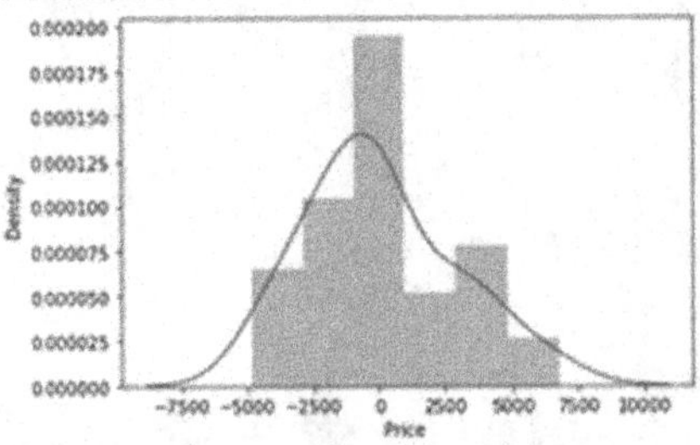

FIGURE 13.11 (*Continued*)
(a) Function used for K-nearest neighbor prediction. (b) Importing KNN algorithm in Python. (c) Predictions and training scores of KNN algorithm. (d) Graph showing the density of price of tickets for KNN predictions.

TABLE 13.1

Performance Accuracy Table of Algorithms

	Random Forest	Linear Regression	Decision Tree	KNN Algorithm
Training Score	0.9345	0.6828	0.1000	0.7738
R-Squared	0.82581	0.5549	0.4838	0.2786
MAE	1,914.4497	1,883.7805	2,725.575	2,181.297
MSE	4,976,585.149	7,119,150.076	11,270,523.625	778,473.767
RMSE	43.75	43.51907	52.20	46.70

13.6.5 Algorithm's Evaluation Table

The training score, R-squared score, MAE, MSE, and RMSE after applying the algorithms to the dataset are shown in Table 13.1.

13.7 Limitations

1. **Random Forest Algorithm**
   This algorithm requires a lot of computational power also resources as the process includes building many numerous trees to merge the outputs. Also, more time is required for the algorithm to train the model.
2. **Decision Tree Algorithm**
   Unstable character of the algorithm means if there is a small change in the data it may lead to drastic changes in the structure of the optimal tree which will lead to false or deviated results.
3. **Linear Regression Algorithm**
   The data we are using must be independent and complexity will rise inversely. It only looks for the mean of the dependent variable.
4. **K-Nearest Neighbor Algorithm**
   The accuracy of this algorithm mainly depends on in what condition data is also it is used for small datasets because for large datasets it is very slow in processing as well as it requires more memory.

13.8 Conclusion

In this chapter, the general data for the dynamic charge modifications withinside the tickets of airlines is presented. This put forward the statistics approximately the peaks and valleys withinside the fares of airlines tickets according to the days, whether it is at the end of the week, or daytime, or during summer or winter vacations, festivals, morning or night. Additionally the gadget gaining knowledge of fashions withinside the artificial intelligence (AI) or field of computation that are already being obtained on unique datasets are measured. Their accuracy and performances are estimation also comparison is done which will show the higher end result. For predicting the prices, price tag outlays perfectly different prediction fashions are examined to get the best predicting accuracy. As the manner of getting prices of different company are developed aiming to make the management of sales as higher as possible. For getting the end result, a more accurate regression evaluation is used. Studies show that the characteristic that affects the expenses of the price tag is to be pondered.

13.9 Future Work

In the future, the approximate count of available tickets can enhance the overall performance of the model. The best weakness of this work is the deficiency of information. Researchers wish to develop it ought to look for elective wellsprings of the chronicled information or be more systematic in gathering information physically over a timeframe.

References

1. T. Wang et al., "A Framework for Airfare Price Prediction: A Machine Learning Approach," in *IEEE 20th International Conference on Information Reuse and Integration for Data Science (IRI)*, 2019, pp. 200–207. doi: 10.1109/IRI.2019.00041.
2. M. Papadakis, "Predicting Airfare Prices", 2014.
3. Swarali M. Pathak, Archana K. Chaudhari, "Comparison of Machine Learning Algorithms for House Price Prediction using Real Time Data," *International Journal of Engineering Research & Technology (IJERT)* vol. 10, issue 12, December 2021.
4. S. Rajankar and N. Sakharkar, "A survey on flight pricing prediction using machine learning," vol. 8, issue 06, pp. 14–19, June–2019.
5. A. V. Jain, A. S. Raval, and R. K. Oza, "Airfare price prediction based on reviews using machine learning techniques," vol. 9, issue 3, March 2020.
6. K. Tziridis, Th. Kalampokas, G. A. Papakostas, and K. I. Diamantaras, Airfare Prices Prediction Using Machine Learning Techniques, EURASIP 2017.
7. Sharma, L., and Garg, P.K. (Eds.). (2021). *Artificial Intelligence: Technologies, Applications, and Challenges* (1st ed.). Chapman and Hall/CRC. doi: 10.1201/9781003140351.
8. B. Mantin and B. Koo, "Dynamic price dispersion in airline markets," *Transportation Research Part E: Logistics and Transportation Review*, vol. 45, no. 6, pp. 1020–1029, 2009.
9. P. Malighetti, S. Paleari, and R. Redondi, "Has Ryanair's pricing strategy changed over time? an empirical analysis of its 2006–2007 flights," *Tourism Management*, vol. 31, no. 1, pp. 36–44, 2010.
10. T. H. Oum, A. Zhang, and Y. Zhang, "Inter-firm rivalry and firm-specific price elasticities in deregulated airline markets," *Journal of Transport Economics and Policy*, vol. 7, no. 2, pp. 171–192, 1993.
11. K. S. Gerardi and A. H. Shapiro, "Does competition reduce price dispersion? New evidence from the airline industry," *Journal of Political Economy*, vol. 117, no. 1, pp. 1–37, 2009.
12. S. A. Rhoades, "The Herfindahl-Hirschman index," *Federal Reserve Bulletin*, vol. 79, p. 188, 1993.
13. G. Francis, A. Fidato, and I. Humphreys, "Airport–airline interaction: the impact of low-cost carriers on two European airports," *Journal of Air Transport Management*, vol. 9, no. 4, pp. 267–273, 2003.

14. International Civil Aviation Organization, "List of lowcost-carriers (LCCs)," cited July 2018. [Online]. Available: https://www.icao.int/sustainability/Documents/LCC-List.pdf.
15. C. Koopmans and R. Lieshout, "Airline cost changes: To what extent are they passed through to the passenger?" *Journal of Air Transport Management*, vol. 53, pp. 1–11, 2016.
16. S. Lee, K. Seo, and A. Sharma, "Corporate social responsibility and firm performance in the airline industry: The moderating role of oil prices," *Tourism Management*, vol. 38, pp. 20–30, 2013.
17. V. Nair and G. E. Hinton, "Rectified linear units improve restricted Boltzmann machines," in the *27th International Conference on Machine Learning*, 2010, pp. 807–814.
18. Sharma, L., and Carpenter, M. Coronavirus, 2021. Available at: https://www.routledge.com/Computer-Vision-and-Internet-of-Things-Technologies-and-Applications/Sharma-Carpenter/p/book/9781032154367
19. Fare prediction. Available: http://arxiv.org/abs/1412.6980 [accessed on 12 June 2021].
20. T. Liu, J. Cao, Y. Tan, and Q. Xiao, "ACER: An adaptive context-aware ensemble regression model for airfare price prediction," in the *International Conference on Progress in Informatics and Computing*, 2017, pp. 312–317.
21. L. Sharma and P. K Garg, "IoT and its applications", *From Visual Surveillance to Internet of Things*, Taylor & Francis, CRC Press, vol. 1, pp. 29.
22. M. S. Ryerson and H. Kim, "Integrating airline operational practices into passenger airline hub definition," *Journal of Transport Geography*, vol. 31, pp. 84–93, 2013.
23. H. Baik, A. A. Trani, N. Hinze, H. Swingle, S. Ashiabor, and A. Seshadri, "Forecasting model for air taxi, commercial airline, and automobile demand in the united states," *Transportation Research Record*, vol. 2052, no. 1, pp. 9–20, 2008.
24. T. Janssen, T. Dijkstra, S. Abbas, and A. C. van Riel, *A Linear Quantile Mixed Regression Model for Prediction of Airline Ticket Prices*, Radboud University, 2014.
25. R. Ren, Y. Yang, and S. Yuan, *Prediction of Airline Ticket Price*, University of Stanford, 2014.
26. T. Wohlfarth, S. Clemenc¸on, F. Roueff, and X. Casellato, "A data-mining approach to travel price forecasting," in the *10th International Conference on Machine Learning and Applications and Workshops*, vol. 1, 2011, pp. 84–89.
27. H.-C. Huang, "A hybrid neural network prediction model of air ticket sales," *Telkomnika Indonesian Journal of Electrical Engineering*, vol. 11, no. 11, pp. 6413–6419, 2013.
28. Sharma, L. (Ed.) *Towards Smart World*, Chapman and Hall/CRC, New York, 2021. doi: 10.1201/9781003056751.
29. Sharma, L., Garg, P. (Eds.) *From Visual Surveillance to Internet of Things*, Chapman and Hall/CRC, New York, 2020. doi: 10.1201/9780429297922.
30. V. Pai, "On the factors that affect airline flight frequency and aircraft size," *Journal of Air Transport Management*, vol. 16, no. 4, pp. 169–177, 2010.
31. L. Breiman, "Random forests," *Machine Learning*, vol. 45, pp. 5–32, 2001.

Part 4

Challenging Issues and Novel Solutions

14

CapsNet and KNN-Based Earthquake Prediction Using Seismic and Wind Data

Sandeep Dwarkanath Pande and Soumitra Das
DYPIT, Pune, India

Pramod Jadhav
Department of Computer Engineering, G. H. Raisoni Institute of Engineering and Technology, Pune, India

Amol D. Sawant
DYPIT, Pune, India

Shantanu S. Pathak
DHI Training and Research Consultancy, Pune, India

Sunil L. Bangare
Department of I.T., Sinhgad Academy of Engineering, Pune, India

CONTENTS

14.1 Introduction

Earthquakes are one of the worst natural hazards that frequently turn into disasters that cause extensive destruction and losses of human lives. Generally, sensor data are used for earthquake prediction. The use of self-adaptive artificial neural networks (ANNs) is a convenient and feasible way to build relationships between different symptom factors

DOI: 10.1201/9781003244165-18

and earthquake incidents. Today's computing world has allowed us to forecast a lot of things with the past data that have been obtained. The data that have been stored for decades are used by data scientists in order to make assumptions and in turn predictions related to the cause and future occurrence of earthquakes. This has also helped to boom most of the machine learning concepts as there is a constant need to apply previously used algorithms with a certain change as the data won't be constant and also the algorithm should perform well with different parameters and with higher accuracy rates. The system being developed would help to predict an estimate of how the seismic activity is and what it would be after a certain period of time [1]. The most suited neural network (NN) termed as CapsNet [2] is used to predict earthquakes using data with multiple possible conjugations, with lesser epochs and earlier than the actual time period.

Predicting each earthquake event in different places can reduce the disaster. The extravagant loss of human life and property can be reduced. Various stations have been deployed through different geographical regions in order to determine the wave activity at that particular instant. By acquiring the information of the seismic activity of that area, we can predict the future occurrence of an earthquake using data analytics tools. Various arrangements such as National Disaster Response Force, medical teams, ambulances, and other necessary disaster mechanisms should be placed well before collateral damage. NN could be a feasible way to solve such problems. An effective, quick, and computationally efficient earthquake prediction system is always crucial for the early prediction of earthquakes from input sensor data. Earthquake prediction is also substantial to access seismic and wind data and generate knowledge from such information to construct the buildings or infrastructures for transports, public or government offices. This promotes the introduction of a new earthquake prediction system which is based on seismic and wind data of the region [3]. The system performs early prediction of the situation of earthquakes on the basis of input seismic and wind data. By learning the patterns from input, it predicts the situation of an earthquake into various classes. The classes predicted are no earthquake, mild earthquake, medium earthquake, strong earthquake, and very strong earthquake.

Early, reliable, and computationally efficient earthquake prediction is also a big challenge. In this chapter, the seismic data obtained from sensors, wind data, and longitude and latitude information are used to predict an earthquake in India. The longitude and latitude information is fed to the K-nearest neighbor (KNN) [4] classifier to obtain one of the five seismic zones. The latest and most suitable NN termed as CapsNet, is then used for quick and reliable earthquake prediction. CapsNet accepts the seismic and wind data for intermediate earthquake prediction. Eventually, it uses the blender network that combines the output of KNN and CapsNet for final earthquake prediction. The proposed method is compared with present cutting-edge earthquake prediction systems concerning performance and results. The comparison reveals proposed approach obtains promising and reliable results. In many parameters, it outperforms other earthquake prediction systems. The proposed method signifies an advantage of low resource overhead and yields accurate results.

The second section of this chapter contains a quick review of recent earthquake prediction studies. In Section 14.3, the preprocessing methods required to prepare data for feeding to the adopted networks are discussed, as well as, it presents a brief description of the KNN and CapsNet architecture. It also explains how to set up CapsNet and KNN for this approach. The experimental results and setup and the conclusion are presented in Sections 14.4 and 14.5 respectively.

14.2 Literature Review

An early warning system in earthquake prediction can minimize the damage. Therefore, various sensors can be deployed at major locations and signals can be retrieved in real time from them. But there are different methods for getting the information from sensors and predicting the earthquake to find out the accurate location of an earthquake. This section discusses a few of the earthquake detection systems developed until now. Sherki et al. [1] have predicted the earthquake through an optical sensor. Furthermore, it uses a noise isolation system to remove various external noises. Distance triangulation is a popular pattern for locating earthquakes, although it is inaccurate since it measures the mean speed of P and S waves. The seismic wave's speeds differ significantly reliant on the type of soil, rocks, and water beneath the ground, but the Azimuth angle is a newer technique. These waves are used to determine the exact location of an earthquake. Three-dimensional accelerometer, digital band-pass filter, microcontroller, global system for mobile communication module, and global positioning system also were part of the system. It turns vibrations into electrical signals along three axes, which are then converted using an analog to digital converter, allowing the epicenter's position to be triangulated. This method is not appropriate anymore as the distance calculation would change according to every sensor making it imperfect. This approach may not provide expected accuracy for a varied and large dataset.

Detection of Electro-Magnetic Emissions Transmitted from Earthquake Regions (DEMETER) satellite observations have been used by Xu et al. [3] to identify anomalies associated with earthquakes. DEMETER is a solar-synchronous spacecraft dedicated to seismic electromagnetic radiation monitoring. These satellite data can be employed to predict earthquakes. The data were utilized to train a backpropagation NN (BPNN) for earthquake prediction. Satellite-based earthquake observations have several benefits, including worldwide coverage, a short revisit duration, high efficiency, and good dynamics. D. Sun and B. Sun [5] provided a fuzzy analysis-based fast model for predicting earthquake damages for structures utilizing a seismic damage index, an average seismic damage index as a return index, and building impact factors as a modification index. However, we are still unable to anticipate earthquakes using this strategy, so a different methodology, the seismic belt, is employed in our work. The earthquake belts are bounded geographical zones on the earth's crystal surface where the majority of earthquakes occur, and data mining algorithms could be used to do the analysis. It is a classification problem to determine whether or not an earthquake will occur in a certain location. The BPNN is a popular way of training ANN and to accomplish this task. Veri et al. [6] used this technique to locate and identify the data point movement patterns. They also attempted to develop a formula on the basis of the study and investigation of prior seismic data movement patterns. The gradient descent down the error surface is used in backpropagation learning. The primary goal of a BPNN is to reduce error. Elman recurrent NN is proposed in research [7] by Deo and Chandra to conduct an experiential investigation over the minimum period required for robust cyclone wind-intensity prediction.

The geomagnetic field was used by many researchers to predict earthquakes. In 2003, a state-space model and filter were designed to reduce electromagnetic wave noise. They had seen electromagnetic radiation in the ultralow frequency range. The distribution characteristics and imaging properties of earth fissures were studied in 2005. The extremely low-frequency signals of the 6.4 San Simone earthquakes from space and on the ground were compared in California. Low-altitude satellites detected electromagnetic, extremely low frequency emissions from earthquake zones in 1992. In 2000, a signal processing method was presented for separating earthquake activity from background noise in a relatively low-frequency electromagnetic wave. However, the work hasn't been able to come up with a precise outcome in their studies. In recent years, satisfactory work has been done on the variation of the earth's geomagnetic field for earthquake prediction; however, some researches on earthquake predictions are still ongoing [8]. The comparison of linear programming boost (LPB), recurrent NN (RNN), pattern recognition NN (PRNN), and random forest (RF) algorithms for earthquake magnitude prediction is described in the work [9]. The LPB performed finest with 65% accuracy for unseen data, while RNN attained 64% accuracy. The PRNN model yielded the least false alarms. Asencio-Cortés et al. [10] compared, generalized linear models (GLM), deep learning (DL), RF, and gradient boosting machines (GBM) for earthquake magnitude prediction.

Artificial earthquake and reaction spectra are generated using NN-based algorithms in the study [11]. Here, five different ANNs are used to replace the empirical model's parameter identification process for generating the Fourier amplitude spectrum. The models are trained to obtain the parameters of the power spectral density and intensity function and generate an acceleration response spectrum with basic information by inverting the fourth ANN. The fundamental problem of this strategy is that the diverse ANNs require a lot of training time, and the ANNs used here are not mutually exclusive. Panakkat and Adeli [12] used eight distinct computed seismicity indicators to estimate earthquake magnitude. For prediction, it utilized three NNs: feed-forward Levenberg–Marquardt backpropagation, RNN, and radial basis function (RBF) NN. Panakkat and Adeli [13] employed a vector that included eight different seismicity markers to forecast the location and timing of future moderate-to-high-risk earthquakes. It was based on NN modeling and used an approximation sense. This method necessitates the computing of a seismicity indicator for each time period, requiring the consideration of the epicentral location's longitude and latitude many times. This method pinpointed the location of the earthquake within 15–39 km. It anticipated a delay of 75–94 days for the mainshock and a delay of 5–16 days for the aftershocks. For discovering long-term patterns in the earthquake, Kanarachos et al. [14] employed deep NN (DNN) with a wavelet and Hilbert transform. Before the Kozani Grevena earthquake, they had an average true positive rate of 83%. In the study [15], the long short-term memory (LSTM) was used to predict the future movement of earthquakes. Using a DL-based model, Bao et al. [16] estimated the magnitude of the P-wave during an earthquake. It correctly predicted 84.21% of samples with an error rate of 0.5. Wang et al. [17] presented an earthquake prediction model based on LSTM. This model had a 63.50% accuracy with one-dimensional input. For the (5 5) subregion, this model attained an accuracy of 86% using two-dimensional input. S. L. Bangare et al. proposed work in a comparable domain [18–21]. Awate et al. [22] employed convolutional NN (CNN), compute unified device architecture (CUDA), and CUDA DNN to diagnose Alzheimer's disease with a low error rate.

Jozinović et al. [23] outlined a CNN-based technique for measuring the intensity of shaking of the earth's surface after an earthquake. It uses 3C acceleration waveforms as an input. A similar CNN-based approach using earthquake data provided by government institutions was used to predict the earthquakes in Ref. [24]. The work further carries out an analytical hierarchy process to examine the vulnerability of the earthquake. Wang et al. [17] proposed an LSTM-based earthquake prediction system using the spatial–temporal relationship of the data. Cui et al. [25] presented a stacked ensemble learning technique for the prediction of casualty in earthquake-disturbed regions for making faster and precise decisions amid rescue operations.

14.3 Methodology

Earthquake prediction is a difficult problem to solve but there could be ways to have a work around that problem. A simple methodology would be illustrated below.

The work done in an earthquake is enormous and is to be taken care of as it would later affect the actual development of the system. This makes it complex so that various techniques need to be combined in order to solve such kinds of problems in an efficient and even more work-friendly environment. A basic method that could be used for this problem is trying to find the zone of certain areas with a simple KNN algorithm. After finding the zonal activity these data would serve as the information source at the user end to try the prediction of the same. The system architecture for the same is shown in Figure 14.1.

14.3.1 KNN Method

KNN is a basic data-mining algorithm which tries to collect all the data in the form of clusters and tries to predict a class of an unknown dataset. This problem can be easily related to the current discussion as the zonal activity can be figured out easily. KNN clusters are not labeled at once but can be easily labeled at the end of the algorithm execution. This suits the problem as the city data or the data provided by the government can be used to serve the dataset for the algorithm. The KNN algorithm works on finding the distances

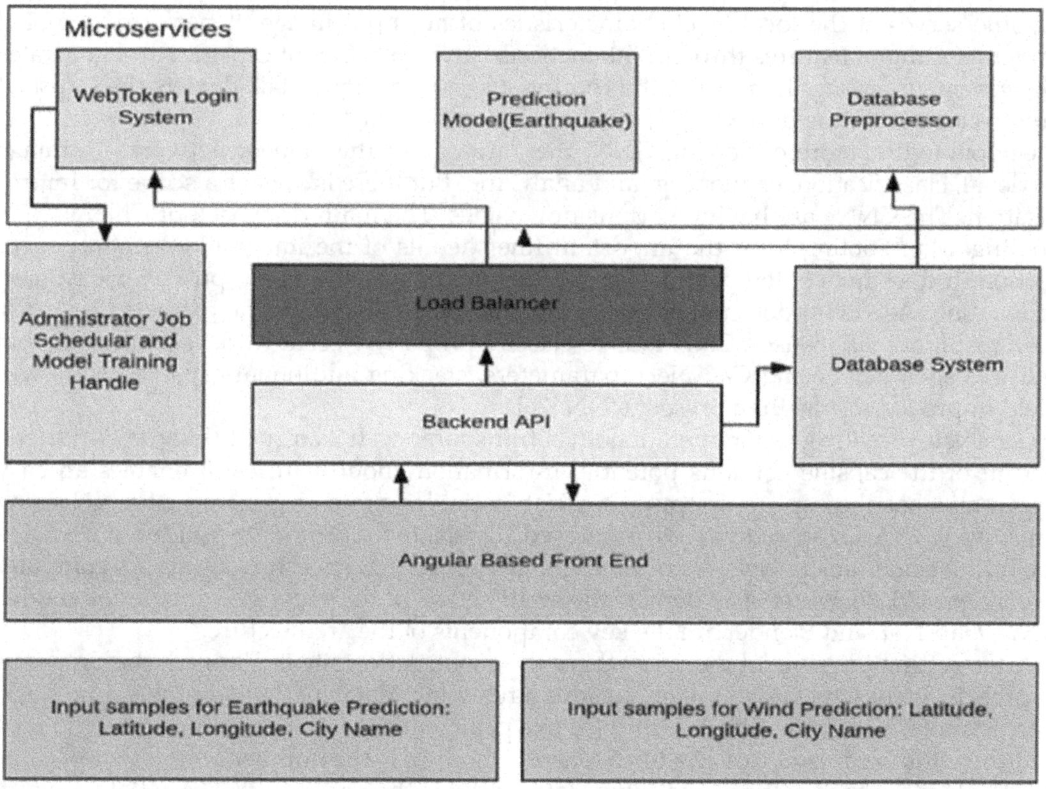

FIGURE 14.1
System architecture.

with each data item from the dataset and then grouping up all the items with near or the same distance parameters. This forms a cluster of the data items which later can be used to predict or classify the label of the unknown item inserted in the model. This is the simplest form of classification which can be used in various machine learning techniques for solving various problems [4].

In this research work, KNN will be able to classify the zones as per the colors represented on the seismic activity map known as BaseMaps data. These map data could be extracted and scaled to fit as per the later prediction input. Each color in the zonal map makes up for a cluster. These clusters can then be used to predict the zone of the specified input from the user. The main advantage of KNN is that it is much simpler than any other model, and the calculations aren't much complex; hence, it decreases the complexity of the task. The only disadvantage of KNN in this approach is that it becomes time-consuming because it has to calculate the distances for each node when the dataset is changed or updated. This would happen very rarely. Therefore, in this context, it is a much more viable and feasible method to implement.

14.3.2 CapsNet

This approach further uses CapsNet [2,26] for generating final output for architectural designs. Sabour et al. [2] introduced the CapsNet architecture in 2017 for applications relating to image processing that overcome the drawbacks of CNN. It is inspired by the concept called inverse graphics. The approach grouped a series of numerous convolutional kernels known as a capsule. The CapsNets contain several such capsules. Every single capsule serves at the local level characteristics of an input image. It performs a global understanding of features through interactions between different capsules and generates an agreement among them using the routing procedure. Figure 14.2 depicts the CapsNet architecture as outlined in Ref. [27] by Shahroudnejad et al.

Undoubtedly before criticizing CNN, they are one of the pioneer networks in image retrieval, classification, captioning, and analyzing. But there is always a scope for improvisation. The CNNs are having various downsides. The main drawback of CNN is Max Pooling. Max Pooling loses the important finer details of the image like location information. It does not contemplate the properties of features like color, pose, velocity, size, orientation. As well it does not consider the relationship between the extracted features. As a result of these causes, CNN fails to classify properly. Recent work recommends that features such as altering CapsNet's parameters, stacking additional capsule layers will yield improved results than present CNN [26].

CapsNets provide greater operation level transparency. It is argued that the vectorized output of the capsule contains potential information about features. It enables an easy explanation of the capsule output. CapsNet improves the transparency of CNN using capsules. Various researchers have employed CapsNets for different application domains including semantic analysis, natural language processing, speech recognition, computer vision, etc. [26,28]. Figure 14.3 depicts the architecture of the suggested prediction model. KNN, CapsNet, and Blender are the key components of the architecture.

After getting the zone of the area where the construction site is planned, the proposed approach fetches the last 5 years seismic and wind data in the preprocessing stage. Furthermore, it identifies the month-wise five peak values for both seismic and wind data and then it is organized in a $12 \times 10 \times 5$ tensor, where 12 is the number of months, 10 is the peak values for each seismic and wind data, and 5 is the year count. These data are also stored in year-wise format for avoiding recomputations. The CapsNet takes this tensor as an input. The CapsNet is the latest and hot variant of CNN that can be used to model

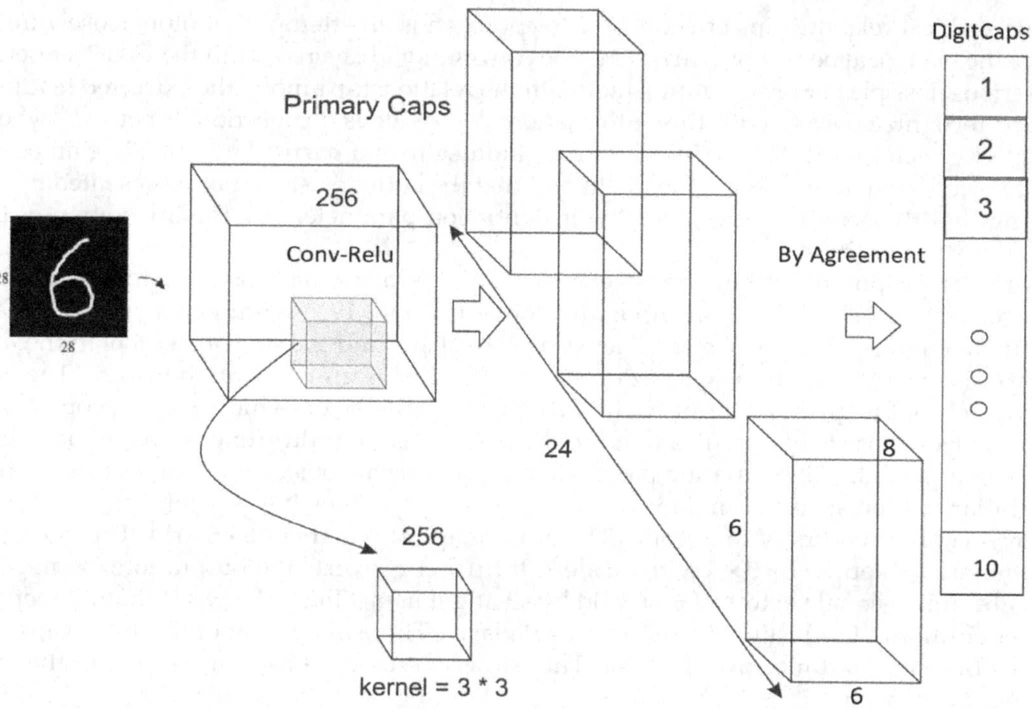

FIGURE 14.2
CapsNet architecture diagram.

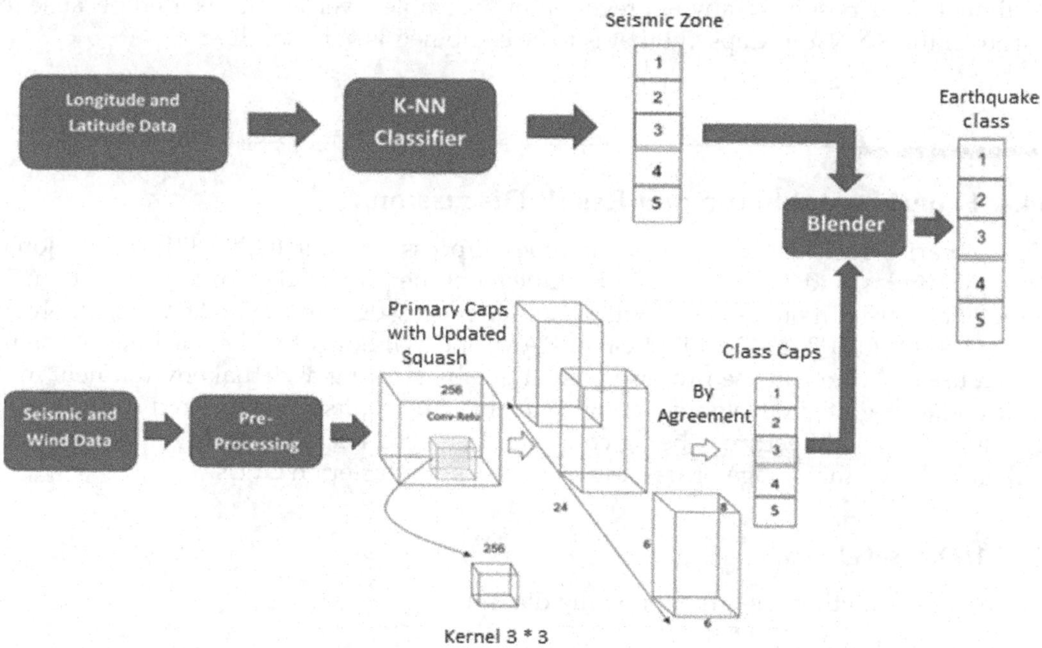

FIGURE 14.3
Proposed prediction architecture.

hierarchical relationships effectively. The approach is an attempt that more closely mimics the biological neural organization. The convolutional layers within the CapsNet detect various possible features in input, maintain the relationship among the extracted features, and recognize objects with this information. A capsule is a collection of neurons whose activity vector indicates the instantiation parameters of a particular entity, like an object or object's part. It represents the likelihood that the entity exists or not represented by the length of the activity vector, and the instantiation parameters by the orientation of the activity vector [29].

In the proposed system, CapsNet's first layer is a normal convolutional layer that receives a 12×10×5 tensor as input and converts it into 192 one-dimensional scalar feature mappings with a 6×6 grid. The stride is kept at 1 and the size of a kernel in the first layer is fixed to 3×3. It makes use of the rectified linear unit activation feature. The first layer output is given as input to the Primary Capsule layer, which is the second layer. This layer consists of capsules that are the basic level of multi-dimensional entities that try to capture the instantiation parameters of the various objects in the input image in a similar fashion as found in an inverse graphics perspective. It is a second convolutional layer configured firstly to output 192 feature maps of scalars of 6×6 grid. The kernel or filter size is kept to be 3×3 with stride 1. It further converts the output into 24 maps of eight-dimensional vectors of 6×6 grid by slicing it using TensorFlow's reshape function. The third and last layer is the class capsule layer. There are a total of five class capsules that output a 16-dimensional vector. There are two stages of routing present in the proposed approach.

The KNN outputs five seismic zones, CapsNet outputs five classes of the probability of earthquake risk. This information is further combined through a blender layer [30], and it is used to provide the five different classes of earthquakes. The classes predicted are no earthquake, mild earthquake, medium earthquake, strong earthquake, and very strong earthquake. Purposefully, any aggregation or ensemble layer is not inserted because the output of the KNN and CapsNet that is to be combined is very small.

14.4 Experimental Setup and Result Discussion

Core libraries such as TensorFlow Graphics Processing Unit (GPU) [31], Flask, Joblib, Numpy, Pandas, and Tqdm are used to implement the proposed technique in Python 3.7. Simulations are run on a system with a Core i3 processor running at 2.5 GHz, an Nvidia GeForce GTX 1080 Ti 2 GB GPU, 4 GB of RAM, and Windows 10. The machine is a setup, which uses GPU to run the proposed technique by creating a virtual environment in an Anaconda-integrated development environment. The results are compared and assessed. The suggested architecture is being configured on GPU because recent implementations [32] have shown that image processing techniques run faster on GPUs.

14.4.1 Datasets Used

The proposed method uses the following datasets.

- **WindData**
  It contains the raw wind speed data as well as other wind-related information.

- **ZoneData**
  A zonal statistic tool calculates a statistic based on values of another dataset for individual zone defined by the zone dataset. For each zone in the input zone dataset, a single output value is produced.
- **BaseMaps**
  It serves as a starting point for overlaying data from layers and visualizing geographic data.

14.4.2 Result Comparison

The proposed technique is compared with the current approaches such as DNN [14], LSTM [15], RNN, BP, RBF [12], RNN, PRNN, RF, LPB [9], and DNN, GLM, RF, GBM [10] in terms of performance. On a dataset with distinct samples, many iterations are conducted. The mean results are then calculated by averaging the results. Table 14.1 demonstrates the comparison of these results to existing methodologies. The proposed method has a good accuracy rate in terms of classification. Due to the right selection of data for training the networks, the suggested strategy outperforms in terms of processing time. Furthermore, the organization of KNN and CapsNet work in parallel on two distinct datasets in two different channels, and their outputs are blended to produce the final output. DNN [14] had an accuracy of 83% for predictions made on April 18, 1995, and a true positive rate of 86% for predictions made on April 19, 1995, before the Kozani Grevena earthquake. For the KiK-net dataset, this model correctly predicted 84.21% of samples. The accuracy of LSTM [15] with one-dimensional input was 63.50%. For the (5 5) subregion, this model attained an accuracy of 86% using two-dimensional input.

The LPB technique introduced in Ref. [9] performed best with 65% accuracy for unseen data, whereas RNN attained 64% accuracy. When compared to the other techniques used in this study, PRNN produced the greatest outcomes. The MAE for the research in Ref. [10] was 0.74. The accuracy of the research in Ref. [12] was 80.85%, whereas the accuracy of the research in Ref. [10] was 85.41%. With an accuracy of 100%, the LSTM model fared the best; however, the computation time of LSTM is a serious issue. The PRNN approach attained an accuracy of 57.97%. The DL-based approaches provided medium performance.

TABLE 14.1
Results Comparison

Approach	Dataset	Performance
DNN [14]	SES of April 18 and 19, 1995	Acc.=83% and 86% (anomaly detection)
	KiK-net	Acc.=84.21% (magnitude prediction)
LSTM [15]	USGS	Acc.=88.57% with 2D input
RNN, PRNN, RF, LPB [9]	CES	Acc.=79% (PRNN and LPB)
	USGS	Predicted location with 15–39 km error
RNN, BP, RBF [12]	SCEC	Acc. 80.85% (RNN)
GLM, DL, RF, GBM [10]	NCEDC	MAE=0.74 (RF), Acc. 85.41%
Proposed (KNN, CapsNet)	WindData, ZoneData, BaseMaps	MAE=0.25, Acc.=87.93

Abbreviations: Acc., accuracy; MAE, mean absolute error; SES, seismic electric signal; KiK-net, Kiban Kyoshin network; SCEC, South California Earthquake Data Center; USGS, United States Geological Survey; CES, Center for Earthquake Studies; NCEDC, National California Earthquake Data Center.

The proposed system attained an acceptable accuracy of 87.93% and an MAE of 0.23. The system combines cutting-edge technology with various NN types and their features to provide benefits such as global coverage, efficacy, and dynamicity.

14.5 Conclusion

This chapter tries to lay out the foundation for the models to predict the zonal levels and seismic data which are needed for earthquake prediction. In addition, this research compares various NNs that can be utilized to handle such a task. The CapsNet employed in this chapter is the best network for predicting earthquakes and other challenges in the same domain. Traditional NNs can obtain superior classification results with far more efficient networks than RBF networks for classification issues; however, the training time of these conventional networks is an issue. This method used an efficient method for feature selection, which further reduced the training time. The longitude and latitude data are entered into the KNN classifier, which produces one of five seismic zones. CapsNet, the most appropriate and up-to-date NN, is then utilized for prediction. For intermediate earthquake prediction, CapsNet accepts seismic and wind data. Finally, for final earthquake prediction, it employs the blender network, which mixes the output of KNN and CapsNet. In terms of performance and outcomes, the suggested approach is compared with current earthquake prediction systems. The comparison shows that the proposed method produces promising and trustworthy outcomes. It outperforms other earthquake prediction algorithms in many ways. The proposed system has acceptable values of MAE and accuracy. The system combines cutting-edge technology with various NN types and their features to provide benefits such as global coverage, efficacy, and dynamicity.

References

1. Y. Sherki, N. Gaikwad, J. Chandle and A. Kulkarni, "Design of real time sensor system for detection and processing of seismic waves for earthquake early warning system," in *International Conference on Power and Advanced Control Engineering (ICPACE)*, 2015, pp. 285–289, doi: 10.1109/ICPACE.2015.7274959.
2. S. Sabour, N. Frosst, and G. E. Hinton. 2017. "Dynamic routing between capsules," *Advances in Neural Information Processing Systems* (October): 3856–3866, *NIPS'17: Proceedings of the 31st International Conference on Neural Information Processing* Systems, December 2017.
3. Sharma, L., and Garg, P.K. (Eds.). (2021). *Artificial Intelligence: Technologies, Applications, and Challenges* (1st ed.). Chapman and Hall/CRC. doi: 10.1201/9781003140351.
4. S. D. Pande and M. S. R. Chetty. 2020. Linear Bezier curve geometrical feature descriptor for image recognition. *Recent Advances in Computer Science and Communications* 13, no. 5 (September): 930–941.
5. D. Sun and B. Sun. 2010. "Rapid prediction of earthquake damage to buildings based on fuzzy analysis," *Seventh International Conference on Fuzzy Systems and Knowledge Discovery*, FSKD 2010, 10–12 August 2010, Yantai, Shandong, China, pp. 1332–1335.

6. J. Veri and T. Y. Wah. 2012. Earthquake prediction based on the pattern of points seismic motion. *International Conference on Advanced Computer Science Applications and Technologies (ACSAT)* (May): 209–212.
7. R. Deo and R. Chandra. 2016. Identification of minimal timespan problem for recurrent neural networks with application to cyclone wind-intensity prediction. *International Joint Conference on Neural Networks* (July): 489–496.
8. J. A. Gazquez, R. M. Garcia, N. N. Castellano, M. Fernandez-Ros, A. J. Perea-Moreno, and F. Manzano-Agugliaro. 2017. Applied Engineering Using Schumann Resonance for Earthquakes Monitoring. *Applied Sciences* 7, no.11 (October): 1113: 1–19.
9. K. M. Asim, F. M. Álvarez, A. Basit, and T. Iqbal. 2017. Earthquake magnitude prediction in Hindukush region using machine learning techniques. *Natural Hazards* 85, no. 1 (September): 471–486.
10. G. Asencio–Cortés, A. Morales–Esteban, X. Shang, and F. Martínez–Álvarez. 2018. Earthquake prediction in California using regression algorithms and cloud-based big data infrastructure. *Computers & Geosciences* 115 (June): 198–210.
11. S. C. Lee and S. W. Han. 2002. Neural-network-based models for generating artificial earthquakes and response spectra. *Computers & Structures* 80, no. 20–21 (August): 1627–1638.
12. A. Panakkat and H. Adeli. 2007. Neural network models for earthquake magnitude prediction using multiple seismicity indicators. *International Journal of Neural Systems* 17, no. 1 (February): 13–33.
13. A. Panakkat and H. Adeli. 2009. Recurrent neural network for approximate earthquake time and location prediction using multiple seismicity indicators. *Computer Aided Civil and Infrastructure Engineering* 24, no. 4 (March): 280–292.
14. S. Kanarachos, S.-R.-G. Christopoulos, A. Chroneos, and M. E. Fitzpatrick. 2017. Detecting anomalies in time series data via a deep learning algorithm combining wavelets, neural networks and Hilbert transform. *Expert Systems with Application* 85 (November): 292–304.
15. T. Bhandarkar, N. Satish, S. Sridhar, R. Sivakumar, and S. Ghosh. 2019. Earthquake trend prediction using long short-term memory RNN. *International Journal of Electrical and Computer Engineering* 9, no.2 (April): 1304–1312.
16. Z. Bao, J. Zhao, P. Huang, S. Yong, and X. Wang. 2021. A deep learning-based electromagnetic signal for earthquake magnitude prediction. *Sensors* 2021, no. 21, 4434 (July): 1–12.
17. Q. Wang, Y. Guo, L. Yu, and P. Li. 2020. Earthquake prediction based on spatio-temporal data mining: An LSTM network approach. *IEEE Transactions on Emerging Topics in Computing* 8, no. 1 (March): 148–158.
18. S. L. Bangare and S. T. Patil. 2015. Implementing tumor detection and area calculation in MRI image of human brain using image processing techniques. *International Journal of Engineering Research and Applications* 5, no. 4, part -6 (April): 60–65.
19. S. L. Bangare and S. T. Patil. 2015. Reviewing Otsu's method for image thresholding. *International Journal of Applied Engineering Research* 10, no. 9 (May): 21777–21783.
20. S. L. Bangare et al. 2018. Regenerative pixel mode and tumor locus algorithm development for brain tumor analysis: a new computational technique for precise medical imaging. *International Journal of Biomedical Engineering and Technology* 27, no.1/2 (January): 76–85.
21. S. L. Bangare et al. 2017. Neuroendoscopy adapter module development for better brain tumor image visualization. *International Journal of Electrical and Computer Engineering* 7, no. 6 (December): 3643–3654.
22. G. Awate, S. L. Bangare, G. Pradeepini, and S. T. Patil. 2018. Detection of alzheimers disease from MRI using convolutional neural network with tensorflow. *arXiv preprint arXiv* (June): 1806.10170.
23. D. Jozinović, A. Lomax, I. Štajduhar, and A. Michelini. 2020. Rapid prediction of earthquake ground shaking intensity using raw waveform data and a convolutional neural network. *Geophysical Journal International* 222, no. 1 (March): 1379–1389.
24. R. Jena, B. Pradhan, S. Naik, and A. Alamri. 2021. Earthquake risk assessment in NE India using deep learning and geospatial analysis. *Geoscience Frontiers* 12, no. 3 (May): 101110. doi: 10.1016/j.gsf.2020.11.007.

25. S. Cui, Y. Yin, D. Wang, Z. Li, and Y. Wang. 2021. A stacking-based ensemble learning method for earthquake casualty prediction. *Applied Soft Computing* 101 (March): 107038. doi: 10.1016/j.asoc.2020.107038.
26. S. D. Pande and M. S. R. Chetty. 2018. Analysis of capsule network (Capsnet) architectures and applications. *Journal of Advanced Research in Dynamical & Control Systems* 10, no. 10 (October): 2765–2771.
27. A. Shahroudnejad, A. Mohammadi, and K. N. Plataniotis. 2018. Improved explainability of capsule networks: Relevance path by agreement. *arXiv preprint arXiv* (February):1802.10204.
28. R. Ma, T. Yu, X. Zhong, Z. L. Yu, Y. Li, and Z. Gu. 2021. Capsule network for ERP detection in brain-computer interface. *IEEE Transactions on Neural Systems and Rehabilitation Engineering* 29 (April): 718–730.
29. S. D. Pande and M. S. R. Chetty. 2021. Fast medicinal leaf retrieval using CapsNet. In: Bhattacharyya S., Nayak J., Prakash K.B., Naik B., Abraham A. (eds) *International Conference on Intelligent and Smart Computing in Data Analytics.* Advances in Intelligent Systems and Computing 1312 (February): 8–16. Springer, Singapore. doi: 10.1007/978-981-33-6176-8_16.
30. Z. Zhou, J. Wu, and W. Tang. 2002. Ensembling neural networks: Many could be better than all. *Artificial Intelligence* 137, no. 1–2 (May): 239–263.
31. M. Abadi, P. Barham, J. Chen, Z. Chen, A. Davis, J. Dean, et al. 2016. Tensorflow: A system for large-scale machine learning. *12th USENIX Symposium on Operating Systems Design and Implementation* (November): 265–283.
32. S. D. Pande and M. S. R. Chetty. 2019. Bezier curve based medicinal leaf classification using capsule network. *International Journal of Advanced Trends in Computer Science and Engineering* 8, no. 6 (December): 2735–2742.

15

Computer-Aided Lung Cancer Detection and Classification of CT Images Using Convolutional Neural Network

Sunil L. Bangare
Department of I.T., Sinhgad Academy of Engineering, Pune, India

Lavanya Sharma
Amity Institute of Information Technology, Amity University, Noida, India

Aditya N. Varade , Yash M. Lokhande, Isha S. Kuchangi, and Nikhil J. Chaudhari
Department of I.T., Sinhgad Academy of Engineering, Pune, India

CONTENTS

DOI: 10.1201/9781003244165-19

15.1 Introduction

Most lung cancers are certainly caused by harmful and life-threatening lifestyles of people in the world. However, early prognosis and treatment of the cancer by interpreting CT images can save lives. Therefore, computer-aided diagnosis can help doctors pinpoint the cancer cells. Computer-related techniques such as deep learning and image processing were also implemented. Diagnosing cancer is one of the most challenging tasks faced by radiologists. An intelligent computerized diagnostic system can be very useful for radiologists to detect, predict, and diagnose lung cancer. Detecting and predicting the type of cancer at the earliest reference point stage can be very useful in improving patient survival. Cancers are the second most significant threat of death on earth, after heart attack as mentioned by Punithavathy et al. [1]. It is the cell growth in tissues of the lungs mainly due to smoking. Cancer remains a delicate subject. Million of people throughout the world are diagnosed with cancer and there is still no final cure. In India, lung cancer is one of the most observed types of cancer in both male and females. According to Cancer Statistics 2020, the projected incidence of patients with cancer in India among males was 1 in 68 males [2]. The survival rate is ensured by a survival probability of only 18% for a survival period of 5 years, as per Mia and Yusuf [3]. The biggest problem with such a high mortality rate is late detection which leads to delayed care. The probability of survival will increase to 60%–70% if lung cancer is diagnosed in early stage. Based on the cell characteristics, this cancer can be categorized into non-small cell and small cell lung cancer [4]. The first one is the most prevalent form of lung cancer, accounting for 90%–95% of all cases, whereas the second one accounts for 15%–18% of all cases [5–7]. It is also classified into stages based on the degree to which the cancer has spread.

15.2 Literature Survey

In the past few days, convolutional neural networks (CNNs) have become a successive method for medical image (MI) analysis. Steva et al. [8] presented the CNN-based system to classify skin lesions, which attains comparable efficiency. So far, the approaches suggested for lung cancer diagnosis have mostly dealt with radiology. CNN is used with a precision of 86.4% in Ref. [9] to identify photographs of pulmonary nodules. Cell-level CNNs have been used in automated pathology activities such as mitosis detection [10] and cell nucleus detection [11]. CNN architectures such as GoogLeNet [12], VGG-Net [13], and ResNet [14] could be trained. Ref. [2] describes the approach that produces the best outcomes in this competition. It uses a patch-based grouping to distinguish tumor patches from standard patches by combining two GoogLeNet architectures, one with and one without rough negative mining. To predict tumor proliferation, a function vector is fed into the SVM classifier. In Ref. [15], author defined the proposed architecture, which consists of several streams of 2-D ConvNets, the outputs of which are merged using a dedicated fusion process to achieve the final classification; however, the morphological variance of the nodules is normally greater than that of a detection algorithm. Narmada et al. [16] suggested a novel method for classifying both small cell and large cell lung cancers, as well as predicting their form and treatment, using CNN. This chapter also touches on the pre-processing and segmentation methods used to improve prediction accuracy. The results of the experiment with Python—Tensor Flow and the Kaggle picture data collection demonstrate that compared to the prior art, the classification and prediction process is superior. Bangare et al. [17–20] have worked on the MI processing and provided solutions to the medical domain issues. Bangare et al. [21] have proposed the detection of Alzheimer's disease by using convolutional neural networks, CUDA, and CuDNN with minimal error rate. Gulati et al. [22,23] have presented IOT and Interface work.

15.3 Proposed System

15.3.1 Flow of Proposed System

To find accurate findings for lung cancer detection, CNN methodology is used. As illustrated in Figure 15.1, the operation is broken down into numerous sections.

This work presents a system that uses a CT scan to detect cancer and displays the results on a web page using the CNN algorithm. For obtaining accurate results, this system makes

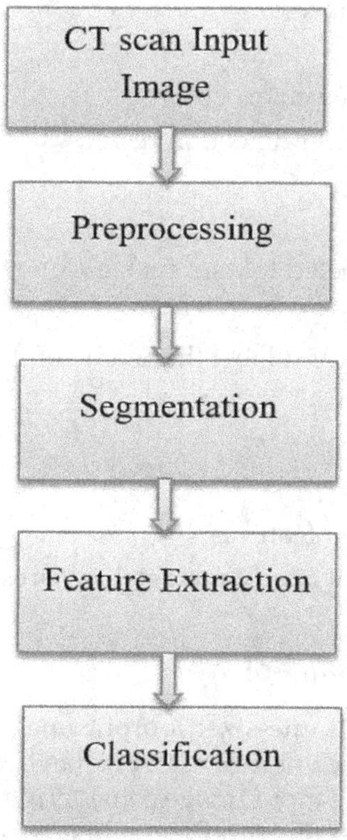

FIGURE 15.1
Flow diagram of proposed system.

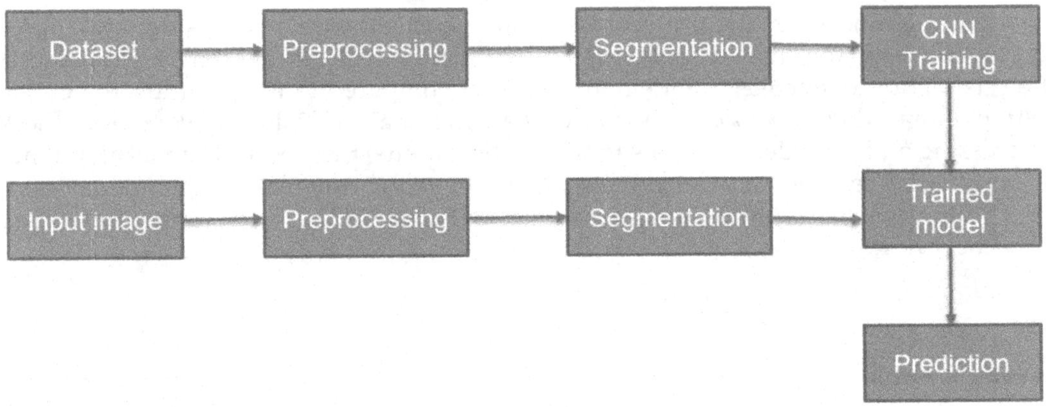

FIGURE 15.2
Architecture of proposed system.

use of a variety of datasets. Feature extraction and reduction models are used in this system. It aids in reducing detection time and increasing accuracy (Figure 15.2).

The steps of the projected algorithm are given below:

Step 1: Upload a CT scan image.
Step 2: Image enhancement.
Step 3: Propagation of feature vector.
Step 4. Suggesting a lung cancer detection result.

Steps of CNN algorithm:

1. Classify dataset under labeled folders such as lung cancer images.
2. Read dataset.
3. Read features of all image and label (her name of dataset folder) of it.
4. Store it in model file.
5. Get input image.
6. Read features of input image.
7. Compare features of stored features.
8. Show label as prediction of nearly matched features.

15.3.2 Morphological Operation

These operations produce the same-sized output image by applying a structuring element (SE) to an input image. The value of each pixel in the output image is certain by comparing the matching pixel in the input image considering its neighbors in a morphological operation. In order to enhance the pixel elements, two methods, dilation and erosion, are used in this proposed work. Dilation increases the number of pixels upon the edges of objects in an image, whereas erosion reduces the number of pixels upon the edges of objects [24,25].

15.3.3 Design of the Proposed System

Use-case diagrams of the system are given in Figures 15.3–15.6.

15.4 Experimental Setup of the Proposed System

15.4.1 Dataset

For experimental analysis, lung cancer CT images of the Iraq-Oncology Teaching Hospital/ National Center for Cancer Diseases (IQ-OTH/NCCD) are used. IQ-OTH/NCCD images were collected at specialist hospitals for 3 months in 2019. This dataset contains 1,095 CT images of

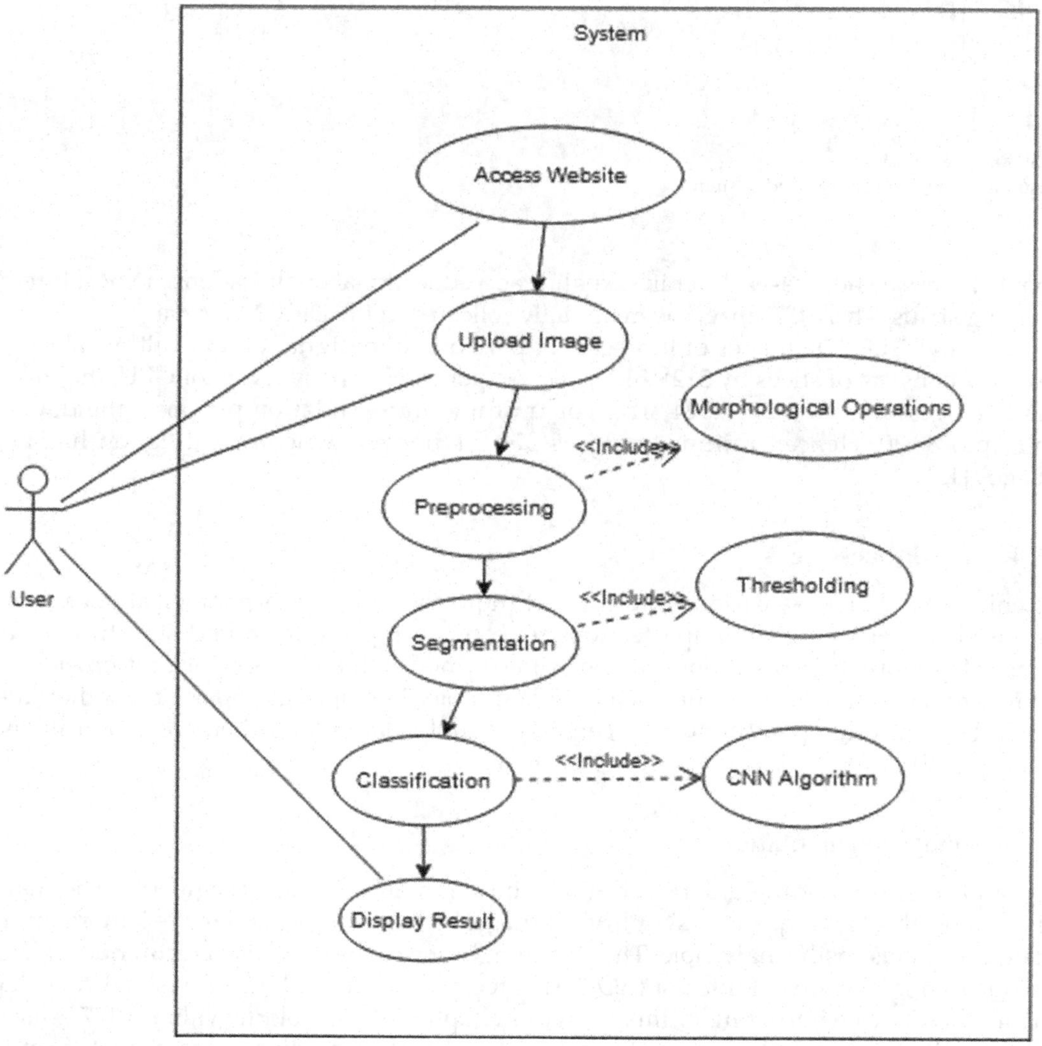

FIGURE 15.3
Use-case diagram of proposed system.

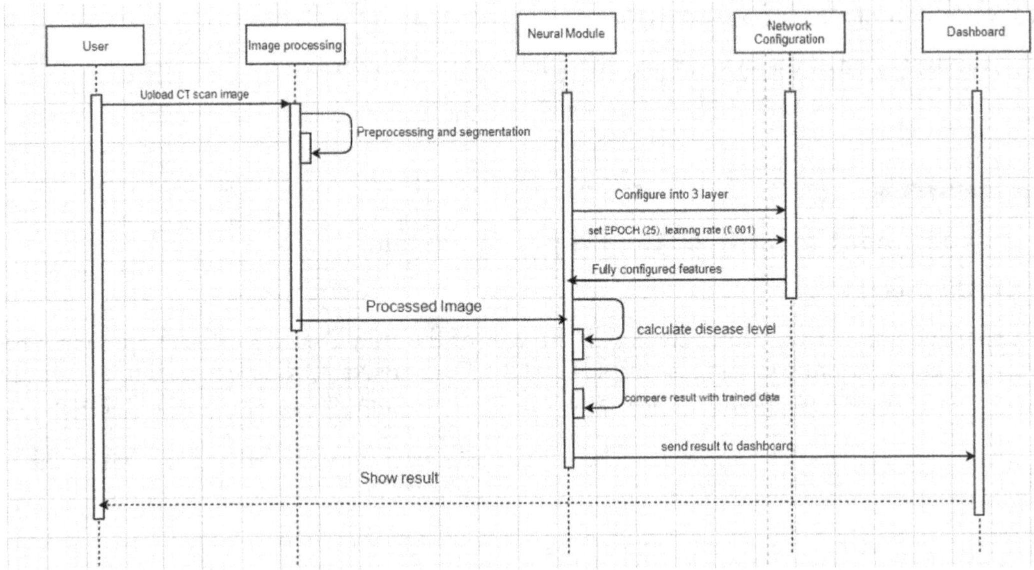

FIGURE 15.4
Sequence diagram of proposed system.

about 110 cases. These cases differ in sexuality, age, educational qualifications, home address, or living status. These CT images were initially collected in the DICOM format.

It also includes CT images of lung cancer patients in malignant and healthy subjects. Each CT consists of slices by 512×512 sizes. As per the memory usage of GPU, the input has been rescaled to a 224×224 size. For training and validation purpose, the data is split into 80:20. Hence, training contains 874 CT images, whereas validation has 221 images [15].

15.4.2 Pre-Processing

To enhance CT images, various filtering techniques are used. Morphological operations are used for removing the imperfections by considering the form and structure of an image. Here opening operation has been performed, which consist of first erosion to eliminate holes and then dilation which fills in hole. The opening operation is then followed by a closing operation which consists of first dilation and then erosion of image (Figure 15.7).

15.4.3 Image Segmentation

It is a process of dividing a digitized image into numerous smaller segments. The main objective of this technique is to simplify the image representation, and make it more informative and thus easily analyzable. Thresholding is one of the most used techniques that is used to extract the area of interest (AOI). This technique is based on a thresholding value to turn a gray-scale into a binary image. In this chapter, a thresholding value of 127 is used for analysis purpose. If the pixel value is less than the thresholding value, then it is set to 0; otherwise, it is set to a maximum value of 1 (Figure 15.8).

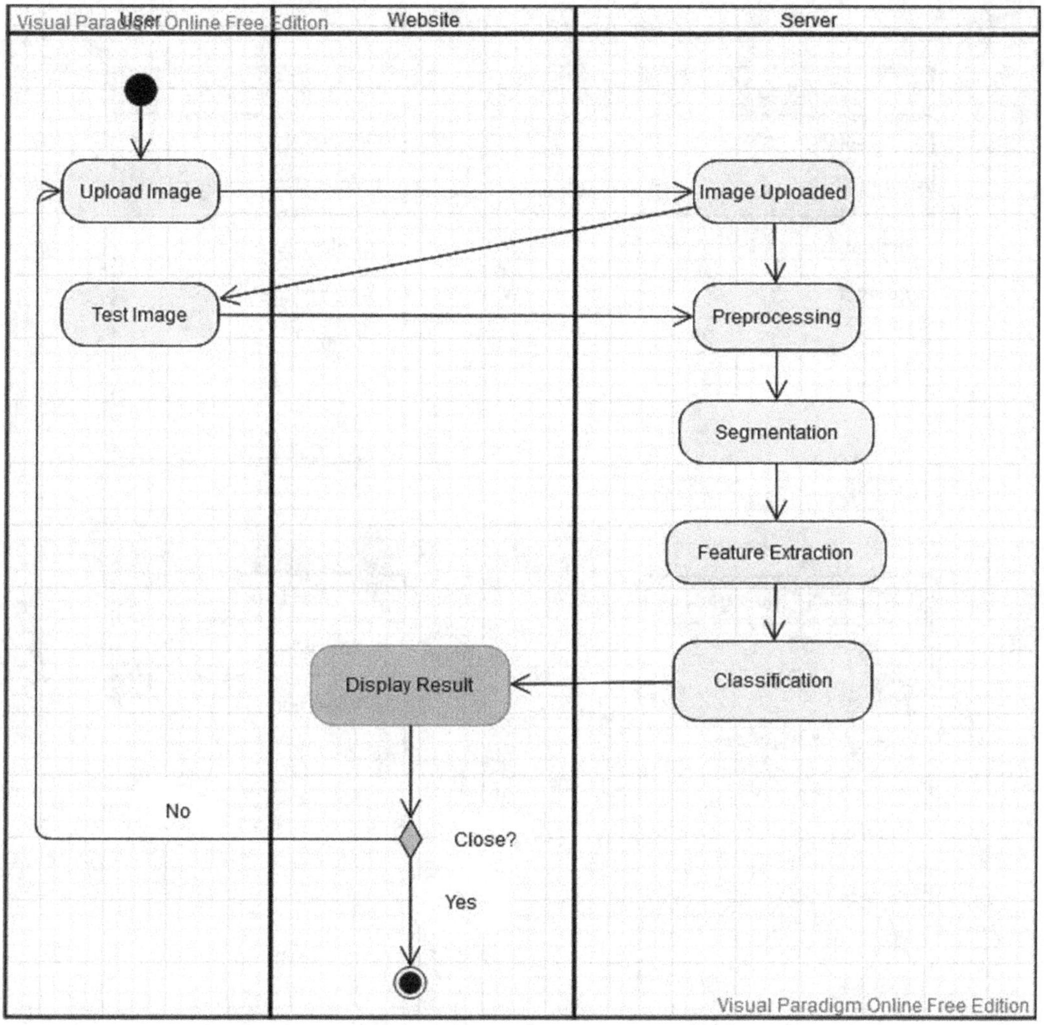

FIGURE 15.5
Activity diagram of proposed system.

15.4.4 Convolutional Neural Network (CNN)

It is used for analysis of visual imagery. It is nothing but an advanced form of multilayer perceptron and is usually designed for two-dimensional structure image. It consists of a multiple convolutional layers and subsampling layers connected to a fully connected layer. The network acquires to optimize the filters or kernels through automated learning, whereas in conventional algorithms, these filters are hand-engineered.

15.4.5 Proposed CNN Model for Lung CT Classification

The main structure of CNN architecture used for feature extraction consists of six convolutional layers (CL), followed by six max-pooling layers (MPLs) and two fully connected layers. The size of kernel in all the convoluting layers is 3×3, whereas every pooling layer

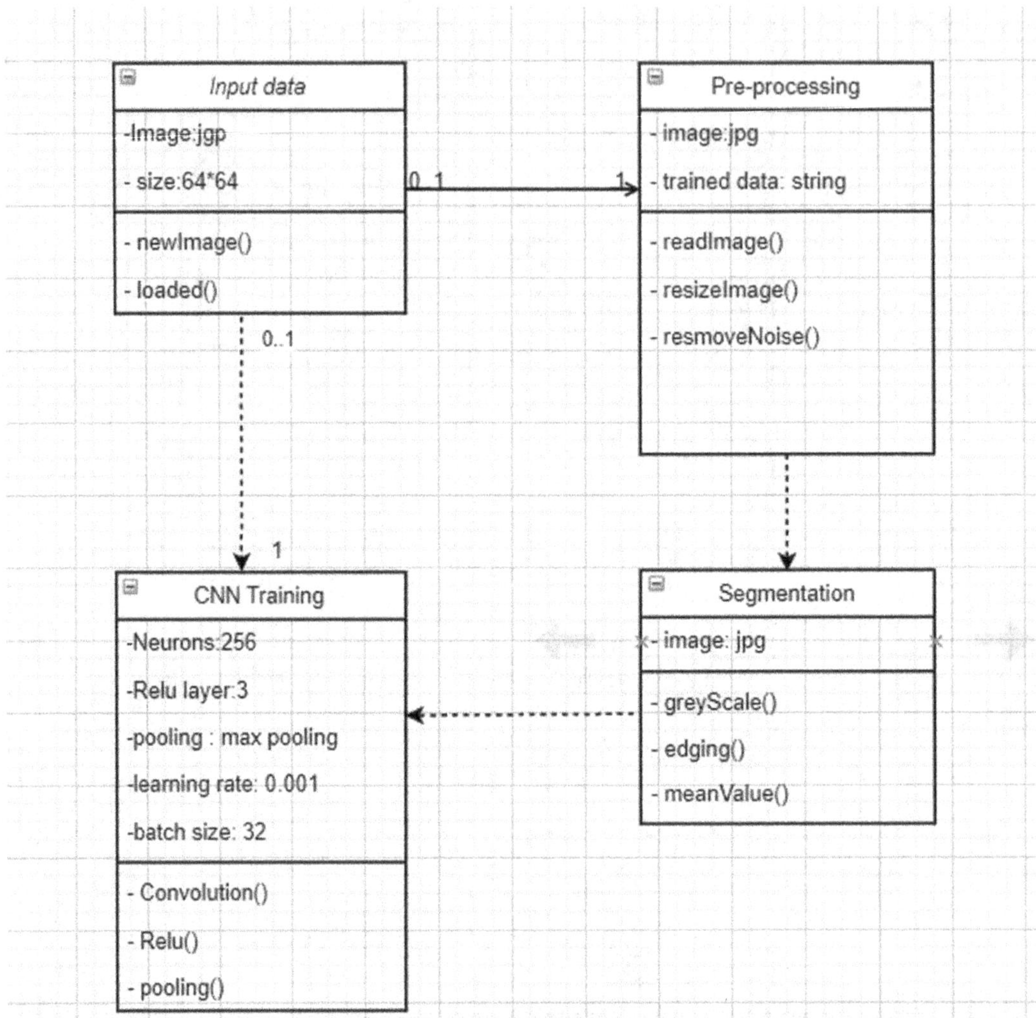

FIGURE 15.6
Class diagram of proposed system.

has a pool size of 2×2. For non-linearity, the ReLu activation function is applied to the output of each CL and fully connected layers. Padding of size 1×1 is done at each convolution layer. Adding to this, the number of kernels that are on feature maps in convoluting layers is 64, 64, 128, 128, 256, and 256, respectively. MPL is used to lessen the output size of the previous CL by applying filters of size 2×2, and it also helps to avoid overfitting. Batch normalization is applied to maintain the mean output close to zero and the output standard deviation close to one. Every layer in the network is followed by batch normalization (Table 15.1).

The model continues by the remaining of the layers until two fully connected layers where all neurons are connected to the previous layer. The first fully connected layer consists of 4,096 neurons for learning along with ReLu activation function and dropout of 0.25. The second dense layer has 256 neurons taking input from the previous dense layer. Softmax classifier is used to classify CT images with malignant and

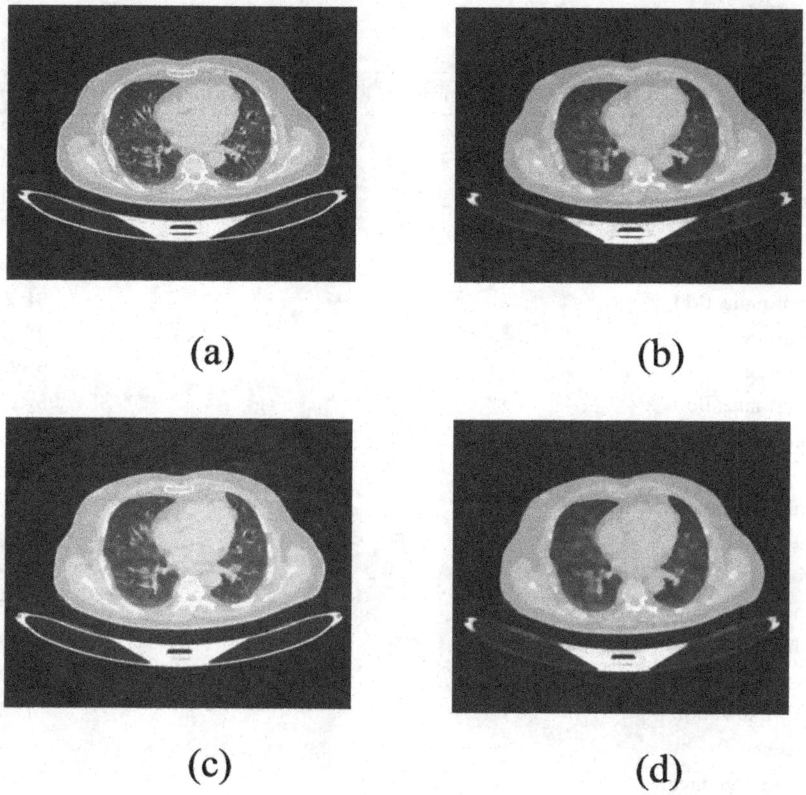

FIGURE 15.7
(a) Original CT image. (b) After closing operation. (c) After opening operation. (d) Opening followed by closing.

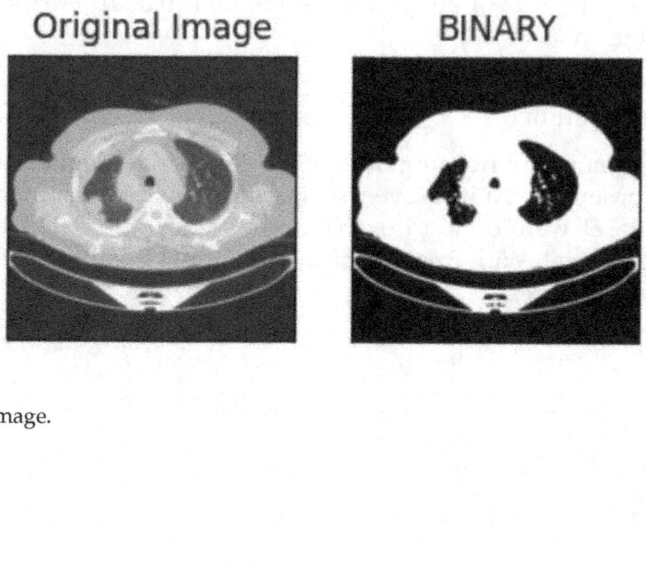

FIGURE 15.8
Original and binary image.

TABLE 15.1

Layers, Kernel, and Output

Layer Type	Number of Kernels	Kernel Size	Output Size
Convolutional	64	3×3	222×222×64
Max pooling			111×111×64
Batch normalization			111×111×64
Convolutional	64	3×3	109×109×64
Max pooling			54×54×64
Batch normalization			54×54×64
Convolutional	128	3×3	52×52×128
Max pooling			26×26×128
Batch normalization			26×26×128
Convolutional	128	3×3	24×24×128
Max pooling			12×12×128
Batch normalization			12×12×128
Convolutional	256	3×3	10×10×256
Max pooling			5×5×256
Batch normalization			5×5×256
Convolutional	256	3×3	3×3×256
Max pooling			1×1×256
Batch normalization			1×1×256
Flattening			1×256
Fully connected layer			1×4,096
Fully connected layer			1×256
Fully connected layer with Softmax			1×2

non-malignant. Here, Stochastic Gradient Descent (SGD) Optimizer is used because any model that uses batch normalization (BN) can be optimized by using this optimizer. In addition to this, batch size, i.e., number of samples through 1 training cycle, is 32, whereas the learning rate of our proposed model is 0.0001. Furthermore, the model is trained on 200 epochs as well.

15.4.6 Data Augmentation

Due to the limited amount of training dataset labeled by radiologists, the initial phase of network was compromised. To improve the classification, data augmentation techniques of the training dataset were done to overcome this issue. Our model further used the enhanced training dataset. With this, the proposed model is well-trained on the learning parameters and data features.

15.5 Performance Evaluation

For quantitative analysis, the given performance metrics are used (Figure 15.9):

		Predicted	
		Negative (N) -	Positive (P) +
Actual	Negative -	True Negatives (TN)	False Positives (FP) **Type I error**
	Positive +	False Negatives (FN) **Type II error**	True Positives (TP)

FIGURE 15.9
Confusion matrix.

Accuracy: it represents the performance metrics of an algorithm

$$\text{Accuracy} = (\text{TP} + \text{TN})/(\text{TP} + \text{FP} + \text{TN} + \text{FN}) \tag{15.1}$$

Precision: it is the percentage of positive instances out of the total predicted positive instances

$$\text{Precision} = \text{TP}/(\text{TP} + \text{FP}) \tag{15.2}$$

Sensitivity (Recall): it indicates the proportion of classified as positive labels have positive class labels

$$\text{Recall} = \text{TP}/(\text{TN} + \text{FN}) \tag{15.3}$$

Specificity: it indicates the amount of classified classes as a negative label has negative class labels

$$\text{Specificity} = \text{TN}/(\text{TN} + \text{FP}) \tag{15.4}$$

***F*1 Score**: positive predictions are actually positives (precision) and don't miss out on positives and predict them negatively (recall)

$$F1\ \text{Score} = 2*(P*R)/(P + R) \tag{15.5}$$

where R represents recall, P is precision, TP is true positive, FP is false positive, TN is true negative, and FN is false negative.

15.6 Experimental Results

The dataset consists of 1,095 total images, which split into 80:20 for training and validation. First, pre-processing and segmentation operations were performed on the whole dataset, and then scans were split into 874 for training and 221 for validation purposes. For testing purposes, nearly 113 scans for malignant and 108 scans for non-malignant class are taken. After splitting those binary images, model is trained with dataset. The model is trained up to 200 epochs, with each epoch consisting of 27 batches, and the proposed system has achieved good results. The graph in Figure 15.10 shows the accuracy achieved by our system. Initially, the accuracy was as low as 50%. The optimization performed at the ending of each epoch consistently increased the accuracy over the training period, resulting in nearly greater than 85% 200th epoch of the training.

The *X*-axis represents epochs, and the *Y*-axis represents accuracy.

After a satisfying accuracy, the model is frozen as a model file with its parameters to keep the learned features and structure of the graph (Figure 15.11).

The confusion matrix can also be obtained through the results predicted labels and actual labels of malignant. Based on the training scans, the number of test scans was tested to obtain this confusion matrix with TP, TN, FP, and FN. From evaluating our model through the confusion matrix, the following values are achieved:

TP:98 TN:93

FP:15 FN:15

The performance metrics (precision, recall, and *F*1-measure/score) can be obtained using the below notations based on the results that are obtained through the confusion matrix. From the above values, the model achieved a sensitivity of 86.11%, accuracy of 86.42%, and specificity of 86.72%. Along with these values, our model has been performed well in a precision of 86.11% and an *F*1 score of 86.11%.

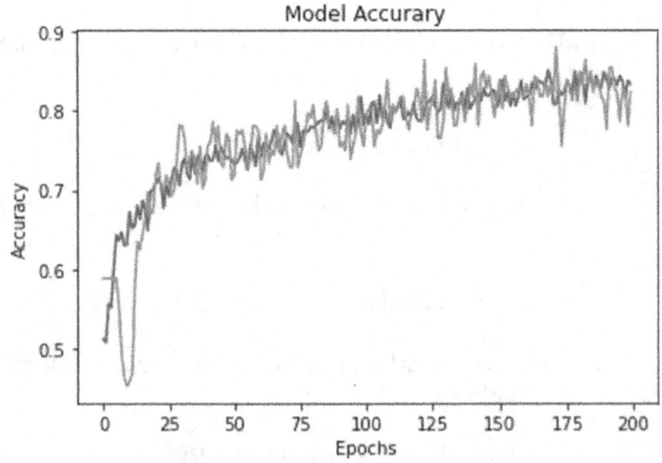

FIGURE 15.10
Accuracy over the training period in epochs.

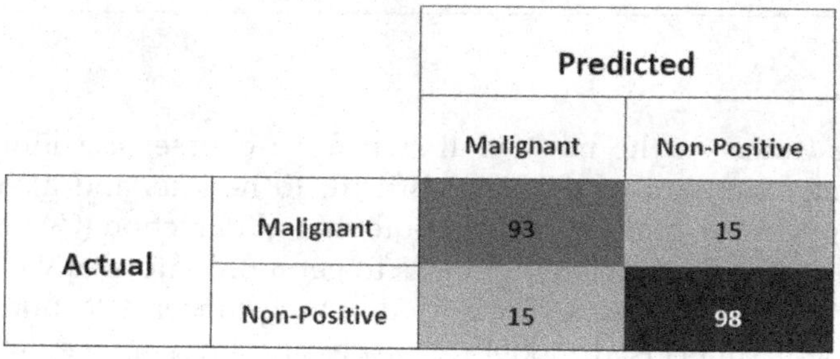

FIGURE 15.11
Confusion matrix results.

15.7 Comparison

For a better illustration of the performance of this method, comparison is shown in Table 15.2. IQ dataset is compared with other works on other datasets, because, on this dataset, any kind of work is not done yet. All the works mentioned above are done using the CNN model proposed by them but on different datasets. Zhao et al. [26] and Masood et al. [27] proposed an approach that made performance up to 80+, but that work was not up to our results even after processing data. Paul et al. [28] proposed a CNN model for private dataset and achieved an accuracy of nearly 78%. Liu et al.'s [29] model has done well on sensitivity, but not in achieving specificity. Rehman et al. [30] and Pradhan et al. [31] acquired better outcomes by using the CNN algorithm, but they don't have much amount of data. Basha et al. [32] and Ma et al. [33] have achieved better outcomes in some factors but didn't achieved the same for others. Our proposed model has achieved good results by achieving sensitivity, accuracy, and specificity by nearly 86%.

TABLE 15.2
Comparison of Results

CAD System	Database	Year	Number of Scans	Accuracy	Sensitivity	Specificity
Zhao et al. [26]	Private	2017	4,581	84.15%	83.96%	84.32%
Masood et al. [27]	LIDC	2018	1,018	86.02%	80.91%	83.22%
Paul et al. [28]	Private	2018	276	76.79%	#	#
Liu et al. [29]	LIDC	2018	1,018	82.51%	96.64%	71.43%
Rehman et al. [30]	JSRT	2020	247	88%	#	#
Pradhan et al. [31]	SPIE	2020	100	83.33%	#	#
Basha et al. [32]	LIDC	2020	1,018	97%	86.5%	80.5%
Ma et al. [33]	Private	2020	1,702	84.1%	91.8#	80.9%
Proposed model	IQ-OTH		1,095	86.42%	86.11%	86.72%

15.8 Discussion

Detection of lung cancer is the most vital step for a course of a lung cancer diagnosis. Manually a huge amount of data are difficult to handle, and also both time- and energy-consuming. On the other hand, computer-aided detection (CAD) systems require segmentation and feature extraction methods. To get more informative information from this large amount of existing data, CNN is used in lung cancer detection and analysis.

CNN is an efficient supervised model for processing huge image data, but still, some deficiencies were still present that needed to be solved. These difficulties were unavailability of resources for heavy models, expensive GPUs, and reliability of data. In these difficulties, dataset is of utmost importance because bad data result in bad learning. In dataset, for the third class, i.e., benign class, a very limited number of scans were available; hence, that class was merged with non-malignant. Initially, the model was overfitting; hence, it introduced data augmentation, and a dropout layer is added to avoid this. Even the complexity was raised of the CNN structure, but the effect of improvement was quite less. In the context of using a CAD system for detection of cancer, the performance can be improved. Various findings are drawn for future perspective. Dataset with more CT images for benign class can make it easy for the model to learn about benign class features. Change in the number of epochs and accurate tuning of other hyperparameters can also help to increase performance.

15.9 Conclusion

This chapter surveys the diverse strategies that can be a part of MI processing and are prominently utilized to come across lung cancer from CT images. This chapter proposed a successful CNN structure for detecting scan whether is cancerous or not using the IQ-OTH dataset. This model was prepared by investigating the effects of learning rate, size of kernel, training or batch size, or dropout on performance. This model perceives a relatively good accuracy of 86% and less prone to overfitting, which is associated with other CNN architectures. The classification model might help to diagnose and treat malignant nodules effectively.

Acknowledgments

We express our gratitude toward IQ-OTH for making such a reliable and accurate dataset available for use. We are thankful to the Department of IT, (STES) Sinhgad Academy of Engineering, Pune, for providing their cooperation and kind support. We are grateful to Dr. K. P. Patil, Principal, Dr. K. M. Gaikwad, Vice Principal, Prof. A. N. Adapanawar, HOD-IT, Dr. G. Pradeepini, Professor-CSE, KLEF, and Dr. S. T. Patil, Professor-CSE, VIT for their motivation, guidance, and support for this work.

References

1. K. Punithavathy, M. M. Ramya, S. Poobal, "Analysis of statistical texture features for automatic lung cancer detection in PET/CT images", *International Conference on Robotics, Automation, Control and Embedded systems (RACE)*, IEEE, Chennai, India, 18–20 February 2015.
2. P. Mathur, K. Sathishkumar, M. Chaturvedi, P. Das, K. L. Sudarshan, S. Santhappan, V. Nallasamy, A. John, S. Narasimhan, F. S. Roselind, ICMR-NCDIR-NCRP Investigator Group, "Cancer Statistics, 2020: report from National Cancer Registry Programme, India", *JCO Global Oncology*, 6, pp. 1063–1075, 2020.
3. B. A. Mia, M. A. Yusuf, "Detection of lung cancer from CT image using image processing and neural network", *International Conference on Electrical Engineering and Information Communication Technology (ICEEICT)*, IEEE, Savar, Bangladesh, May 2015.
4. G. P. Pratap, R. P. Chauhan, "Detection of lung cancer cells using image processing techniques", *1st IEEE International Conference on Power Electronics, Intelligent Control and Energy Systems*, IEEE, Delhi, India, 2016.
5. A. Chaudhary, S. S. Singh, "Lung cancer detection on CT images using image processing", *Computing Sciences 2012 International Conference*, IEEE, Phagwara, India, 2012.
6. N. Hadavi, Md. J. Nordin, A. Shojaeipour, "Lung cancer diagnosis using CT-scan images based on cellular learning automata", *International Conference on Computer and Information Sciences (ICCOINS)*, IEEE, Kuala Lampur, Malaysia, 2014.
7. M. Anshad, S. S. Kumar, "Recent methods for the detection of tumor using computer aided diagnosis", *International Conference on Control, Instrumentation, Communication and Computational Technologies (ICCICCT)*, IEEE, Kanyakumari, India, 2014.
8. M. Keshani, Z. Azimifar, R. Boostani, "Lung nodule segmentation using active contour modeling", *MVIP*, IEEE, Isfahan, Iran, 2016.
9. R. M. Haralick, K. Shanmugam, I. Dinstein, "Textural features for image classification", *IEEE Transactions on Systems, Man and Cybernetics*, 3(6), pp. 610–621, 1973.
10. A. A. A. Setio, F. Ciompi, G. Litjens, P. Gerke, C. Jacobs, S. J. van Riel, M. M. Wille, "Pulmonary nodule detection in CT images: false positive reduction using multi-view convolutional networks", *IEEE Transactions on Medical Imaging*, 35(5), pp. 1160–1169, 2016.
11. J. Xie, R. Girshick, "Unsupervised deep embedding for clustering analysis", *ICML'16: Proceedings of the 33rd International Conference on International Conference on Machine Learning*, ACM, New York, 2016.
12. M. Buty, Z. Xu, M. Gao, "Characterization of lung nodule malignancy using hybrid shape and appearance features", *Medical Image Computing and Computer-Assisted Intervention – MICCAI*, 2016.
13. Y. Zhou, L. Xie, E. K. Fishman, A. L. Yuille, "Deep supervision for pancreatic cyst segmentation in abdominal CT scans", CoRR abs/1706.07346, 2017.
14. M. Buty, Z. Xu, M. Gao, U. Bagci, A. Wu, D. J. Mollura, "Characterization of lung nodule malignancy using hybrid shape and appearance features", *MICCAI*, pp. 662–670. Springer, Cham, 2016.
15. H. Alyasriy, "The IQ-OTHNCCD lung cancer dataset", *Mendeley Data*, V1, 2020. doi: 10.17632/bhmdr45bh2.1.
16. K. Narmada, G. Prabakaran, R. Madhan Mohan, "A study on lung nodule segmentation and classification using supervised machine learning techniques", *International Journal of Computer Sciences and Engineering*, 6(12), pp. 497–503, 2018.
17. S. L. Bangare, S. T. Patil, "Implementing tumor detection and area calculation in MRI image of human brain using image processing techniques", *International Journal of Engineering Research and Applications*, 5(4 Part 6), pp. 60–65, 2015. ISSN: 2248-9622.
18. S. L. Bangare, S. T. Patil, "Reviewing Otsu's method for image thresholding", *International Journal of Applied Engineering Research*, 10(9), pp. 21777–21783, 2015.

19. S. L. Bangare et al., "Regenerative pixel mode and tumor locus algorithm development for brain tumor analysis: a new computational technique for precise medical imaging", *International Journal of Biomedical Engineering and Technology*, 27(1/2), pp. 76–85, 2018.
20. S. L. Bangare et al., "Neuroendoscopy adapter module development for better brain tumor image visualization", *International Journal of Electrical and Computer Engineering*, 7(6), pp. 3643–3654, 2017.
21. G. Awate, S. L. Bangare, G. Pradeepini, S. T. Patil, "Detection of Alzheimer's disease from MRI using convolutional neural network with tensorflow". arXiv preprint arXiv:1806.10170.
22. K. Gulati et al., "Use for graphical user tools in data analytics and machine learning application", *Turkish Journal of Physiotherapy and Rehabilitation*, 32(3), pp. 3540–3546, 2021. ISSN 2651-4451, e-ISSN 2651-446X.
23. L. Sharma, P. K. Garg (Eds.), *Artificial Intelligence: Technologies, Applications, and Challenges* (1st ed.). New York: Chapman and Hall/CRC, 2021. doi: 10.1201/9781003140351.
24. L. Sharma (Ed.), *Towards Smart World*. New York: Chapman and Hall/CRC, 2021. doi: 10.1201/9781003056751.
25. L. Sharma, P. Garg (Eds.), *From Visual Surveillance to Internet of Things*. New York: Chapman and Hall/CRC, 2020. doi: 10.1201/9780429297922.
26. Q. Song, L. Zhao, X. Luo, X. Dou, "Using deep learning for classification of lung nodules on computed tomography images", *Journal of Healthcare Engineering*, 2017, 8314740. doi: 10.1155/2017/8314740.
27. A. Masooda, B. Shenga, P. Lic, X. Houd, X. Weid, J. Qine, D. Feng, "Computer-assisted decision support system in pulmonary cancer detection and stage classification on CT images", *Journal of Biomedical Informatics*, 79, pp. 117–128, 2018.
28. R. Paul, S. H. Hawkins, M. B. Schabath, R. J. Gillies, L. O. Hall, D. B. Goidof, "Predicting malignant nodules by fusing deep features with classical radiomics features", *Journal of Medical Imaging*, 5, p. 011021, 2018.
29. G. Han, G. Zheng, M. Wang, S. Huang, "Automatic recognition of 3D GGO CT imaging signs through the fusion of hybrid resampling and layer wise fine tuning", *Medical & Biological Engineering & Computing*, 2018. doi: 10.1007/s11517-018-1850z.
30. M. Z. Rehman, N. Md. Nawi, A. Tanveer, H. Zafar, H. Munir, S. Hassan, "Lungs cancer nodules detection from CT scan images with convolutional neural networks", *International Conference on Soft Computing and Data Mining*, pp. 382–391, Springer, Cham, 2020.
31. A. Pradhan, B. Sarma, B. K. Dey, "Lung cancer detection using 3D convolutional neural networks", *2020 International Conference on Computational Performance Evaluation*, IEEE, Shillong, India, 2020, 978-1-7281-6644-5/20.
32. Z. Basha, B. L. Pravallika, D. Vineela, S. L. Prathyusha, "An effective and robust cancer detection in the lungs with BPNN and watershed segmentation", *International Conference for Emerging Technology (INCET)*, IEEE, 2020. doi: 10.1109/INCET49848.2020.9154186.
33. Y. Han, Y. Ma, Z. Wu, F. Zhang, D. Zheng, X. Liu, L. Tao, Z. Liang, Z. Yang, X. Li, J. Huang, X. Guo, "Histologic subtype classification of non-small cell lung cancer using PET/CT images", *European Journal of Nuclear Medicine and Molecular Imaging*, 2020. doi: 10.1007/s00259-020-04771-5.

16

Real-Time Implementations of Background Subtraction for IoT Applications

Belmar Garcia-Garcia and Thierry Bouwmans
Lab. MIA, University of La Rochelle, La Rochelle, France

Kamal Sehairi
Lab. LTSS, University of Laghouat, Laghouat, Algeria

El-Hadi Zahzah
Lab. L3i, University of La Rochelle, La Rochelle, France

CONTENTS

16.1 Introduction

The Internet of Things (IoT) applications for smart homes and cities require detecting and recognizing moving objects such as humans, animals, and vehicles to optimize home and traffic surveillance. These applications employ background subtraction algorithms, increasingly used in video surveillance tasks. In practice, most of these applications require real-time processing with lightweight algorithms for global video systems and embedded

DOI: 10.1201/9781003244165-20

systems. Even if background subtraction methods offer a good compromise between robustness and computational time requirement, they often require much processing time and computing resources to be used directly in real-time applications. Thus, different strategies have been developed over time to deal with those computational and memory limitations. For global video systems, two main computing strategies are employed: fog computing and edge computing. Both of them under the same objective carry out tasks of the cloud in a local network. Edge computing is present on the physical devices connected and close to the sensors, while fog computing occurs on the data processed by an IoT gateway or fog node. Once the strategy is decided, different kinds of implementations can be employed as Graphics Processing Unit (GPU) implementations or specific implementations on dedicated cards with digital signal processing (DSP) technology, very large-scale integration (VLSI) technology, or field programmable gate array (FPGA) technology. Real-time strategies are also concerned with the use of a specific language adapted to the chosen architecture. These languages are often C/C+ for central processing unit (CPU) implementations and CUDA or Fermi for GPU implementations. For embedded systems, the strategy used is lightweight algorithms. For all the above mentioned, in this document we present an analysis of real-time implementations of background subtraction methods, surveying GPU implementations, embedded implementations, and design of specific architectures (FPGA, DSP, and VLSI). The following sections of this chapter are structured as follows: In Section 16.2, a short preliminary overview about moving objects detection by background subtraction is presented. Section 16.3 reviews GPU implementation approaches. In Section 16.4, embedded implementations are reviewed for smart cameras and systems. In Section 16.5, the design of specific architectures is surveyed while Sections 16.6 and 16.7 concern parallel implementations and specific programming languages, respectively. Low complexity strategies are analyzed in Section 16.8. Section 16.9 reviews fog computing and edge computing strategies. At the end of this chapter, perspectives and conclusions are presented.

16.2 Background Subtraction: A Short Preliminary Overview

Background subtraction algorithms offer the best compromise between robustness and real-time performance, making them the most popular in the scientific literature. Throughout this section, we will briefly describe the background subtraction process as well as the most representative methods reported so far. Under the assumption that the background of the scene is static and only the objects of interest are in motion, the background subtraction approach can be described (see Figure 16.1). The background initialization computes the first image(s) of the video for the construction of the background model. Background modeling describes the technique used to describe the background. Background maintenance updates the background model over time. Foreground detection refers to the pixel classification mechanism in both background and foreground (objects of interest).

In the scientific literature, there is a big amount of supervised, unsupervised, and semi-supervised methodologies for the background model representation. Readers can refer to valuable books [1,2] and texts with a more detailed analysis on this topic [3–7]. The wide variety of models are based mainly on signal processing, machine learning, and mathematical treatment, usually classified in (i) unsupervised models such as crisp models [8–10], statistical models [11–14], fuzzy models [15–17], Dempster-Schafer models

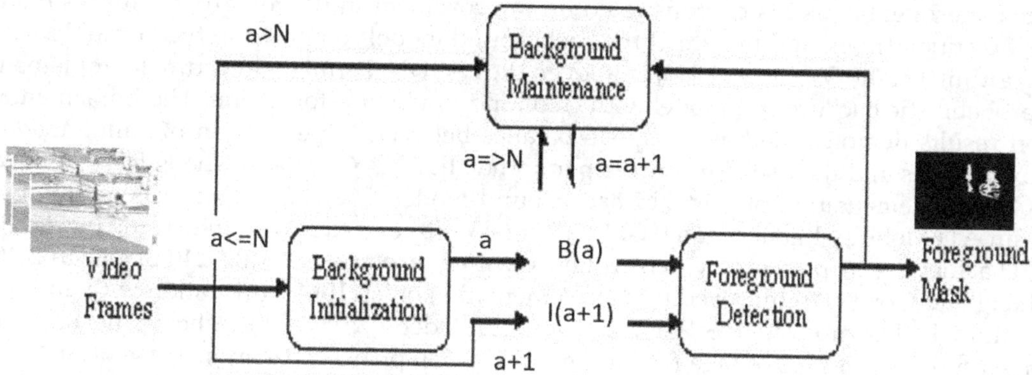

FIGURE 16.1
Stages of a background subtraction algorithm.

[18], subspace learning models [19–23], and robust learning models [24–27], (ii) supervised models such as neural networks models [28–30], and (iii) semi-supervised models such as graph signal processing-based models [31–34]. Most of the time, researchers develop and test on PCs these models, and then provide CPU implementations. However, their aim mainly focuses on the robustness against challenges met in the videos paying less attention to the real-time and memory requirements, and their implementation in embedded systems, portable devices, and IoT applications. In the following, we survey attempts to reach real-time achievements for those diverse cases.

16.3 GPU Implementations

The first way to reach real-time processing is to exploit General Purpose Graphics Processing Unit (GPGPU)'s abilities instead of conventional CPU abilities. Indeed, the advent of specialized multi- and many-core chip multiprocessors such as Compute Unified Device Architecture (CUDA) provided by Nvidia company and Cell Broadband Engine Architecture (CBEA) by IBM offer attractive alternatives to accelerate background subtraction algorithms. In 2005, Griesser et al. [35,36] implemented in a GPU a background subtraction algorithm based on an extended collinearity criterion methodology. This implementation takes less than 4 ms per frame. In 2008, Fabian and Gaura [37] proposed a parallel implementation of the original MOG [13] in a CUDA device for vehicle traffic surveillance and object tracking. This method processes 20 frames per second (fps) for images of size 320 * 240. In 2009, Liyanage [38] also proposed a GPU implementation of the original MOG [13] obtaining similar performances. In 2009, Fauske et al. [39] implemented three background subtraction algorithms using the CUDA technology for parallel processing. The performance observed in CUDA architectures shows execution times below 10 ms per frame and up to 96 times faster than its serial implementation. In 2009, Schreiber and Rauter [38] proposed a background subtraction algorithm in the YCbCr color space, without training, based on a compressed non-parametric model. This method achieved a rate of 635 fps for the background subtraction algorithm, in a video of 352 * 288 pixels of resolution and was implemented on a GeForce 9800 GT GPU graphics card Nvidia.

The experimental results demonstrated an improvement in the foreground masks quality, execution times, and update of the background model compared to the original MOG algorithm. In 2010, Wong et al. [40] employed the NVIDIA Fermi architecture to implement a probabilistic background model with color and gradient information. The implementation results demonstrated an adequate balance between the execution of simultaneous subprocesses and the use of resources per kernel. In 2008, Gong and Cheng [41] provided a GPU implementation of a 1-SVM background model. The experiment was carried out on an ATI Mobility Radeon X1400 GPU, a Lenovo T60 computer with Intel Centrino T2500 CPU achieving 16 fps for the video sequence with a resolution of 320 * 240 pixels. In 2010, Cheng and Gong [42] improved this implementation on an IBM workstation with an Intel Pentium 4 CPU running at 3.4 GHz and an ATI Radeon X1950 GPU. The results showed an execution of 81.5 fps, image resolution of 320 * 240 pixels, with an increase of up to 40 times higher than its implementation in the CPU of the same workstation. In 2011, Cheng and Gong [43] processed motion signals by combining SVM and optical flow algorithms. The experiments were performed on a laptop computer processing QVGA size images at 39.3 fps. In 2006, Fukui et al. [44,45] designed a GPU implementation of moving objects detection under intensity changes and shadows elimination based on the CIELAB color space. This implementation requires about 0.34 ms/frame whereas the CPU implementation needs about 864.53 ms/frame. In 2009, Yamamoto and Iwai [46,47] used the Margined Sign Correlation to propose a background subtraction algorithm implemented in a GPGPU, resulting in 0.13 seconds for each frame. In 2010, Culibrk and Crnojevic [48] proposed to parallelize the operations done for each pixel in the background modeling neural networks on NVIDIA GPU. Several modifications to the original algorithm are made for an efficient implementation. This approach achieves real-time processing of standard definition video in real-time, never before reported for a probabilistic approach with frame rates of 10–15 fps for image sequences of 160 * 120.

In 2010, Poremba et al. [49] evaluated the acceleration of the original MOG algorithm on the CUDA and CBEA architectures by taking the largest images without losing real-time performance. As a result, an acceleration of 2.77 times in CEBA versus 17.82 times in CUDA was obtained. In 2010, Pham et al. [50] implemented the Zivkovic-Heijden GMM on GPU to achieve real-time performance. At a rate of 50 fps in most experiments, with low and high-resolution video, an increase in speed of at least 10 times was achieved using a low-performance GeForce 9600GT GPU. In 2010, Gupta et al. [51] designed a GPU implementation of a change detection method using a locally stable monotonic change invariant feature. In 2013, Kumar et al. [52] provided a parallel implementation of the original MOG [13], labeling objects using the connected component labeling (CCL) technique and a post-processing stage through morphological operations. In this work, speeds 15 times higher for the original MOG, 2 times for blob labeling, and 250 times for binary masks were achieved compared to a sequential implementation on a Xeon processor at a rate of 22.3 fps in HD videos. In 2012, Li et al. [53] proposed a three-level GPU accelerated MOG implementation by modifying the third level in order to reduce processing times and thus increase bandwidth, and in turn, by using memory address saving and memory coalescing techniques for the second level. This method outperformed a conventional CPU 24 and 23 times, with rates of 145 fps for VGA quality videos and 505 fps for QVGA videos, respectively. In 2013, Gule et al. [54] implemented on GPU the video surveillance and monitoring algorithm with a performance 21.88 times higher than a conventional CPU, which allowed the integration of a greater number of video surveillance cameras. In 2013, Popa et al. [55] proposed the implementation of the original MOG [13] algorithm on a GPU using compression techniques in the MJPEG format. The compression process was implemented in the frequency domain through a discrete cosine transform (DCT) with a reduced number of coefficients, achieving an efficiency of 170 times higher than the sequential implementation in a CPU. The experimental results also showed real-time segmentation in 122 video streams at 25 fps. In 2014, Zhang et al. [56] proposed the computation of the original MOG algorithm [13] in a GPU by optimizing its parameters such as window optimization, overlapping operations applied to communication and computing, memory coalescing, optimization in flow control and register usage, and shared memory. This implementation in full HD videos at 60 Hz and 1080 resolution improves performance over sequential implementation by 57 times, 97 times, and 101 times without impact to the foreground mask quality. In 2015, Szwoch [57] designed a GPU-optimized parallel implementation of the MOG algorithm using CUDA and OpenCL. This work bases its optimization on configuring parameters of the GPU architecture, for example, data transfers, access to device memory, workgroup debugging. In addition, the MOG algorithm is improved, and different GPU performance states were tested including desktop PC graphic cards, ultrabooks, and the Tegra mobile processor. The total processing speed of CUDA implementation was 62.46 fps versus 62.15 fps for OpenCL with image sequences of 1920 * 1080.

In 2015, Qin et al. [58] proposed a methodology to calculate image edge information through optimized Gabor wavelets and the ViBe method. To boost this method, a GPU parallel optimization was employed providing a processing time of 22.4 ms for an image sequence of 640 * 480. In 2015, Karahan and Sevilgen [59] presented an implementation of pixel-based adaptive segmenter (PBAS) on CUDA. The performance is improved by using Nsight profiler metrics such as consumed time for functions, usage of GPU compute and memory, occupancy, and divergence. On a Tesla K20 graphic processing unit, 646 fps is needed for a video with a 320 * 240 resolution when GPU memory transfer is taken into account. As a result, about 17 times speedup is achieved with regards to the single thread version. In 2015, Wilson and Tavakkoli [60] proposed the design of a GPU-based architecture through the implementation of a non-parametric background model within a framework in order to facilitate the parallelization process required to accelerate the computation and execution of algorithms in real time. Wilson and Tavakkoli [60] targeted the Kepler CUDA architecture from Nvidia Corporation. This heterogeneous computing model is made up of five kernels within the GPU. Except for the training kernel, the rest of the kernels run simultaneously once per image and per pixel. Experimental tests demonstrated 38 times superiority for 320 * 240 low-resolution images and 82 times for 720 * 480 high-resolution frames compared to sequential implementation on an Intel Xeon E5-2620 v3 CPU device. In 2015, Boghdady et al. [61] exploited GPU parallelism to accelerate the SG algorithm achieving real-time processing. In 2016, Song et al. [62] proposed a framework for the implementation of a parallel-connected component labeling algorithm and the parallel computation of a background subtraction algorithm. The PCCL algorithm clusters the pixels classified as foreground in order to obtain well-defined and separated blobs, while the background model was based on color histogram techniques and implemented within the memory of a GPU device. Remarkable performance in precision and processing speed was achieved using an Nvidia CUDA development kit. The average processing speed in the 640 * 480 resolution is 56.32 fps. For the 768 * 576 it is 34.14 fps. In 2018, Choi et al. [63] proposed to CPU-GPU heterogeneous experimentation of an algorithm based on the background subtraction approach using depth videos. In 2019, Cuzzocrea and Mumolo [64] implemented algorithms on GPUs applied to IoT and big data using histogram techniques for object recognition and detection.

In 2020, Janus et al. [65] worked with the RGB-D format by implementing the MOG [13] and an improved version of the PBAS algorithm on a GPU device. Three GPUs models were employed for the experimental tests in this research work: efficient NVIDIA RTX 2070 (Turing architecture), embedded NVIDIA Jetson TX2 (Maxwell architecture), and mobile NVIDIA GeForce GTX (Pascal architecture). These GPUs were configured to work together with the RGB-D sensors: RealSense D435 and D415. The performance in real-time processing speed and foreground detection accuracy is similar to the research articles mentioned so far in this section of this chapter.

Note: In 2016, Song et al. [60] emphasized that real-time segmentation is not possible with background subtraction algorithms such as MOG and codebook models processing high-resolution video. This means that the GPU thread is suitable only for the parallelism of low complexity algorithms, and not for the computation of HD videos or high complexity operations. As evidence, the GPU performance was not enough to compute in real-time segmentation operations from the Gaussian probability density function in HD videos.

16.4 Embedded Implementations

Embedded computing and VLSI technology have enabled the expansion of smart cameras on the market. Video cameras can be considered stand-alone units that integrate sensing, communication, and processing elements into an independent platform. This fact motivates the distribution of a large number of video cameras communicated via wireless. However, this in turn leads to a host of challenges due to their characteristics of being battery-powered, wireless, and embedded devices. Challenges such as power delivery, limited bandwidth, and memory resources must be addressed. Thus, features such as simplicity and efficiency are highly desirable in background subtraction algorithms to be implemented in embedded systems. Therefore, in addition to the expected robustness of the algorithm, its portability in embedded systems must be taken into account during design.

In 2007, Valentine et al. [66] designed an adaptive multimodal background model methodology that computes a certain number of modes for each pixel in the image. This algorithm bases its operation on the temporal information observed in each pixel in order to discriminate the foreground from the background. Furthermore, Valentine et al. [66] also developed an implementation of the proposed algorithm in an embedded system optimizing storage resources and execution times. In 2007, Apewokin et al. [67–69] evaluated the multimodal mean (MM) model. Experiments show that MM provides similar accuracy to the original MOG [13] by reducing the storage resources of an eBox-2300-embedded device by 18% and observing an improvement in 6 times less operating times. Due to the temporal adaptive nature of the MM algorithm, the values of the new pixels must be previously observed for a while before being incorporated into the background. Pixel values at the threshold of two colors tend to be unstable leading to false positives in foreground detection. In 2008, Choi et al. [70] addressed this issue through edge detection operations. With this proposal, false positives were reduced by 70% at a low computational cost. In 2008, Valentine et al. [71] developed a hybrid background modeling algorithm. First, a background model called "BigBackground" is designed, represented by a 15-color palette. Its function is to identify large homogeneous areas with little activity in the scene. Second, depending on the behavior of the pixel, that is, stable or variant over time, the "BigBackground" or MM algorithm is applied, respectively. The hybrid nature of this algorithm achieved an acceleration of up to 58% in computation times while maintaining the segmentation quality for the foreground. In 2009, Williford et al. [72] proposed an improvement to the MM algorithm by designing the spatial multimodal mean (SMM) method. With this new idea, the computation times of the MOG model are optimized and the best results were observed in scenarios with dynamic backgrounds. The SMM algorithm is similar to the MOG background model; however, SMM models each pixel as a random distribution. In addition, Williford et al. [72] demonstrated how to integrate the SMM model into an MTI system for real-time applications without depth calculations. For frame sizes of 3380 * 400 pixels, times of 149 ms against 361 ms for MM and SMM were obtained, respectively. For similar "speedups" between MM and MOG, SMM improved the runtime by 2.5. Better efficiency in computational overhead could be achieved by using GPGPUs. In 2010, Valentine et al. [73] developed a new proposal that combines elements such as the chromatic distribution of the scene, the MM algorithm, and a color palette based on background matching. Since the adaptation and updating of complex models are avoided, this proposal obtained a 45% better performance in processing time and 58% in storage requirements compared to MM. In 2012, Azmat et al. [74] presented the background model called temporal multimodal media in order to model the static objects that appear over time, and that in the long term these can present occlusions. To solve this, Azmat et al. [75,76] worked on the definition of additional fields from the data structures that describe the modes of each pixel: the quantification of observations, the temporal record in which a pixel appears for the first time, the temporal record in which a pixel appears for the last time, and the use of flags to record the frequency of appearance of certain pixels in the image. In this way, Azmat et al. [75,76] filtered out spurious objects by characterizing the static objects of the scene based on their temporal attributes. In 2014, Azmat et al. [75,76] developed the model called spatiotemporal multimodal media (STM3), designed to work on a spatiotemporal environment through multilayer background analysis and the characterization of the dynamic foreground in the scene. The basic essence of this algorithm lies in modeling the foreground objects that have been incorporated into the background, along with the foreground and background of the scene. As a result, objects that even exhibit spatial displacement are successfully modeled. In summary terms, the STM3 algorithm offers adequate analysis and modeling for spatiotemporal situations than the classical background subtraction methodologies. In 2017, Azmat et al. [77] implemented the MM algorithm on a 12-watt GPU with Thermal Design Power exploiting memory access pattern resources and fine-grained data parallelism. The performance achieved in VGA video quality (640 * 480) was 392 fps. This result corresponds to a speed up 20 times higher compared to its implementation in its integrated CPU. The Hardware used is an Nvidia ION GPU and Intel Atom CPU. In addition, it was shown that MM exceeds 5 to 6 times the efficiency of the MOG algorithm when both are executed on the GPU chip. In 2008, Casares and Velipasalar [78] designed a lightweight algorithm, which is a background subtraction model for object detection with low computational overhead to be implemented in smart cameras. Notable improvements were made to deal with dynamic backgrounds, lighting changes, movement of trees caused by wind/rain, wave motion of water, among others. Experimental tests support the robustness of the algorithm for grayscale videos, without saturating the memory resources of the embedded system. Selective update and automatic adaptation rate stand out in the design of the LW model. Pixels are sometimes treated differently and their neighborhood is taken into account for segmentation. In this way, through the information of neighboring pixels and the reliability of the pixel location, LW is able to distinguish between salient and non-salient movement. Experiments show that LW gives either similar or better performance than MOG [13], PCA, and codebook with less memory and computational requirements. In 2009, Casares and Velipasalar [79,80] integrated adaptive memory into the LW algorithm. With this methodology, memory resources decrease throughout the scene. Each pixel is processed differently, that is, memory requirements are lower for static background pixels than for foreground or dynamic backgrounds. This method is named ALW [79]. In 2010, Casares and Velipasalar [81,82] used feedback tracking concepts to design a resource-efficient methodology for foreground detection. This method uses feedback during the object tracking stage in the video sequence rather than processing images sequentially and independently. A saving of 56% in energy and computing resources and 48.7% in execution times were achieved through this proposal.

In 2011, Cuevas and Garcia [83,84] proposed a fast non-parametric background model for smart cameras. This method consists mainly of two stages: a Bayesian method for classification, based on the approaches of Sheikh and Shah [85] and Zhang and Yang [86], and a tracking step. Using this strategy, the foreground model mixes the pixel's spatial and color information, instead of taking into account only the color space as in a conventional background model. The reduction of the analysis area in the image and the updating of the spatial information are carried out by a particle filter. The experimental tests showed a saving in memory and computing resources of 93.1% in indoor environments and 90.4% in outdoor environments by significantly reducing the number of pixels under analysis. In 2012, Wang et al. [87] designed a foreground segmentation algorithm based on contour detection rather than full object detection. With this method called resource-aware fast foreground detection (RaFFD), the compute and storage overhead of the embedded system is optimized. RaFFD handles contour detection and updating of non-static backgrounds with acceptable processing times for real-time applications. Its implementation was on a CITRIC platform with an integrated smart camera. RaFFD showed a 68% decrease in hardware resources and 2 times more performance in execution speed in relation to the ALW [79] and LW [78] algorithms. Additionally, even in challenging situations such as sudden lighting changes or camera jitter, the segmentation quality yielded 6% false alarms and 95% accuracy. In 2012, Shen et al. [88,89] developed a method for detecting objects in video sequences based on compression sensing concepts and mainly contributed to the dimensional reduction of the data. This method named CS-MOG employs the original MOG [13] with compressive measurements instead of the color pixel information. Its accuracy is comparable to that of the original MOG. Its implementation on an embedded system with an integrated camera proved to be 5 times more efficient in execution speed with less consumption in storage and processing resources compared to its MOG counterpart. In 2012, Tsai et al. [90] contributed through a block-based background subtraction algorithm efficient in processing and fast in execution. In a grayscale environment, the blocks are analyzed, and subsequently the most representative background pixels of the block are selected by MOG for the construction of the model. This method saves up to 94% memory consumption improving also the objection detection speed. Its memory consumption is 94.2% lower than ALW [79]. This method was tested using the TI DaVinci-embedded system to demonstrate its robustness reaching 25 fps at 640 * 480 image sizes. In 2012, Tsai et al. [91] improved the previous method by using color information and processed 26 fps for an image size of 768 * 576. In 2012, Tsai et al. [92] designed a strategy that contemplates adaptive multilayer non-static background and static background modeling. This proposal first analyzes a homogeneous region to later model the dependence between adjacent pixels within the image. The tests were carried out on the TI TMS320DM6446 DaVinci platform applied to 768 * 576 resolution frames reaching up to 18 fps in performance. In 2012, Salvadori et al. [93]

optimized the conventional MOG methodology for implementation in low-performance embedded systems. Modifications such as the granulation of the mean value, the variance rounded to whole numbers, and novel weight optimization strategies lightened the computational load required by a classic MOG. The results observed in a microcontroller with low memory resources are comparable to the performance of the MOG implemented in a conventional CPU. In 2014, Salvadori et al. [94] further improved their previous proposal by adding characteristics to the learning rate and generalizing their model to any method based on a Gaussian mixture in order to extend its application in more real and practical situations. Thus, Salvadori et al. [94] addressed sub-integer implementations of the MOG model to allow its deployment on low-end processors that typically do not have an integrated floating-point unit. Two methods similar to MOG were designed: (i) iGMM-l uses a linear function to round the values of the weights of the mixture. (ii) iGMM-s uses a staircase function for the same purpose. In other words, two low complexity arithmetic techniques were designed to update the parameters of a Gaussian distribution in order to optimize computing resources. Fundamental parameters such as the mean and variance present in the update rates represent a large amount of data in the original MOG and therefore receive a rounding treatment to reduce their complexity. The behavior of the weights is modeled by a linear or staircase function as well as by stochastic methods. All of these enhancements substantially optimize hardware resources without losing robustness in foreground segmentation quality. In 2021, Li et al. [95] designed a background subtraction model for embedded systems called embedded background subtraction (EBGS). The fundamental element of this technique corresponds to an additive-increase/multiplicative-decrease filter aimed at optimizing the algorithm while improving the quality of the foreground masks. This filter consists of a combination of random and codebook background models. EBGS proved robust to environmental challenges such as dynamic backgrounds, lighting variations, and jitter while running in real time within the embedded system.

16.5 Specific Architectures

In 2013, Cottini et al. [96] designed a very low power vision prototype with a resolution of 64 * 64 pixels, in order to implement an improved version of the moving average algorithm. The hardware consists of two clocked comparators devices and two switched-capacitor low-pass filter platforms as programmable devices with a 12% fill factor. The energy efficiency characteristics correspond to a 3.3 V power supply, 33 W at a rate of 13 fps and 620 W/frame.pixel. In 2014, Ratnayake and Amer [97] implemented the MOG background model within a field-programmable gate array (FPGA) device. Using parameter reduction techniques of the Gaussian distribution function, power consumption, memory, and storage resources were optimized on the FPGA platform. The chosen model corresponds to a Virtex-5 which allowed HD image processing at rates of 30 fps. The orders of magnitude achieved are high in relation to the hardware requirements. In 2016, Calvo et al. [98] designed Internet Protocol modules within reconfigurable FPGA hardware for vision applications using background models. All four Internet Protocol module designs are based on architectures provided by Xilinx FPGA. Without additional hardware resources to the embedded system, acceptable results were achieved in low-density frames of QCIF and CIF quality.

16.5.1 Digital Signal Processor

In 2006, Arth et al. [99] proposed a portable system called TRICam for traffic monitoring applications. Using up to three Texas Instruments TMTMS320C6414 DSP, Arth et al. [99] employed the approximated median \cite{P1C1-Median-1} to detect the vehicle. The average time required for a region of interest is about 2.27 ms. In 2007, Boragno et al. [100] developed an embedded system based on a DSP device and its application to the monitoring of vehicles parked in restricted areas. Using the Philips Trimedia DSPs (PNX1302), Boragno et al. [100] used the LBP-based background model [101] and provided experiments on the i-LIDS dataset. But, no information about memory and time requirements is provided. In 2007, Qiao et al. [102] designed a portable system for real-time monitoring implemented in DSP hardware. The background model employed is the multimodal mean [67]. In 2007, Sheu et al. [103] proposed a DSP implementation for a multimodal background maintenance (MBM) [104]. In 2012, Saptharishi et al. [105] presented a discriminative foreground/background separation algorithm called detector. Its implementation including the classification and tracking feedback on a TI DM6437 DSP achieves 12 fps for resolution of 720 * 480 and 10 fps for 1080p resolution.

16.5.2 Very Large Scale Integration

In 2009, Peng et al. [106–108] presented a VLSI design for MBM [104] with labeling and noise reduction. The MBM contains two principal features related to static and dynamic pixels. Obtaining the background reference model is carried out from the main statistical parameters of the Gaussian distribution function. For division and exponential operations of this distribution, three lookup tables were used. Several additional properties of the Gaussian distribution were taken into account to significantly reduce the complexity of lookup tables as well as hardware costs. MBM required 14.4 K logic gates for its real-time execution in high-density 720 * 480 pixel frames with a maximum working frequency of 100 MHz. The device used is built under TSMC 0.18 um technology.

16.5.3 Field-Programmable Gate Array

According to the nature of the background subtraction algorithm, FPGAs can be classified as follows:

- **Median Algorithms**
  In 2010, Hiraiwa et al. [109] presented an FPGA framework designed for real-time foreground detection in video sequences. The architecture is divided into work modules to facilitate the implementation of different object detection algorithms. This platform also allows remote observation of detection. Hiraiwa et al. [109] implemented the running average, an approximate and selective running Gaussian, and the conditional sigma–delta background estimation. With morphological post-processing based on component-connected labeling, the system reaches 30 fps.
- **MOG Algorithms**
  The original algorithm of Stauffer and Grimson [13] is the most investigated one to be applied in FPGA. Its operation is based on calculating a large number of Gaussian parameters per pixel and image, making it inappropriate for practical applications where low computational complexity is required. In the original

publication by Stauffer and Grimson [13] the implementation in an SGI O2 computer is reported, for images of 160 * 120 at a rate that oscillates between 11 and 13 fps. In the literature, several hardware processors based on FPGA have been proposed for MOG algorithms [110–123]. In 2004, Jiang et al. [124] proposed architecture to address the excessive computational overhead and memory resource demands of the MOG algorithm. The design was configured on an FPGA platform, Xilinx Virtex 2 1000 model, and VHDL hardware description language. Among the most relevant parameters under study are bandwidth optimization, memory saving, and fixed-point operations. Simulation analysis in a C++ environment demonstrated bandwidth reduction of up to 60%. High-resolution image processing of 1024 * 1024 pixels was achieved at a rate of 38 fps. In 2005, Appiah et al. [110,111] combined various properties of the W4 and MOG algorithms to meet the challenge of the low contrast of foreground objects against the background of the scene. The new algorithm implemented in a Xilinx Virtex II FPGA platform and Handel-C programming environment was called GW4. The performance achieved a rate of 30 fps in 640 * 480 images. In 2005, Appiah and Hunter [112] combined the MOG algorithm with a temporal low-pass filtering algorithm [125]. The operating frequencies observed in the experimental tests ranged from 57 MHz to 65 MHz, with a standard quality video processing capacity of 640 * 480 at 210 fps. In 2010, Appiah et al. [113] improved this method by adding a connected component labeling process. In 2005, Jiang et al. [126] proposed on an FPGA platform a robust circuit designed to process up to 39.8 Mps. This amount of pixels corresponds to 1014 * 1024 resolution images running at 38 fps. In 2008, Kristensen et al. [115] developed the VirtexII xc2pro30-embedded platform with an operating frequency of 83 MHz, an automatic monitoring system for surveillance applications. In 2009, Jiang et al. [116] optimized the memory requirements of an embedded system compared to the work reported in [115]. The scheme consisted of reducing the number of pixels to process to 7.68 Mps. In 2010, Genovese et al. [117] implemented the classic MOG algorithm in an OpenCV programming environment. Encoding compiled on a Virtex5 xc5vlx50 FPGA device achieved processing rates of up to 22 fps on full HD images. In 2011 Genovese and Napoli [120] proposed a hybrid system based on work previously developed by Genovese et al. [117] and a denoising element that works through opening, closing, dilation, and erosion operations. Subsequently, In 2012 Genovese and Napoli [121] made improvements to the MOG under an OpenCV programming environment. The circuit developed includes the debugging of some equations of the original MOG algorithm as well as the bandwidth optimization of the signals. In 2013, Genovese et al. [118] implemented the MOG algorithm within an embedded system. The innovations of this work are divided into design methodologies for efficient use of ROM memory and the computation of truncated binary multipliers. The chosen hardware corresponds to the FPGAs Stratix IV and Virtex6 which provided a high performance above 45 fps for high-resolution images of 1080p. In 2013, Genovese and Napoli [119] proposed two designs of the MOG algorithm for implementation in the OpenCV programming language. The contributions were focused on an accelerated initialization of the background model and the mathematical optimization of the equations for their efficient execution in hardware. Real-time foreground object segmentation was achieved in 1920 * 1080 size frames. The first circuit is built on the Virtex6 vlx75t-embedded system, which corresponds to FPGA architecture. With only 3% of its logical elements, rates of 91 fps were achieved in full HD images. The second prototype was designed to be implemented on "UMC-90nm CMOS" manufacturing technology achieving high processing rates above 60 fps in HD frames. In 2019, Morantes-Guzman et al. [127] proposed a highly flexible MOG design compared to previous work exploiting the Register Transfer Level description and numerical representation by a fixed point. The proposed model makes use of a ZedBoard development kit from the Xilinx Company. The MOG algorithm designed in OpenCV source code is transferred to a Vivado-HLS representation for its implementation in hardware. The advantages offered by the FPGA Artix-7 included in the development kit were fully exploited, for example, the execution of processes in parallel and the use of the Zynq processing block allowed to accelerate the design and configuration of the proposed architecture thanks to system on chip manufacturing technology. The system processed 768 * 576 size frames in real time at a low computational cost.

16.6 Parallel Implementations

In 2012, Yang and Chen [128] designed a framework for the parallel implementation of a background subtraction algorithm within a GPU device. From two consecutive frames, the most relevant characteristics are extracted using the scale-invariant feature transform technique, which is later computed by a global motion compensation (GMC) and affine transformation operations. Subsequently, a background subtraction module is designed for the segmentation of foreground objects with the ability to compensate for dynamic backgrounds. Thus, the architecture is composed of three blocks: a foreground segmentation block, a dynamic background update block, and a GMC block. The computing capabilities of a CPU and GPU were harnessed together for better performance, reinforcing the concept of parallel computing. In 2011, Szwoch [129] exploited the parallelism provided by multicore processors in an OpenMP encoding environment for the real-time implementation of the codebook algorithm. Tests on previously recorded videos demonstrated the robustness of algorithmic processing in multi-threaded configuration. Parameters such as the number of cores, the frame size, fps, encoding methods, and the nature of the scenes were tested. The number of cores running simultaneously turned out to be a determining factor in the performance of the proposed architecture. The main reason for using the OpenMP programming environment is due to the efficiency of dynamic task coding compared to static task coding. The tasks concerning multi-threading are executed with better efficiency and performance under a dynamic programming environment. The experimental tests were carried out on the Linux operating system and the GCC 4.3.2 compiler of the OpenMP environment. In 2012, Szwoch et al. [130] tested and evaluated different parallel implementations of the MOG algorithm in addition to the previous ones of the codebook algorithm. Thus, different experimental tests were carried out to evaluate the results such as the design of a thread management platform, pixel processing by specific threads, and methodologies for efficient scheduling of tasks. These approaches were implemented on a workstation accelerated by 12 processing cores. One of the most obvious results was the low demand for computing resources of the MOG algorithm compared to its codebook counterpart. In high-quality frames, it was not possible to achieve the original video fps rate, even increasing the number of threads with their corresponding physical cores working concurrently. In images of reduced dimensions, acceptable results were obtained by activating two to six threads; however, even increasing this amount, the execution times remained invariant.

The codebook algorithm turned out to be poor in performance when real-time execution is needed. In addition, MOG provided quite acceptable detection results—indoors and outdoors—, successfully compensating for abrupt lighting variations. In 2015, Szwoch [131] implemented the MOG algorithm on GPU in OpenCL and CUDA. Different improvements to the algorithm were reported by varying the processing and performance rates to observe its behavior, including desktop PC graphic cards, ultrabooks, and the Tegra mobile processor. Experiments demonstrated that employing GPU devices for parallel background subtraction in high-resolution video streams provides a significant performance boost compared with serial CPU implementations. As a result, previously tested devices will be able to run the MOG algorithm with sufficient computational power to process 1920 * 1080 resolution streams in online mode with the progress of technology.

16.7 Programming Languages

For implementation either on CPU and GPU, programming languages are required. For CPU implementations, the most current languages used are C and C++. For GPU implementations, CUDA is the natural language. However, for testing and prototyping new algorithms, MATLAB® is currently used by researchers but these implementations cannot achieve real-time performance. In 2010, Wu and Yu [132] provided an implementation in C++ for SG and KDE on a standard PC. Wu and Yu [132] recorded long video sequences from a web camera with 20 fps. The processing time of SG is less than 50 ms and thus ensures real-time processing. KDE achieved real-time processing only through the early-break technique and the use of a lookup table. Wu and Yu [132] verified this by obtaining satisfactory results in the modeling of multimodal backgrounds and challenges in lighting variations. In 2008, Fabian and Gaura [37] proposed architecture for the MOG algorithm to run inside a low-end GPU (GeForce 8400 GS) on 640 * 480 resolution digital videos. Parallelization faced difficulties when adapting the CCL algorithm on the CUDA environment due to its high memory demand since it operates at the pixel level in each frame. Also, the strong dependency between sub-blocks while merging slowed down parallelization. To solve the problem of merging blocks in CUDA, a parallel model of the CCL algorithm was designed for its implementation in concurrent mode. The speedups obtained for CCL, MOG, and morphologic operations (erosion, dilation, etc.) correspond to 11 times, 10 times, and 260 times, respectively. These results contrast significantly with sequential execution on the GeForce GTX 280 platform. In 2009, Fauske et al. [39] described three implementations corresponding to the codebook algorithm [133], SOBS [134], and the statistical approach of Horprasert et al. [135]. These implementations used the CUDA technology for parallel processing. In these high-level systems, real-time implementation requires a processing time of at least 33 ms per image. Each of the CUDA implementations provided by Fauske et al. [39] has run times below 10 ms per frame. The highest speed achieved by these experiments is 96 times with respect to their sequential execution. This data corresponds to the statistical approach of Horprasert et al. [135] equivalent to 0.779 ms per image, or 42 digital cameras at 30 fps operating simultaneously. However, this enormous speedup performance reduces the quality of segmentation. In general terms, the SOBS method obtained the best score although it only achieves a rate of 30 fps with three cameras working simultaneously, equivalent to 9.8 ms per frame. In 2009, Schreiber and Rauter [136] proposed a non-parametric description for the values of each pixel in the YCbCr space. This method is similar to a codebook [133]. The main difference lies in keeping the code list always organized in order to perform smarter searches. A processing time of 1.6 ms was achieved in small frames of 352 * 288 pixels and 2.6 ms including memory transfer processes on the GPU. The experiments show five times better performance on GPU compared to the CPU version. Specifically, tests on the NVidia GTX260 GPU on a Q6600 CPU showed twice the performance, and approximately 8 times better than a conventional CPU. In 2010, Pham et al. [50] provided an optimized version of the Zivkovic-Heijden GMM [137] on GPUs to achieve real-time performance. The test platform was a low-performance GeForce 9600GT GPU with a speed increase of up to 10 times. High and low definition frames were processed at 50 fps with satisfactory results. In 2010, Wong et al. [40] used gradient and color elements [138] to propose a probabilistic methodology. Using the NVIDIA Fermi architecture and heterogeneous computing, this implementation offers design features suitable for real-time implementation. But, the authors did not provide a quantitative evaluation of their poster.

16.8 Low-Complexity Strategies

In 2011, Chen et al. [139] developed an efficient hardware resource management system for video requirements. The first stage consists of computing a low-resource, low-performance algorithm in combination with another algorithm of better performance and high memory demand. In this step, the best-performing algorithm has little intervention to save resources. Then, a dynamic programming allocation (DPA)'s inference algorithm is employed to determine to which frames further expensive processing should be applied. This DPA algorithm is based on dynamic programming to optimize value of information (VoIDP) designed by Krause and Guestrin [140]. VoIDP operates by applying graphic models to a video sequence in such a way that it allows combining characteristics of high and low-performance algorithms to work together. In addition, Chen et al. [139] demonstrated how, based on an algorithm previously designed for short chains, it can be efficiently executed in longer digital videos. Experiments are provided for background subtraction. As a few algorithm, Chen et al. [139] used the two Consecutive Frame Difference, and ZHGMM [137] as the expensive algorithm. In 2014, Lee and Lee [141] developed a foreground detection model using the SSIM algorithm. The implicit operations in the SSIM method are used to provide robustness to the built background model. Thus, through a spatial similarity analysis, typical challenges such as camouflage problems between the background of the scene and the objects to be detected, highly multimodal backgrounds, and shadow detection are addressed. Pixels with low similarity to the background model are classified as foreground. Therefore, by discarding the background pixels and working only the foreground pixels, the computational demand and complexity are significantly reduced.

16.9 Fog Computing and Edge Computing

In 2019, Liu et al. [142] designed a background subtraction system for video surveillance applications. The system operates at the edge layer level, that is, the video sequences are stored in an edge layer device and are subsequently sent to the fog layer for processing by a PCA background subtraction algorithm. The results go back to the edge layer where object detection

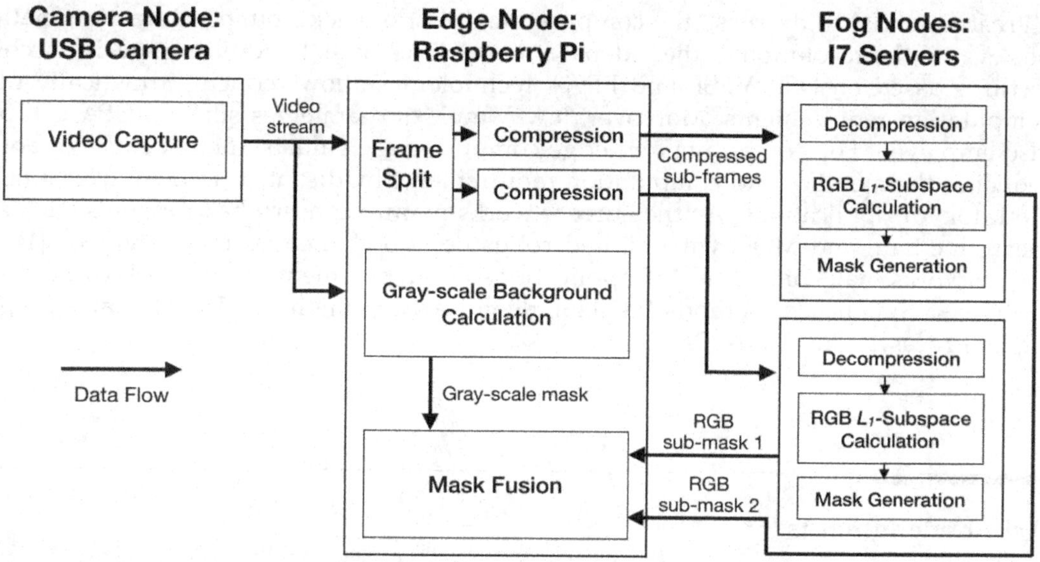

FIGURE 16.2
Edge-to-fog computing architecture. (Image from Liu et al. [142].)

is finally performed. Since the PCA algorithm works in RGB color space, better accuracy is obtained compared to a grayscale space. In addition, the idea of migrating between the edge and fog levels optimizes processing costs since the computational overhead is divided into different layers. Figure 16.2 represents the proposed edge–fog architecture which has an analogy to the client–server network model. The basic mechanism consists of broadcasting the video through an edge device to the nodes in the fog layer in order to reduce the computing load. The edge device splits the frames into subframes and sends them to different nodes in the fog layer for processing by a PCA background subtraction model in the RGB color space. The PCA algorithm segments each of the received subframes. Simultaneously, the segmentation of the subframes in a grayscale environment is carried out in the edge layer. Finally, the segmentation of each of the subframes in the fog layer is sent back to the edge device and is combined with the binary masks generated locally in the edge layer and thus showing the complete foreground object detection result of the video sequences.

In 2020, Siddharth and Aghila [143] addressed the challenge of implementing real-time foreground segmentation algorithms in the fog computing layer. The methodology includes the random projection (RP) for the reduction of high-resolution images and the structural similarity index measure (SSIM) algorithm to discriminate between the frames with presence and absence of movement. The primary purpose is to save memory and bandwidth in fog computing. Using videos from the CDnet 2014 dataset for experimental tests, RP SSIM optimized cloud storage of 67% at a very low computational cost.

16.10 Conclusion

In this chapter, we reviewed and classified the different strategies developed over time to reach efficient real-time moving object detection. First, GPU implementations show a great potential to decrease the computational cost of background subtraction methods even if the resolution of the videos increases over time. In addition, specific architectures based on DSP, VLSI, and FPGA technologies allow reducing drastically the computation requirements. Moreover, low-complexity strategies such as DPA can be also employed. Fog computing and edge computing are valuable perspectives to both reducing the memory and computation requirements by dispatching them over a network for IoT applications. As the current works mainly concern basic models such as mean, median, and MOG with limited robustness to challenges met in videos [144], future works may concern in IoT applications the deployment of more robust models such as sample-based methods [145,146], deep learning methods [147–149], and graph signal processing methods [31–33].

Acknowledgments

The authors thank the support of CONACYT, which provides the scholarship for the post-doc of Dr. Belmar Garcia-Garcia.

References

1. T. Bouwmans, F. Porikli, B. Horferlin, A. Vacavant, *Handbook on Background Modeling and Foreground Detection for Video Surveillance*, CRCPress, Taylor and Francis Group, Boca Raton, FL (2014).
2. T. Bouwmans, N. Aybat, E. Zahzah, *Handbook on Robust Low-Rank and Sparse Matrix Decomposition: Applications in Image and Video Processing*, Taylor and Francis Group, Boca Raton, FL (2016).
3. T. Bouwmans, B. Höferlin, F. Porikli, A. Vacavant, Traditional approaches in background modeling for video surveillance, in T. Bouwmans, B. Höferlin, F. Porikli, A. Vacavant (eds) *Handbook Background Modeling and Foreground Detection for Video Surveillance*, Taylor and Francis Group (2014).
4. T. Bouwmans, B. Höferlin, F. Porikli, A. Vacavant, Recent approaches in background modeling for video surveillance, *Handbook Background Modeling and Foreground Detection for Video Surveillance*, Taylor and Francis Group (2014).
5. T. Bouwmans, Traditional and recent approaches in background modeling for foreground detection: An overview, *Computer Science Review* 11 (2014) 31–66.
6. T. Bouwmans, C. Silva, C. Marghes, et al., On the role and the importance of features for background modeling and foreground detection, *Computer Science Review* 28 (2018) 26–91.
7. L. Maddalena, A. Petrosino, Background subtraction for moving object detection in RGB-D data: A survey, *Journal of Imaging* 4 (2018) 71.
8. B. Lee, M. Hedley, Background estimation for video surveillance. In: *Image & Vision Computing New Zealand* (IVCNZ '02), Auckland, NZ (2002) 315–320. http://hdl.handle.net/102.100.100/199802?index=1.
9. P. Graszka, "Median mixture model for background-foreground segmentation in video sequences", *WSCG 2014* (2014) 104–114.

10. S. Roy, A. Ghosh, "Real-time adaptive histogram min-max bucket (HMMB) model for background subtraction", *IEEE T-CSVT*, 2017.
11. A. Elgammal, L. Davis, "Non-parametric model for background subtraction", *ECCV 2000*, pp. 751–767, 2000.
12. R. Caseiro, P. Martins, J. Batista, "Background modelling on tensor field for foreground segmentation", *BMVC 2010*, pp. 1–12, 2010.
13. C. Stauffer, E. Grimson, "Adaptive background mixture models for real-time tracking", *IEEE Conference on Computer Vision and Pattern Recognition*, CVPR 1999, pp. 246–252, 1999.
14. S. Varadarajan, P. Miller, H. Zhou, "Spatial mixture of Gaussians fordynamic background modelling", *AVSS 2013*, pp. 63–68, 2013.
15. F. E. Baf, T. Bouwmans, B. Vachon, "Fuzzy integral for moving object detection", *IEEE International Conference on Fuzzy Systems, FUZZ-IEEE 2008*, pp. 1729–1736, 2008.
16. F. E. Baf, T. Bouwmans, B. Vachon, "Type-2 fuzzy mixture of Gaussians model: Application to background modeling", *ISVC 2008*, pp. 772–781, 2008.
17. F. E. Baf, T. Bouwmans, B. Vachon, "Fuzzy statistical modeling of dynamic backgrounds for moving object detection in infrared videos", *IEEE-Workshop OTCBVS 2009*, pp. 60–65, 2009.
18. O. Munteanu, T. Bouwmans, E. Zahzah, R. Vasiu, The detection of moving objects in video by background subtraction using Dempster-Shafer theory, *Transactions on Electronics and Communications* 60(1) (2015) 1–9.
19. N. Oliver, B. Rosario, A. Pentland, "A Bayesian computer vision system for modeling human interactions", *International Conference on Vision Systems, ICVS 1999*, January 1999.
20. D. Farcas, T. Bouwmans, "Background modeling via a supervised subspace learning", *IVPCV 2010*, pp. 1–7, 2010.
21. D. Farcas, C. Marghes, T. Bouwmans, Background subtraction via incremental maximum margin criterion: A discriminative approach, *Machine Vision and Applications* 23(6) (2012) 1083–1101.
22. C. Marghes, T. Bouwmans, "Background modeling via incremental maximum margin criterion", *ACCV 2010 Workshop Subspace*, 2010.
23. C. Marghes, T. Bouwmans, R. Vasiu, "Background modeling and foreground detection via a reconstructive and discriminative subspace learning approach", *IPCV 2012*, July 2012.
24. E. Candes, X. Li, Y. Ma, J. Wright, Robust principal component analysis? *International Journal of ACM* 58(3), 2011.
25. A. Sobral, T. Bouwmans, E. Zahzah, "Double-constrained RPCA based on saliency maps for foreground detection in automated maritime surveillance", *AVSS* 2015.
26. S. Javed, A. Mahmood, T. Bouwmans, S. Jung, "Motion-aware graph regularized RPCA for background modeling of complex scenes, scene background modeling contest", *International Conference on Pattern Recognition, ICPR 2016*, December 2016.
27. S. Javed, A. Mahmood, T. Bouwmans, S. Jung, Spatiotemporal low-rank modeling for complex scene background initialization, *IEEE Transactions on Circuits and Systems for Video Technology* 28 (2016) 1315–1329.
28. G. Ramirez-Alonso, M. Chacon-Murguia, Self-adaptive SOM-CNN neural system for dynamic object detection in normal and complex scenarios, *Pattern Recognition* 48 (2015) 1137–1149.
29. J. Ramirez-Quintana, M. Chacon-Murguia, "Self-organizing retinotopic maps applied to background modeling for dynamic object segmentation in video sequences", *IJCNN 2013*.
30. A. Schofield, P. Mehta, T. Stonham, A system for counting people in video images using neural networks to identify the background scene, *Pattern Recognition* 29 (1996) 1421–1428.
31. J. Giraldo, S. Javed, T. Bouwmans, "Graph moving object segmentation", *IEEE Transactions on Pattern Analysis and Machine Intelligence*, 2021.
32. J. Giraldo, S. Javed, M. Sultana, S. Jung, T. Bouwmans, "The emerging field of graph signal processing for moving object segmentation", *International Workshop on Frontiers of Computer Vision, IW-FCV 2021*, Daegu, South Korea, February 2021.
33. J. Giraldo, T. Bouwmans, "GraphBGS: Background subtraction via recovery of graph signals", *International Conference on Pattern Recognition, ICPR 2020*, Milan, Italy, January 2021.

34. J. Giraldo, T. Bouwmans, "Semi-supervised background subtraction of unseen videos: minimization of the total variation of graph signals", *IEEE International Conference on Image Processing, ICIP 2020*, Abu Dhabi, UAE, October 2020.
35. A. Griesser, S. De Roeck, A. Neubeck, "GPU-based foreground-background segmentation using an extended colinearity criterion", *Vision, Modeling, and Visualization, VMV 2005*, 2005.
36. A. Griesser, "Real-time GPU-based foreground-background segmentation", Technical Report 269, 2005.
37. T. Fabian, J. Gaura, "Parallel implementation of recursive background modeling technique in CUDA for tracking moving objects in video traffic surveillance", *Annual Doctoral Workshop on Mathematical and Engineering Methods in Computer Science*, Brno University of Technology, Czechia, November 2008.
38. J. Liyanage, *Background Subtraction Using Gaussian Mixture Models*, University of Central Florida, Orlando (2009).
39. E. Fauske, L. Eliassen, R. Bakken, "A comparison of learning based background subtraction techniques implemented in CUDA", *NAIS 2009*, pp. 181–192, 2009.
40. M. Wong, K. Leman, N. Pham, R. Chang, T. Chua, G. Feng, "Fermi in action: Robust background subtraction for real-time video analysis", *NVIDIA Research Summit 2010*, 2010.
41. M. Gong, L. Cheng, "Real time foreground segmentation on GPUs using local online learning and global graph cut optimization", *International Conference on Pattern Recognition, ICPR 2008*, Tampa, FL, December 2008.
42. L. Cheng, M. Gong, "Real time background subtraction from dynamics scenes", *International Conference on Computer Vision, ICCV 2009*, Kyoto, Japan, September 2009.
43. L. Cheng, M. Gong, D. Schuurmans, T. Caelli, Real-time discriminative background subtraction, *IEEE Transaction on Image Processing*, 20(5) (2011) 1401–1441.
44. S. Fukui, Y. Iwahori, R. Woodham, "GPU based extraction of moving objects without shadows under intensity changes", *IEEE Congress on Evolutionary Computation, CEC 2008*, pp. 4166–4173, June 2008.
45. S. Fukui, Y. Iwahori, H. Itoh, H. Kawanaka, R. Woodham, "Robust background subtraction for quick illumination changes", *PSIVT 2006*, pp. 1244–1253, 2006.
46. A. Yamamoto, Y. Iwai, M. Yachida, "Real-time object detection with adaptive background model and margined sign correlation", IEICE Technical Report, Vol. 108, No. 327, PRMU2008-136, pp. 177–184, November 2008.
47. A. Yamamoto, Y. Iwai, "Real-time object detection with adaptive background model and margined sign correlation", *Asian Conference on Computer Vision, ACCV 2009*, Xi'an, China, September 2009.
48. D. Culibrk, V. Crnojevic, "GPU-based complex-background segmentation using neural networks", *International Conference on Machine Vision and Image Processing, IMVIP 2010*, Limerick, Ireland, September 2010.
49. M. Poremba, Y. Xie, M. Wolf, "Accelerating adaptive background subtraction with GPU and CBEA architecture", *IEEE Workshop on Signal Processing Systems, SiPS 2010*; San Francisco, CA, October 2010.
50. V. Pham, P. Vo, H. Vu Thanh, B. Le Hoai, "GPU implementation of extended Gaussian mixture model for background subtraction", *IEEE International Conference on Computing and Telecommunication Technologies, RIVF 2010*, Vietnam National University, November 2010.
51. R. Gupta, S. Reddy, S. Panda, S. Sharma, A. Mittal, "Foreground-background separation on GPU using order based approaches", *Indian Conference on Computer Vision, Graphics and Image Processing, ICVGIP 2010*, December 2010.
52. P. Kumar, A. Singhal, S. Mehta, A. Mittal, Real-time moving object detection algorithm on high-resolution videos using GPUs, *Journal of Real-Time Image Processing* 11 (2013) 93–109.
53. Y. Li, G. Wang, X. Lin, "Three-level GPU accelerated Gaussian mixture model for background subtraction", *Proceedings of SPIE 8295, Image Processing: Algorithms and Systems X; and Parallel Processing for Imaging Applications II*, 829514 (2 February 2012); https://doi.org/10.1117/12.906385.

54. P. Gule, D. Emeksiz, A. Temizel, M. Teke, T. Temizel, "Real-time multi-camera video analytics system on GPU", *Real-Time Image, RTI 2013*, March 2013.
55. S. Popa, D. Crookes, P. Miller, Hardware acceleration of background modeling in the compressed domain", *IEEE Transactions on Information Forensics and Security* 8 (2013) 1562–1574.
56. C. Zhang, H. Tabkhi, G. Schirner, "A GPU-based algorithm-specific optimization for high-performance background subtraction", *ICPP 2014*, 2014.
57. G. Szwoch, Performance evaluation of parallel background subtraction on GPU platforms, *Elektronika: konstrukcje, technologie, zastosowanian* 56(4) (2015) 23–27.
58. L. Qin, B. Sheng, W. Lin, W. Wu, R. Shen, GPU-accelerated video background subtraction using gabor detector, *Journal of Visual Communication and Image Representation* 32 (2015) 1–9.
59. S. Karahan, F. Sevilgen, "CUDA implementation of the pixel based adaptive segmentation algorithm", *Signal Processing and Communications Applications Conference, SIU 2015*, May 2015.
60. B. Wilson, A. Tavakkoli, "An efficient non parametric background modeling technique with CUDA heterogeneous parallel architecture", *ISVC 2015*, Las Vegas, NV, December 2015.
61. R. Boghdady, C. Salama, A. Wahba, "GPU-accelerated real-time video background subtraction", *IEEE International Conference on Computer Engineering and Systems, ICCES 2015*, Cairo, Egypt, pp. 34–39, 2015.
62. W. Song, Y. Tian, S. Fong, K. Cho, W. Wang, W. Zhang, GPU-accelerated foreground segmentation and labeling for real-time video surveillance, *Sustainability* 8 (2016) 916.
63. Y. Choi, J. Kim, J. Kim, Y. Chung, D. Park, S. Lee, CPU-GPU heterogeneous implementations of depth-based foreground detection, *IEICE Electronics Express* 15(4) (2018) 1–9.
64. A. Cuzzocrea, E. Mumolo, A novel GPU-aware histogram-based algorithm for supporting moving object segmentation in big-data-based IoT application scenarios. *Information Sciences* 496 (2019) 592–612.
65. P. Janus, T. Kryjak, M. Gorgon, "Foreground object segmentation in RGB-D data implemented on GPU", Preprint, February 2020.
66. B. Valentine, S. Apewokin, L. Wills, S. Wills, A. Gentile, "Midground object detection in real world video scenes", *AVSS 2007*, 2007.
67. S. Apewokin, B. Valentine, L. Wills, S. Wills, A. Gentile, "Multimodal mean adaptive backgrounding for embedded real-time video surveillance", *Embedded Computer Vision Workshop, ECVW 2007*, June 2007.
68. S. Apewokin, B. Valentine, D. Forsthoefel, L. Wills, S. Wills, A. Gentile, "Embedded real-time surveillance using multimodal mean background modeling", *Advances in Pattern Recognition, Embedded Computer Vision*, Part II, pp. 163–175, Springer, London, 2009.
69. S. Apewokin, "Efficiently mapping high performance early vision algorithms onto multicore embedded platforms", Thesis, Georgia Institute of Technology, May 2009.
70. J. Choi, S. Apewokin, B. Valentine, D. Wills, L. Wills, Edge noise removal in multimodal background modeling techniques, *Image Processing: Machine Vision Applications* 6813 (2008) 68130K.
71. B. Valentine, J. Choi, S. Apewokin, D Wills, L. Wills, "Bypassing BigBackground: An efficient hybrid background modeling algorithm for embedded video surveillance", *International Conference on Distributed Smart Cameras, ICDSC 2008*, pp. 1–8, September 2008.
72. J. Williford, C. Dalal, M. Shim, "Spatial multi modal mean background model for real-time MTI", *Proceedings of SPIE*, Vol. 7338, 2009.
73. B. Valentine, S. Apewokin, L. Wills, S. Wills, "An efficient, chromatic clustering-based background model for embedded vision platforms", *Computer Vision and Image Understanding, CVIU 2010*, Vol. 114, No. 11, pp. 1152–1163, 2010.
74. S. Azmat, L. Wills, S. Wills, "Temporal multi-modal mean", *IEEE Southwest Symposium on Image Analysis and Interpretation*, pp. 73–76, Santa Fe, NM, April 2012.
75. S. Azmat, "Multilayer background modeling under occlusions for spatio-temporal scene analysis", PhD Thesis, Georgia Institute of Technology, USA, 2014.
76. S. Azmat, L. Wills, S. Wills, "Spatio-temporal multimodal mean", *IEEE Southwest Symposium on Image Analysis and Interpretation, SSIAI 2014*, San Diego, CA, April 2014.

77. S. Azmat, L. Wills, S. Wills, Parallelizing multimodal background modeling on a low-power integrated GPU, *Journal of Signal Processing Systems* 88(1) (2017) 43–53.
78. M. Casares, S. Velipasalar, "Light-weight salient foreground detection for embedded smart cameras", *IEEE International Conference on Distributed Smart Cameras, ICDSC 2008*, pp. 1–7, Palo Alto, CA, September 2008.
79. M. Casares, S. Velipasalar, A. Pinto, "Light-weight salient foreground detection for embedded smart cameras", *Computer Vision and Image Understanding*, 2010.
80. M. Casares, S. Velipasalar, "Light-weight salient foreground detection with adaptive memory requirement", *International Conference on Acoustic, Speech and Signal Processing, ICASSP 2009*, pp. 1245–1248, Taipei, Taiwan, April 2009.
81. M. Casares, S. Velipasalar, Adaptive methodologies for energy-efficient object detection and tracking with battery-powered embedded smart cameras, *IEEE Transactions on Circuits Systems and Video Technologies* 2(10) (2011) 1438–1452.
82. M. Casares, S. Velipasalar, "Resource-efficient salient foreground detection for embedded smart cameras by tracking feedback", *International Conference on Advanced Video and Signal Based Surveillance*, AVSS 2010, Boston, USA, September 2010.
83. C. Cuevas, N. Garcia, "Efficient moving object detection for lightweight applications on smart cameras", *IEEE Transactions on Circuits and Systems for Video Technology*, Vol. 23, No. 1, pp. 1–14, January 2013.
84. C. Cuevas, N. Garcia, "Moving object detection for real-time high-quality lightweight applications on smart cameras", *IEEE International Conference on Consumer Electronics, ICCE 2011*, pp. 479–480, 2011.
85. Y. Sheikh, M. Shah, "Bayesian object detection in dynamic scenes", *IEEE Conference on Computer Vision and Pattern Recognition, CVPR 2005*, June 2005.
86. X. Zhang, J. Yang, Foreground segmentation based on selective foreground model, *Electronics Letters* 44(14) (2008) 851.
87. Q. Wang, P. Zhou, J. Wu, C. Long, "RaFFD: Resource-aware fast foreground detection in embedded smart cameras", *Globecom 2012*, pp. 499–504, 2012.
88. Y. Shen, W. Hu, J. Liu, M. Yang, B. Wei, C. Chou, "Efficient background subtraction for real-time tracking in embedded camera networks", *ACM International Conference on Embedded Network Sensor Systems, Sensys 2012*, pp. 295–308, November 2012.
89. Y. Shen, W. Hu, M. Yang, J. Liu, C. Chou, B. Wei, "Efficient background subtraction for tracking in embedded camera networks", *International Conference on Information Processing in Sensor Networks, IPSN 2012*, April 2012.
90. W. Tsai, M. Sheu, S. Yang, C. Gao, T. Chen, "Efficient block-based foreground object detection for outdoor scenes on embedded SoC platforms", *International Conference on Image, Vision and Computing, ICIVC 2012*, 2012.
91. W. Tsai, M. Sheu, C. Lin, H. Liao, "A robust background modeling and foreground object detection using color component analysis", *IEEE International Conference on Systems, Man, and Cybernetics, SMC 2012*, pp. 263–267, October 2012.
92. W. Tsai, M. Sheu, C. Lin, "Region-based background subtraction for complex sense on embedded platforms", *International Conference on Intelligent Information Hiding and Multimedia Signal Processing*, pp. 351–354, July 2012.
93. C. Salvadori, D. Makris, M. Petracca, J. Rincon, S. Velasti, "Gaussian mixture background modelling optimisation for micro-controllers", *International Symposium on Visual Computing, ISVC 2012*, pp. 241–251, 2012.
94. C. Salvadori, M. Petracca, J. Rincon, S. Velastin, D. Makris, "An optimisation of Gaussian mixture models for integer processing units", *Real-Time Image Processing*, February 2014.
95. S. Li, J. Wu, C. Long, Y. Lin, "A full-process optimization-based background subtraction for moving object detection on general-purpose embedded devices", *IEEE Transactions on Consumer Electronics*, 2021.

96. N. Cottini, M. Gottardi, N. Massari, R. Passerone, Z. Smilansky, A 33 W 64*64 pixel vision sensor embedding robust dynamic background subtraction for event detection and scene interpretation, *IEEE Journal of Solid-State Circuits* 48(3) (2013) 850–863.
97. K. Ratnayake, A. Amer, Embedded architecture for noise-adaptive video object detection using parameter-compressed background modeling, *Journal of Real-Time Image Processing* 13(2) (June 2017) 397–414. https://doi.org/10.1007/s11554-014-0418-x.
98. E. Calvo, P. Brox, S. Sanchez-Solano, Low-cost dedicated hardware IP modules for background subtraction in embedded vision systems, *Journal of Real-Time Image Processing* 12 (2016) 681–695.
99. C. Arth, H. Bischof, C. Leistner, "TRICam—An embedded platform for remote traffic surveillance", *CVPR Workshop on Embedded Computer Vision, CVPRW 2006*, June 2006.
100. S. Boragno, B. Boghossian, J. Black, D. Makris, S. Velastin, "A DSP-based system for the detection of vehicles parked in prohibited areas", *AVSS 2007*, pp. 1–6, 2007.
101. M. Heikkila, M. Pietikainen, "A texture-based method for modeling the background and detecting moving objects", *IEEE Transactions on Pattern Analysis and Machine Intelligence, PAMI 2006*, Vol. 28, No. 4, pp. 657–662, 2006.
102. Q. Qiao, Y. Peng, D. Zhang, Real-time surveillance of people on an embedded DSP-platform, *Journal of Ubiquitous Convergence Technology* 1(1) (2007) 125–132.
103. W. Sheu, "Foreground object detection based on multi-model background maintenance and its DSP implementation", Master Thesis, Electrical Engineering, June 2007.
104. T. Tsai, W. Sheu, C. Lin, "Foreground object detection based on multi-model background maintenance", *IEEE International Symposium on Multimedia Workshops, ISMW 2007*, pp. 151–159, 2007.
105. M. Saptharishi, A. Lipchin, D. Lisin, "Discriminative focus of attention for real-time object detection in video", *IEEE Workshop on Signal Processing Systems, SiPS 2012*, pp. 85–90, October 2012.
106. D. Peng, "VLSI design for foreground object segmentation with labeling and noise reduction mode in video surveillance application", Thesis, July 2009.
107. D. Peng, C. Lin, W. Sheu, T. Tsai, "Architecture design for a low-cost and low-complexity foreground object segmentation with multi-model background maintenance algorithm", *ICIP 2009*, pp. 3241–3244, Cairo, Egypt, 2009.
108. T. Tsai, D. Peng, C. Lin, W. Sheu, "A low cost foreground object detection architecture design with multi-model background maintenance algorithm", *VLSI Design Symposium, VLSI 2008*, August 2008.
109. J. Hiraiwa, E. Vargas, S. Toral, "An FPGA based embedded vision system for real-time motion segmentation", *International Conference on Systems, Signals and Image Processing, IWSSIP 2010*, pp. 360–363, 2010.
110. K. Appiah, A. Hunter, T. Kluge, "GW4: An FPGA-driven image segmentation algorithm", *WSEAS International Conference on Signal, Speech and Image Processing, SSIP 2005*, Corfu Island, Greece, August 2005.
111. K. Appiah, A. Hunter, T. Kluge, "GW4: A real-time background subtraction and maintenance algorithm for FPGA implementation", *WSEAS Transactions on Systems*, pp. 1741–1751, 2005.
112. K. Appiah, A. Hunter, "A single-chip FPGA implementation of real-time adaptive background model", *IEEE Conference on Field-Programmable Technology, FPT 2005*, National University of Singapore, Singapore, December 2005.
113. K. Appiah, A. Hunter, P. Dickinson, H. Meng, "Accelerated hardware video object segmentation: From foreground detection to connected components labelling", *Computer Vision and Image Understanding, CVIU 2010*, 2010.
114. H. Jiang, V. Öwall, H. Ardo, "Real-time video segmentation with VGA Resolution and memory bandwidth reduction", *International Conference on Video and Signal Based Surveillance AVSS 2006*, 2006.
115. F. Kristensen, H. Hedberg, H. Jiang, P. Nilsson, V. Öwall, An embedded real-time surveillance system: Implementation and evaluation, *Journal of VLSI Signal Processing* (2007).

116. H. Jiang, H. Ardö, V. Öwall, A hardware architecture for real-time video segmentation utilizing memory reduction techniques, *IEEE Transactions on Circuits and Systems for Video Technology* 19(2) (2009) 226–336.
117. M. Genovese, E. Napoli, N. Petra, "OpenCV compatible real time processor for background foreground identification", *International Conference on Microelectronics, ICM 2010*, Cairo, Egypt, December 2010.
118. M. Genovese, E. Napoli, D. De Caro, N. Petra, Antonio G. M. Strollo, FPGA Implementation of Gaussian Mixture Model Algorithm for 47fps Segmentation of 1080p Video, *Journal of Electrical and Computer Engineering* 2013, Article ID 129589, 8 pages, 2013. https://doi.org/10.1155/2013/129589.
119. M. Genovese, E. Napoli, ASIC and FPGA implementation of the Gaussian mixture model algorithm for real-time segmentation of high definition video, *IEEE Transactions on Very Large Scale Integration (VLSI) Systems* 22 (2013) 537–547.
120. M. Genovese, E. Napoli, FPGA-based architecture for real time segmentation and denoising of HD video, *Journal of Real-Time Image Processing* 8 (2013) 389–401.
121. M. Genovese, E. Napoli, "FPGA implementation of OpenCV compatible background identification circuit", *International Symposium on Computational Modelling of Objects Represented in Images: Fundamentals, Methods and Applications, CompIMAGE 2012*, pp. 75–80, September 2012.
122. S. Minghua, A. Bermak, S. Chandrasekaran, A. Amira, "An efficient FPGA implementation of Gaussian mixture models-based classifier using distributed arithmetic", *IEEE International Conference on Electronic Circuits and Systems*, pp. 1276–1729, December 2006.
123. S. Arivazhagan, K. Kiruthika "FPGA Implementation of GMM algorithm for background subtractions in video sequences", *International Conference on Computer Vision and Image Processing, CVIP 2016*, 2016.
124. H. Jiang, V. Owall, "Controller synthesis in hardware accelerator design for video segmentation", *SSoCC*, 2004.
125. A. Makarov, "Comparison of Background extractionbased intrusion detection algorithms", *IEEE ICIP 1996*, 1996.
126. H. Jiang, H. Ardo, V. Owall, "Hardware accelerator design for video segmentation with multimodal background modeling", *International Symposium on Circuits and Systems, ISCAS 2005*, Vol. 2, pp. 1142–1145, May 2005.
127. L. Morantes-Guzman, C. Alzate, L. Castano-Londono, D. Marquez-Viloria, J. Vargas-Bonilla, "Performance evaluation of SoC-FPGA based floating-point implementation of GMM for real-time background subtraction", *WEA 2019*, October 2019.
128. Y. Yang, W. Chen, "Parallel algorithm for moving foreground detection in dynamic background", *International Symposium on Computational Intelligence and Design, ISCID 2012*, Vol. 2, pp. 442–445, 2012.
129. G. Szwoch, "Performance evaluation of the parallel codebook algorithm for background subtraction in video stream", *International Conference on Multimedia Communications, Services and Security, MCSS 2011*, Krakow, Poland, June 2011.
130. G Szwoch, D. Ellwart, A. Czyzewski, Parallel implementation of background subtraction algorithms for real-time video processing on a supercomputer platform, *Journal of Real-Time Image Processing* (2012).
131. G. Szwoch, Performance evaluation of parallel background subtraction on GPU platforms, *Elektronika: konstrukcje, technologie, zastosowanian* 56(4) (2015) 23–27.
132. H. Wu, Y. Yu, "Real-time background subtraction in C++", Technical Report No. ECE-2010-06, Department of Electrical and Computer Engineering, Boston University, May 2010.
133. Kyungnam Kim, T. H. Chalidabhongse, D. Harwood and L. Davis, "Background modeling and subtraction by codebook construction," in *2004 International Conference on Image Processing*, 2004. ICIP '04., 2004, pp. 3061–3064 Vol. 5. doi: 10.1109/ICIP.2004.1421759.
134. L. Maddalena, A. Petrosino, "A self-organizing approach to detection of moving patterns for real-time applications", *Advances in Brain, Vision, and Artificial Intelligence*, LNCS 2007, Vol. 4729, 2007.

135. T. Horprasert, D. Harwood, L. Davis. "A statistical approach for real-time robust background subtraction and shadow detection". *IEEE International Conference on Computer Vision, FRAME-RATE Workshop*, September 1999.
136. D. Schreiber, M. Rauter, "GPU-based non-parametric background subtraction for a practical surveillance system", *IEEE Workshop on Embedded Computer Vision in conjunction with ICCV 2009*, Kyoto, Japan, pp. 870–877, October 2009.
137. Z. Zivkovic, F. Heijden, Recursive unsupervised learning of finite mixture models, *IEEE Transaction on Pattern Analysis and Machine Intelligence* 5(26) (2004) 651–656.
138. L. Li, R. Luo, W. Huang, K. Leman, W. Yau, "Adaptive background subtraction with multiple feedbacks for video surveillance", *International Symposium on Visual Computing, ISVC 2005*, pp. 380–387, 2005.
139. D. Chen, M. Bilgic, L. Getoor, D. Jacobs, Dynamic processing allocation in video, *IEEE Transactions on Pattern Analysis and Machine Intelligence* 33 (2011) 2174–2187.
140. A. Krause, C. Guestrin, Optimal value of information in graphical models, *Journal of Artificial Intelligence Research* 35 (2009) 557–591.
141. S. Lee, C. Lee, Low-complexity background subtraction based on spatial similarity, *EURASIP Journal on Image and Video Processing* 2014 (2014) 1–16.
142. Y. Liu, Z. Bellay, P. Bradsky, G. Chandler, B. Craig. "Edge-to-fog computing for color-assisted moving object detection", *Big Data: Learning, Analytics and Applications*, May 2019.
143. R. Siddharth and G. Aghila. A light weight background subtraction algorithm for motion detection in fog computing, *IEEE Letters of the Computer Society* 3(1) (2020) 17–20.
144. B. Garcia-Garcia, T. Bouwmans, A. Rosales-Silva, Background subtraction in real applications: Challenges, current models and future directions, *Computer Science Review* 35 (2020) 100204.
145. O. Barnich, M. Van Droogenbroeck, "ViBe: A powerful random technique to estimate the background in video sequences", *International Conference on Acoustics, Speech, and Signal Processing, ICASSP 2009*, pp. 945–948, April 2009.
146. A. Hossain, I. Hossain, D. Hossain, N. Thu, E. Huh, Fast-D: When non-smoothing color feature meets moving object detection in real-time, *IEEE Access* 8 (2020) 186756–186772.
147. T. Bouwmans, S. Javed, M. Sultana, S. Jung, Deep neural network concepts in background subtraction: A systematic review and a comparative evaluation, *Neural Networks* 117 (2019) 8–66.
148. M. Mandal, S. Vipparthi, "An empirical review of deep learning frameworks for change detection: Model design, experimental frameworks, challenges and research needs", *IEEE Transactions on Intelligent Transportation Systems*, 2021.
149. T. Minematsu, A. Shimada, H. Uchiyama, R. Taniguchi, Analytics of deep neural network-based background subtraction, *Journal of Imaging* 4 (2018) 78.

17

The Role of Artificial Intelligence in E-Health: Concept, Possibilities, and Challenges

Priyanka Gupta, Hardeo Kumar Thakur, and Alpana
Department of Computer Science and Technology, Manav Rachna University, Haryana, India

CONTENTS

17.1 Introduction

The global population is changing, with a significant increase in the number of aged people, many of whom have particular diseases, frailty, and co-morbidity. This group has highly specialized health demands, yet they also have a high degree of autonomy for increasingly lengthy periods. The unique demographic characteristics of the countries necessitate the adoption of different strategies, resulting in a holistic scenario.

DOI: 10.1201/9781003244165-21

Healthcare must be available but must also ensure patient security and empowerment. New digital and information and communications technologies (ICT) allow for the development of new services (e-health) that offer numerous benefits to the patient or older person (mainly, safety, high availability, and increasing quality of service). To make the above process ease, artificial intelligence (AI) is introduced to the world (Nicoletti 2021). Nathanson depicted the AI as the ability to learn, comprehend, or manage new circumstances and the capacity to use data or abilities to control one's own environmental elements (Nathanson 2020). The emergence of AI allows us to think about ways to implement more procedures and process automation. It will enable personal data to be incorporated (Cabestany et al. 2018).

17.2 Artificial Intelligence (AI)

AI has been utilized to help in the advancement of PC frameworks (Zhang et al. 2018), including its use in making decisions, detecting objects, and solving complex issues. AI has four key benefits: (i) higher accuracy predictions, (ii) it helps us to make decisions timely and accurately, (iii) it is needed to tackle complex problems, and (iv) it is significant to perform calculations for human resources. AI is helpful because it deals with complex problems and gives better performance in difficult situations. Let's focus on how AI can be applied in the healthcare industry. AI's impact on different sectors has occurred since the 1950s, including marketing, banking, gaming businesses, and the music industry. Now, AI's most significant economic impact will be in the healthcare sector (Nathanson 2020). As healthcare continues to become important, AI advances, and we may expect these trends to continue.

What caused the garden to devote so much time and attention to AI in the healthcare industry? When we looked for the main reasons for the surge in AI in the healthcare industry, we discovered that it comes from two different perspectives. The first of these reasons is a large amount of medical data readily available (Sharma 2021). Now, we have more medical data than we know what to do with. It always seems that if we're ever hospitalized, our history will be recorded in accounting books. The more data we have, the simpler it is to create AI. Deep learning and machine learning systems deal with enormous amounts of data. AI is becoming easier to employ since more data is now available. Another essential reason to lead to the development of AI is the introduction of complex algorithms. Now what happens in machine learning is merging. Learning cannot handle high-dimensional data, particularly the medical or healthcare data of very high quality (Sharma 2020). The data is high-speed; there are thousands and thousands of attributes. To process and analyze data of this dimension is hard to do with machine learning. Still, as soon as deep learning and neural networks were introduced, this became much easier because deep learning and neural networks focus on solving complex problems that involve high-dimensional data. So, the development of deep learning and neural networks also played a significant role in the impact of AI in healthcare. That's why AI is impacting the healthcare industry in such a considerable manner. There are a couple of use cases of AI in healthcare shared with this chapter to clarify the role of AI in e-healthcare.

TABLE 17.1

Types of Artificial Intelligence

Reactive Machine Artificial Intelligence (RMAI) No memory E.g., deep blue super chess player developed by IBM	**Limited Memory Artificial Intelligence (LMAI)** Limited amount of memory E.g., cancer diagnosis, robot dentists, or robot machines used in various medical procedures, and virtual nursing
Theory of Mind Able to understand and store E.g., chalk bot, Sophia Robot	**Self-Conscious** Capable of making decisions on their own

17.3 Types of Artificial Intelligence

One of the types of AI is what is known as reactive machine artificial intelligence (RMAI). It is the first version of AI to be developed. It has no memory, and it has no ability to learn new things. It simply responds to the current scenario. For example, the deep blue super chess player developed by IBM in the 1980s was a breakthrough. As a result, the next type of revolution is now being referred to as limited memory. Limited memory artificial intelligence (LMAI) means AI with a limited amount of memory. That is something which is currently in use. AI with a limited amount of memory is referred to as limited memory AI. As a result of past experience, dataset, and events, this LMAI will be applied to future events in accordance with a set of guidelines. Most cancer diagnosis, robot dentists, or robot machines used in various medical procedures, and virtual nursing assistants are examples of LMAI. AI of the next generation is the "theory of mind." It is defined as follows: machines are able to understand the needs as well as the expectations of humanity and machines (Hintze 2016). Table 17.1 shows the types of AI.

And then take appropriate action. There are limited number of examples such as chalk bot. A further case in point is Sophia Robot. In this particular field, of course, this is the culmination of AI's development. What next is self-conscious AI when machines are completely aware of themselves? They exist and are capable of making decisions on their own (Hintze 2016).

17.4 AI in E-Health

AI is capable of performing real-time operations and behave with the goal of achieving a predetermined result (Chang 2020). It is intended to be a simulation of the human mind. It is a computer-based system that is meant to tackle a specific problem. AI has consistently demonstrated that it is vastly superior to any single human mind. The role of AI in the e-healthcare domain is helping in detection of diseases in their early stages (Triantafillou and Minas 2020), for example, cancer. If cancer is determined in stages 2 and 3, the person can't survive, but if it is determined in the first stage, then the chance of survival of that particular patient increases. For example, Google AI is better at detecting breast cancer than doctors. Another upcoming application of AI is Fitbit applications. They recommend nutritional diets based on a person's daily activities in order to maintain the person healthy enough. There are many other applications based on AI concepts, which make individuals' life easy today. In short, AI domain helps a person to recover faster from diseases. But, there are some limitations also in AI approaches.

17.5 Limitations of Artificial Intelligence (AI)

There are some challenges with respect to the AI being implemented in healthcare. Some of the challenges are that there is a massive amount of money required to do this research for predicting various things with respect to healthcare. Another challenge is that the model should be perfect; if there is a minute change of 0.01%, the modeling agency has a huge impact. For example, if the person has cancer and the model predicts that the person does not have cancer, then this becomes a massive mistake though the person is being skipped, and vice versa, if the person is not having cancer and the model predicts that the person has cancer, then this also becomes a major mistake. Then, the person has to undergo some other tests to determine whether he/she has cancer or not. Considering this, the model accuracy should be very high to detect diseases correctly. To check the feasibility of the model, it can be validated with performance metrics such as confusion matrix and receiver operating characteristic (ROC) curve.

17.5.1 The Confusion Matrix

Confusion matrix is a representation of a classification model's performance in binary format for the given known true values. Table 17.1 measures the results of a test, taking place in two dimensions. The matrix depicts an attributed class for each row, and a corresponding actual scenario for each column (the truth). Averaging the percentages of true positives, false positives, false negatives, and true negatives gives the confusion matrix (Indhumathy 2020).

When both true positive and true negative results are present, the conditions have been met for these measurements, which enables accurate predictions to be made about the algorithm's future performance. Using fresh data to see how the method will perform is significant, since both Tp (true positive) and Tn (true negative) are involved (Indhumathy

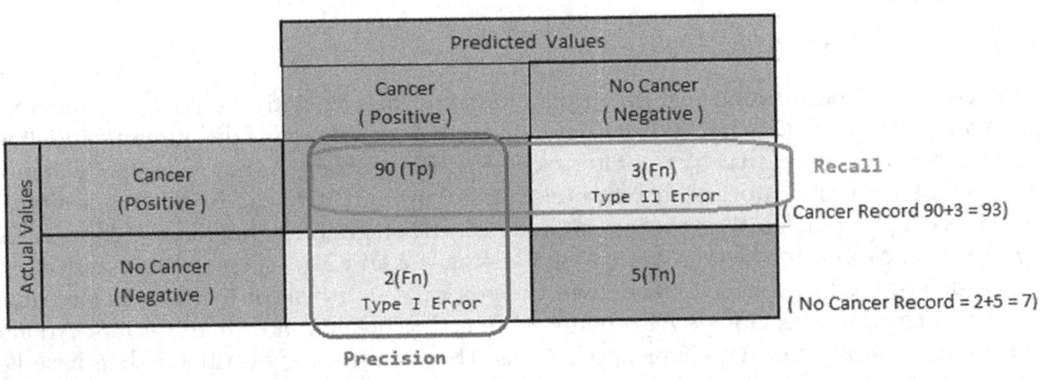

FIGURE 17.1
Confusion matrix.

2020; Haghighi et al. 2018; Levy 2020). For example, if in a dataset comprising 100 items, 93 are cancer and 7 are non-cancer, its prediction accuracy is 90 out of 93 cancer cases. Three cancer records are inadequately expected [3-Fn], which is illustrated in Figure 17.1.

The Recall is a measure of how many positives the model can identify from the data. It is sometimes referred to as Sensitivity or Total Recall Rate (TPR – True Positive Rate). Mathematically, it is formulated as follows:

$$\text{Recall} = \frac{T_p}{T_p + F_n} = \frac{90}{90+3} = 0.967 \tag{17.1}$$

$$\text{Recall} = 97\%$$

The precision of a prediction is defined as the ratio of correct positive predictions to total positive predictions (Indhumathy 2020). Type 1 and Type 2 errors based on the problem statement need to be considered valiantly, which can have a huge impact. Mathematically, it is formulated as follows:

$$\text{Precision} = \frac{T_p}{T_p + F_p} = \frac{90}{90+2} = 0.978 \tag{17.2}$$

$$\text{Precision} = 98\%$$

When it comes to sensitivity and specificity, there is a crucial error: false prediction. The trouble with centering on sensitivity and specificity for prediction is that you just are preparing on the incorrect thing (Indhumathy 2020; Haghighi et al. 2018; Levy 2020); you are conditioning on the real underlying state.

17.5.2 ROC Curve

The ROC bend is utilized to look at the genuine (True) positive rate (TPR) against the false positive rate (FPR) at various limit levels. Affectability is sometimes used to allude to the pace of genuine positives (or likelihood of location in AI) (Indhumathy 2020; Haghighi et al. 2018; Levy 2020). Specificity is used to calculate the FPR. By analyzing ROC curves, Guo at el. discovered the sensitivity and specificity of biomarkers in predicting overall patient survival (Guo et al. 2016).

17.6 Literature Survey

There are many literatures available on the applications of AI in healthcare. Some of the literatures are described briefly in Table 17.2.

Dendral is the world's first developed problem-solving software or expert system as a result of research conducted in the 1960s and 1970s. It was based on organic chemistry (Clancey and Shortliffe 1984). In the 1980s and 1990s, microcomputers and network connectivity became increasingly widespread. Miller et al. proposed AI technologies that must be developed to compensate for the absence of accurate data and to augment medical

TABLE 17.2

Literature Survey

Author	Year	Description	Techniques	Results
Clancey and Shortliffe	1984	Used AI to identify bacteria causing severe infections	Image analysis	65%
Lindsay, Buchanan, and Feigenbaum	1993	Produced the first problem-solving program, or expert system	Knowledge engineering	Molecular wait, and structure
Miller	1994	Medical diagnostic decision support systems that better practitioners' thinking	MDDS	Support system that focuses on patients' records
Koomey et al.	2010	Improvements in computational power leading in quicker data collecting and data processing	Power conversion with high efficiency and durability	Approx. in every 18 months, the computational performance and power consumption of compute nodes roughly double
Pivovarov and Elhadad	2015	Interpreting mounting wealth of health information	i2b2-SMART platform	Translation across hospitals data to EHRs
Bloch-Budzier	2016	NHS utilizing Google innovation to treat patients	Google technology	Detect life-threatening illnesses
Pakdemirli and Wegner	2020	The study outlines considerable advance within the academic domain with respect to AI-driven calculations	Noted continuous progress of AI technology in healthcare field	Discover that how numerous numbers of papers distributed in medical AI
Shaikh, Krishnan, and Thanki	2021	Breast cancer detection and diagnosis systems has considerably improved	Digital imaging technology	Computer-aided cancer detection system

expertise. This is advantageous for specific application areas where information can be known with a high degree of precision. It's an important part of this process. According to the author, decision support systems could supplement physicians' and other health-care practitioners' thinking (Lindsay, Buchanan and Feigenbaum 1993; Miller 1994). In 2010, Koomey et al. showed how the advancement within the field of medical and technology happens (Koomey et al. 2010). Data gathering and processing are becoming more efficient as computing power improves continually. Increased use of AI employs robotic technology and machine learning algorithms in the health industry, as well as innovations in high-tech equipment (Koomey et al. 2010). Pivovarov and Elhadad (2015) explain that a clinical instrument with a single goal in mind is to assist physicians at the front-line of patient care. Emergency clinics records and Electronic Health Record (EHR) information can be simplified with the i2b2-SMART stage. The electronic capacity to store understanding data has created to exceptionally about 80% within the past decade. In a 2016 article, Sarah Bloch-Budzier wrote that the National Health Service (NHS) is utilizing Google technologies to treat patients. Specialists and medical caretakers at the NHS got to leaf through understanding notes to construct a picture of a patient's test reports. Presently, they are able to get a long-time information at their fingertips. When any results are abnormal, an alert will sound (Bloch-Budzier 2016).

17.7 Current Era of Artificial Intelligence and Its Impact on E-Health

AI has increased substantially over a long period of time, according to Pakdemrili et al. (2020). Similar patterns appear to be taking place in 2020, in spite of the fact that they were able to include information only from the start of the year. The five most regularly highlighted nations are in North America, Europe (Italy and Germany), and China. The outlines considerable in advance within the scholarly domain with regard to AI-driven calculations within the areas of oncology, radiology, neuroradiology, and ophthalmology (Pakdemirli and Wegner 2020). In 2021, Shaikh et al. introduced different image methods for breast cancer detection in their book. They showed how the design and development of breast cancer detection and diagnosis systems have considerably improved, and how different imaging technologies such as mammography, ultrasound, and magnetic resonance imaging (MRI) are extensively utilized for breast cancer therapy. Breast cancer detection and diagnosis have improved considerably as a result of the screening of these pictures. Digital imaging technology is currently employed, which may assist in the development of a computer-aided cancer detection system. Ultrasound, MRI, and nuclear medicine have all made major advances in the detection and diagnosis of breast cancer (Shaikh, Krishnan and Thanki 2021).

17.8 Artificial Intelligence Techniques

Now, consider how AI is used to train medical assistants. With the developing request for medical collaborators (Roy 2021), the necessity for AI is expanding. Machine learning and profound learning are subsets of AI, and its procedures are utilized in healthcare as often as possible.

17.8.1 Machine Learning

Machine learning (ML) may be a subset of fake insights that empowers machines to memorize and make strides naturally without explicit programming (Bhargavi and Jyothi 2020). In simpler words, the machine learning methodology is best described as a process. Take a lot of feedback to the machine learning process and then the process will try to understand and interpret the data by using ML models and algorithms, and it will finally solve a problem and predict an outcome based on the data (Chen and Decary 2020). Basically, machine learning is a statistical technique for fitting models to data and allowing models to 'learn' through data training (Davenport and Kalakota 2019). Machine learning could be a developing field of AI that centers on expansive information sets. Utilizing machine learning, computers use algorithms to recognize patterns in data (Niranjanamurthy 2021). Classification algorithms are frequently used to categorize subjects while analyzing various datasets with classifiers and training the machine to apply the complex algorithms, generate the model, and then use these models to generate predictions (from an original test set to a new set of data). Figure 17.2 describes the machine learning process.

In machine learning (ML), various algorithms are used to analyze the acquired data and construct a model that predicts the risk of having a disease like a heart attack. Utilizing a

FIGURE 17.2
Machine learning process.

bed can now forecast whether or not a person is at risk of having a heart attack. Another example of an Apple Watch that saved a person's life and averted death for this individual. This Apple Watch saved his life by providing critical warnings regarding his blood pressure and high heart rate (Buettner and Schunter 2019).

17.8.2 Types of Machine Learning

ML is commonly utilized in other shapes of AI innovations such as natural language processing (NLP), voice innovation, and robots (Chen and Decary 2020). The main ML algorithms must be learned by health Reacher's because ML algorithms take the data and interpret it, in order to correctly predict the resulting results within an acceptable range. If new data is fed into these algorithms, then they 'learn' and 'optimize' their processes in order to achieve the maximum possible efficiency. As a result, the algorithms achieve 'intelligence' (Chitra and Abirami 2019). ML is ordinarily categorized into three sorts of learning calculations: supervised, unsupervised, and semi-supervised, which is briefly described below (Wulfovich and Meyers 2020), as shown in Figure 17.3.

17.8.2.1 Supervised Learning

Supervised learning: A collection of labeled data are used to train the algorithm to anticipate the result, after which it generates the new dataset from which the prediction will be made (Wulfovich and Meyers 2020). In supervised learning, instances of human illnesses are used to teach the machine. To be successful, the ML algorithm must be delivered along with a dataset containing necessary inputs and outputs. Afterward, the process of identifying the appropriate inputs and outputs must be developed, which is an extremely challenging undertaking (Wakefield 2021). It is possible that the operator may know all the correct answers, though. The program examines the data and discovers patterns. The patterns are observed and used to create predictions. The operator uses the data to choose the most effective algorithm for predictions, and they keep trying new algorithms until both performance and accuracy are high (Wakefield 2021). This algorithm has found widespread use in healthcare, where it is applied to calculate quantitative classification data that can be used to categorize variable inputs (for example, by using medical imaging to diagnose cancer tumors, their subtypes, and occurrences) (Nathanson 2020). In addition, predictive analytics is found within continuous output (such as EHR data) (Chen and Decary 2020). Classification, regression, and forecasting all arise from supervised learning techniques. When using a classification scheme, output variables are grouped together and denoted by their kind, but when using a regression scheme, the numerical values of the output variables are represented, and future predictions are made on the basis of the data from the past and present (Nathanson 2020). Classification is used when the output

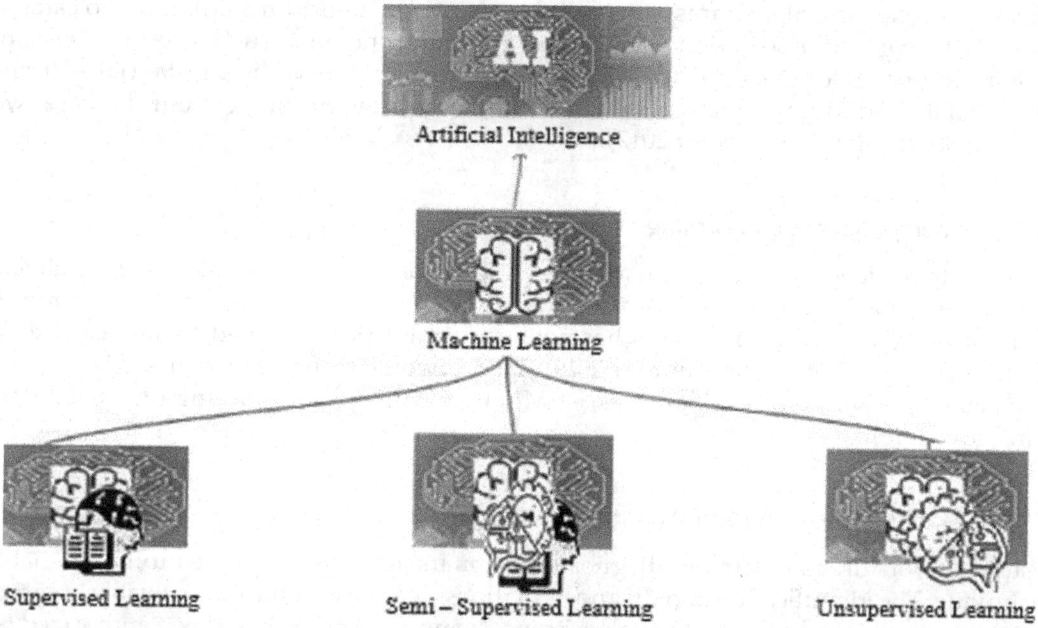

FIGURE 17.3
Machine learning.

variable is a category, such as "red" or "green," or "illness" or "no disease." A classification model seeks to reach a conclusion about the observed values by creating a categorization system. Classification models evaluate the value of various outputs by referencing the influence of one or several inputs (Jain 2021). In this way, for example, the words "spam" and "not spam" are used when filtering emails, and the terms "male" and "female" are applied when examining gender data. Classification does the following: first, it predicts the categorical class labels (known as labels) or class data (model) from training data, and then it applies that model to fresh data to classify it. Different categorization models are available. These categorization models, in the order of their usage frequency, are logistic regression, decision tree, random forest, gradient-boosted tree, multilayer perceptron, one-vs-rest, and Naive Bayes (Raschka 2014). When working with a real or continuous variable such as "salary" or "weight," a regression is appropriate. Nonlinear regression is linear regression in its most basic form. It attempts to locate the best hyperplane that covers the data points and that can be fitted to the data. Forecasting a person's age is part of linear regression (Jain 2021). Forecasting is the process of generating predictions about the future based on information that is acquired over time, as well as information about the past and present (Wakefield 2021).

17.8.2.2 Unsupervised Learning

Unsupervised learning is training a computer on data that are neither categorized nor labeled, and then letting the algorithm to act on that data without explicit instructions. In order to find comparable bits of information that aren't programmed into the computer, the machine's objective is to group the various information items based on similarities, patterns, and variations (Sharma and Garg 2019, 2021, Veeramani 2019). In addition to supervised learning algorithms, unsupervised learning algorithms fall into two categories: clustering and association. To find the groups in the data, such as on the basis of purchasing behavior (Mishra 2017). Identifying rules that describe substantial sections of your data and discovering rules that can explain a huge portion of your data are two facets of learning issue (Mishra 2017).

17.8.2.3 Semi-Supervised Learning

In order to achieve the goal, semi-supervised learning uses both labeled and unlabeled data (Chitra and Abirami 2019). When data has been appropriately tagged, it is referred to as labeled data; however, when it has not been appropriately tagged, it is referred to as unlabeled data (Chitra and Abirami 2019). These two learning approaches can be used in conjunction with one another to assist ML algorithms in categorizing unlabeled data (Grzegorz 2018).

17.8.3 Machine Learning in E-Health

Using ML for illness detection, diagnosis, and management has proven to be a reliable technique. The identification and diagnosis of illnesses have become more reliable in recent years as a result of advances in ML algorithms. Many algorithms have been authorized by the Food and Drug Administration (FDA) of the United States for use in healthcare settings where they are considered to be safe. In accordance with Mesko, a detailed analysis of FDA-approved algorithms has been presented, with the majority of them having been established in the disciplines of radiology, pathology, cardiology and cancer, as well as endocrinology and dermatology, among other subjects (Chen and Decary 2020). Beyond illness detection and diagnosis, several hospitals in the United States and Canada have begun to employ ML and predictive analytics for hospital management reasons, in addition to disease detection and diagnosis (for example, predicting adverse events, mortality rates, and the number of patients who visit the emergency department). Consequently, hospitals are able to prepare for predicted occurrences many days in advance as a result of the predictability of the situation (Amin and Ali 2018).

17.8.4 Deep Learning

Deep learning is a subset of ML, and it is playing an increasingly important role in the treatment of healthcare problems. Incorporating innovative DL-based models produces unparalleled outcomes, with prediction/detection rates. These outcomes have been on par with or even better than those achieved by current state-of-the-art methods using intuitive and non-intuitive disease descriptions (Gupta 2021). According to the findings, advanced deep learning algorithms are most suited for large healthcare data sets and surpass traditional statistical models in terms of prediction power while maintaining interpretability (Gupta 2021).

17.8.4.1 Deep Learning in E-Health

Let's take a closer look at what deep learning is all about. Because deep learning is a more advanced notion in ML, it deserves to be highlighted. Essentially, planning is a more advanced discipline of training learning that makes use of the notion of neural networks in order to solve difficult problems that demand high-dimensional data and automated feature extraction. Planning is used to address more complex and high-dimensional data problems, and the fundamental feature of deep learning is that it makes use of neural networks in order to map input or output into something useful. So, in order to address an issue in planning, it is necessary to employ neural networks, or artificial neural networks to be more specific, to solve the problem. The operator needs to provide a large amount of data as input. Data is being changed, and the operator will apply weights and biases to the data. It is necessary to go through a few transformations in order to get a size increase. To forecast the stage of breast cancer in the early stages, deep learning can be applied. The problem statement is to investigate a breast cancer data set and create a neural network classifier that predicts the stage of breast cancer as either malignant or benign in nature. Malignant cells are those that are cancerous, whereas benign cells are those that are not cancerous or normal cells. This means that the neural network error set will comprise observations and samples of both malignant and benign cells, resulting in a situation in which both cancerous and non-cancerous cells are present. This data collection, for which you will employ artificial neural networks in order to classify them into two distinct categories on a new data set that contains cells. In one, there will be no malignant cells, whereas the other will have cells that are cancerous.

17.9 Healthcare Data and Databases

Future pharmacovigilance (or pharmacoepidemiology) primarily depends on automated databases being utilized. Data misclassification can be a cause of considerable inaccuracy. EHR data can be utilized to detect whether or not an illness or injury has occurred (Lanes et al. 2015). While Big Data, continuous medical education, and continuous medical learning and teaching are not mutually linked, data from all three fields has yet to be completely explored. Dr. John O'Neill believes that basic healthcare education should be provided to staff with a basic understanding of lifelong learning. The outcomes of their study are extremely valuable for healthcare stakeholders and professionals, especially for establishing an excellent route for academia and practice (Au-Yong-Oliveira et al. 2021).

17.10 Ideas, Possibilities, and Challenges

AI will play a major part in healthcare services of the future. ML is largely accepted as a much-needed advancement in medical care. Early diagnosis of diseases helps in advanced treatment recommendations. It is quite conceivable that almost all radiology and pathology images will be examined by a machine in the near future. AI techniques are becoming increasingly capable of discovering complicated patterns in data. Among the sources of information are EHRs, medical literature (including clinical trials), insurance claims data, pharmacy records, data entered by patients or recorded on activity trackers, and so on. A significant push has been made toward the creation and refining of customized medical procedures as a result of the convergence of these two factors (Nicholson Price II 2017). There are a lot of expectations and ambitions regarding the application of AI technology in the medical profession. First and foremost, there is the prospect of constructing predictive models, which have obvious benefits in terms of disease prevention. Second, there is the ability to implement early diagnosis in order to ensure that the most suitable care is provided in the shortest amount of time possible. In conclusion, the affirmation of chatbot-based settings promises to ensure that patients receive the correct information while also accompanying them through their treatment processes (Guarda 2019). Systems with powerful AI capabilities support intelligent virtual assistants, which are now embedded in smartphones or dedicated home speakers, such as Microsoft Cortana and Apple Siri, but are also increasingly present in people's home devices, such as voice assistants Amazon Alexa or Google Assistant, which are supported by powerful AI features. These products represent the most advanced and enthralling frontier in the application of AI to make everyday living easier now. Voice assistants have a distinct advantage, particularly when used in conjunction with healthcare applications, because they allow for greater integration of AI tools into daily medical applications, thereby improving treatment efficiency, avoiding costs, and reducing the risk of false diagnose The potential for AI in healthcare is that the combination of these factors can clearly aid in more targeted pre-operative planning as well as the post-operative stages, among other things (Pfeifle 2019; Davenport and Kalakota 2019).

17.11 Conclusions

This chapter presents an intriguing scenario with the goal of easing some of the issues that many actual healthcare services are currently experiencing. The AI e-health concept provides technological solutions. It also conjuncts with massive communication systems. Now, AI is implemented in order to provide a patient-centered service that promotes patient empowerment. It also strengthens the relationship between doctors and patients. In this chapter, the general role of AI has been discussed, and various methodology has been presented in which the technology was used to find a solution for a more correct and intelligent way to manage the treatments.

References

Amin, M. Z., and A. Ali. 2018. *Performance Evaluation of Supervised Machine Learning Classifiers for Predicting Healthcare Operational Decisions*. Wavy AI Research Foundation: Lahore.

Au-Yong-Oliveira, M., A. Pesqueira, M. J. Sousa, F. Dal Mas, and M. Soliman. 2021. "The potential of Big Data research in healthcare for medical doctors' learning." *Journal of Medical Systems* 45, 1–14.

Bhargavi, P., and S. Jyothi. 2020. "Object detection in Fog computing using machine learning algorithms." In *Architecture and Security Issues in Fog Computing Applications*, by S. J. P. Bhargavi, 90–107. IGI Global: Hershey, PA.

Bloch-Budzier, S. 2016. *NHS using Google technology to treat patients*. https://www.bbc.com/news/health-38055509.

Buettner, R., and M. Schunter. 2019. "Efficient machine learning based detection of heart disease." *2019 IEEE International Conference on E-health Networking, Application & Services (HealthCom)*. IEEE, Bogota, Colombia, pp. 1–6.

Cabestany, J., D. Rodriguez-Martín, C. Pérez, and A. Sama. 2018. "Mixed design of integrated circuits and system." *Artificial Intelligence Contribution to eHealth Application. 25th International Conference.* IEEE, Gdynia, Poland, pp. 15–21.
Chang, A. 2020. "The role of artificial intelligence in digital health." In *Digital Health Entrepreneurship*, by S. Wulfovich, 71–81. Springer: Cham.
Chen, M., and M. Decary. 2020. "Artificial intelligence in healthcare: An essential guide for health leaders." *Healthcare Management Forum* 33(1), 10–18.
Chitra, P., and S. Abirami. 2019. "Smart pollution alert system using machine learning." In *Integrating the Internet of Things Into Software Engineering Practices*, by P. Chitra and S. Abirami, 219–235. IGI Global: Hershey, PA.
Clancey, W. J., and E. H. Shortliffe. 1984. *Readings in Medical Artificial Intelligence: The First Decade.* Addison-Wesley Longman Publishing Co., Inc.: Boston, MA.
Davenport, T., and R. Kalakota. 2019. "The potential for artificial intelligence in healthcare." *Future Healthcare Journal* 6(2), 94.
Grzegorz, M. 2018. *Supervised and unsupervised machine learning–types of ML.* https://www.netguru.com/blog/supervised-machine-learning.
Guarda, P. 2019. "'Ok Google, am I sick?': Artificial intelligence, e-health, and data protection regulation." *BioLaw* 15, 359–375.
Guo, W., Q. Wang, Y. Zhan, and X. Chen. 2016. "Transcriptome sequencing uncovers a three-long noncoding RNA signature in predicting breast cancer survival." *Scientific Reports* 6(1), 1–10.
Gupta, C. 2021. "A call for deep learning in healthcare." *Turkish Journal of Computer and Mathematics Education (TURCOMAT)* 12(12), 2711–2713.
Haghighi, S., M. Jasemi, S. Hessabi, and A. Zolanvari. 2018. "PyCM: Multiclass confusion matrix library in Python." *Journal of Open Source Software* 3, 729.
Hintze, A. 2016. *Understanding the four types of artificial intelligence.* https://www.govtech.com/-computing/understanding-the-four-types-of-artificial-intelligence.html.
Indhumathy, C. 2020. *Confusion matrix—Clearly explained.* https://towardsdatascience.com/confusion-matrix-clearly-explained-fee63614dc7.
Jain, S. 2021. *Regression and classification|supervised machine learning.* https://www.geeksforgeeks.org/regression-classification-supervised-machine-learning/.
Koomey, J., S. Berard, M. Sanchez, and H. Wong. 2010. "Implications of historical trends in the electrical efficiency of computing." *IEEE Annals of the History of Computing* 33(3), 46–54.
Lanes, S., J. S. Brown, K. Haynes, and M. F. Pollack. 2015. "Identifying health outcomes in healthcare databases." *Pharmacoepidemiology and Drug Safety* 24(10), 1009–1016.
Levy, D. G. 2020. *In machine learning predictions for health care the confusion matrix is a matrix of confusion*, pp. 11–15. https://www.fharrell.com/post/mlconfusion/.
Lindsay, R. K., B. G. Buchanan, and E. A. Feigenbaum. 1993. "DENDRAL: A case study of the first expert system for scientific hypothesis formation." *Artificial Intelligence* 61, 209–261.
Miller, R. A. 1994. "Medical diagnostic decision support systems—Past, present, and future: A threaded bibliography and brief commentary." *Journal of the American Medical Informatics Association* 1(1), 8–27.
Mishra, S. 2017. *Unsupervised learning and data clustering.* https://towardsdatascience.com/unsupervised-learning-and-data-clustering-eeecb78b422a.
Nicholson Price II, W. 2017. *Artificial intelligence in health care: Applications and legal implications.* https://repository.law.umich.edu/articles/1932/.
Nicoletti, B. 2021. "Partnerships in Banking 5.0." In *Banking 5.0*, by B. Nicoletti, 359–368. Palgrave Macmillan: Cham.
Niranjanamurthy, M. 2021. *Intelligent Data Analysis for COVID-19 Pandemic.* Springer Nature: Singapore.
Pakdemirli, E., and U. Wegner. 2020. "Artificial intelligence in various medical fields with emphasis on radiology: Statistical evaluation of the literature." *Cureus* 12(10), e10961.

Pfeifle, A. 2019. "Alexa, what should we do about privacy: Protecting privacy for users of voice-activated devices." *Washington Law Review* 93, 421.

Pivovarov, R., and N. Elhadad. 2015. "Automated methods for the summarization of electronic health records." *Journal of the American Medical Informatics Association* 22(5), 938–947.

Raschka, S. 2014. *Predictive modeling, supervised machine learning, and pattern classification*. https://-sebastianraschka.com/Articles/2014_intro_supervised_learning.html.

Sharma, L., and Garg, P.K. (Eds.). 2019. *From Visual Surveillance to Internet of Things: Technology and Applications* (1st ed.). Chapman and Hall/CRC: New York. https://doi.org/10.1201/9780429297922.

Sharma, L., and Garg, P.K. (Eds.). 2021. *Artificial Intelligence: Technologies, Applications, and Challenges* (1st ed.). Chapman and Hall/CRC: New York. https://doi.org/10.1201/9781003140351.

Shaikh, K., S. Krishnan, and R. Thanki. 2021. *Artificial Intelligence in Breast Cancer Early Detection and Diagnosis*. Springer: Cham.

Sharma, L. (Ed.). 2021. *Towards Smart World*. Chapman and Hall/CRC: New York. https://doi.org/10.1201/9781003056751.

Sharma, L., and P. Garg (Eds.). 2020. *From Visual Surveillance to Internet of Things*. Chapman and Hall/-CRC: New York. https://doi.org/10.1201/9780429297922.

Triantafillou, G., and M. Y. Minas. 2020. "Future of the artificial intelligence in daily health applications." *The European Journal of Social & Behavioural Sciences*. https://doi.org/10.15405/ejsbs.278.

Veeramani, G. 2019. *Introduction to machine learning and how it works*. https://www.cloudiqtech.com/machine-learning-an-introduction/.

Wakefield, K. 2021. *A guide to the types of machine learning algorithms and their applications*. https://-www.sas.com/en_gb/insights/articles/analytics/machine-learning-algorithms.html.

Wulfovich, S., and A. Meyers. 2020. "Introduction to digital health entrepreneurship." In *Digital Health Entrepreneurship*, by S. Wulfovich and A. Meyers, 1–6. Springer: Cham.

Zhang, Y., Lang, P., Zheng, D., Yang, M., & Guo, R. 2018. "A secure and privacy-aware smart health system with secret key leakage resilience." *Security and Communication Networks*. hindawi. Article ID 7202598.

Index